Triple-Helical Nucleic Acids

Springer
New York
Berlin
Heidelberg
Barcelona
Budapest
Hong Kong
London
Milan
Paris
Santa Clara
Singapore
Tokyo

Valery N. Soyfer Vladimir N. Potaman

Triple-Helical Nucleic Acids

With 121 Figures

Springer

Valery N. Soyfer, Ph.D.
Laboratory of Molecular Genetics
Department of Biology
George Mason University
Fairfax, VA 22030-4444
USA

Vladimir N. Potaman, Ph.D.
Center for Genome Research
Institute for Biosciences and Technology
Texas A&M University
Houston, TX 77030
USA

The cover illustration depicts a triple-helical model of DNA, derived from Arnott fiber DNA coordinates. Teresa Larsen, of The Scripps Research Institute, built the model using NAB by Thomas J. Macke, and rendered the computer graphic image using custom software by David S. Goodsell. © 1995, T. Larsen, TSRI.

Library of Congress Cataloging in Publication Data
Soyfer, Valeriĭ.
Triple-helical nucleic acids / Valery N. Soyfer & Vladimir N. Potaman.
p. cm.
Includes bibliographical references and index.
ISBN 0-387-94495-8 (alk. paper)
1. DNA. I. Potaman, Vladimir N. II. Title.
QP624.S69 1995
574.87′3282—dc20 95-12214

Printed on acid-free paper.

Acquiring editor: Robert C. Garber.
Production coordinated by Chernow Editorial Services, Inc., and managed by Terry Kornak; manufacturing supervised by Joe Quatela.
Typeset by Best-set Typesetter Ltd., Hong Kong.
Printed and bound by Braun-Brumfield, Inc., Ann Arbor, MI.
Printed in the United States of America.

9 8 7 6 5 4 3 2 1

ISBN 0-387-94495-8 Springer-Verlag New York Berlin Heidelberg

To Nina, Marina, and Vladimir Soyfer,
and Olga and Gosha Potaman

Preface

Alexander Rich, Gary Felsenfeld, and David Davis published the first observation of triple-helical structures in nucleic acids in 1957. Great changes in the field occurred in the mid-1980s. For us personally, the pioneering thoughts and suggestions of Dr. Maxim Frank-Kamenetskii had particular importance and greatly influenced our interest in triplexes. He presented the first model of H DNA and began his efforts to understand triplex structures, especially their physical chemistry and the role of DNA supercoiling. We were inspired by Maxim, with whom we have collaborated for several years.

Ten years have flown by like one day in our lives, and the science of triplexes has changed dramatically during this decade. Because of outstanding work by Peter Dervan, Claude Helen, Robert Wells, Jeremy Lee, Valentin Vlassov, and many others, the field of DNA triplexes has become a center of interest for specialists in nucleic acids specifically, and, more widely, molecular biologists. Although the biological role of triplexes is not yet clear, their importance in many key processes of life is now obvious. We understand that without a definite clarification of their biological role, any description of triplexes is incomplete. However, the enormous volume of information that is now available motivates us to provide a systematic description of this field in book form. This book discusses the structure and stability of triplexes, the factors involved in their appearance, the methods used for their investigation, and, of course, the current understanding of their biological role. New information discovered in the search for the potential role of triplexes in antisense regulation of gene expression, as well as numerous attempts to apply triplexes in gene therapy and other medical applications, have widened hopes for the great potential presented by triplex studies. Although none of these new ideas has yet resulted in therapeutic use, there is widespread anticipation that triplex nucleic acids will enjoy numerous applications.

Many of our friends and colleagues helped make this book possible. Without their encourgement discussions, and generosity in sending us copies of reprints, photos, and unpublished materials, this book could not have been prepared so quickly. Drs. Valery Ivanov from the Moscow Institute of Molecular Biology, Valentin Vlassov from the Novosibirsk Institute of Bioorganic Chemistry, Sergei Mirkin from the University of

Chicago, and Oleg Voloshin from the National Institutes of Health were the first reviewers of the manuscript; it was they who first encouraged us to try to find a publisher. Since then, the moral support of Dr. Robert Garber, Senior Editor at Springer-Verlag, and his kind openness in considering our manuscript, cannot be overestimated. His continuing support helped us avoid mistakes and expedite the publication of this book.

We are very happy to express our deep gratitude and appreciation to Alex Rich, Joshua Lederberg, Gary Felsenfeld, Maxim Frank-Kamenetskii, Vadim Demidov, Lyudmila Shlyakhtenko, B. Montgomery Pettitt, Albino Bacolla, Paul Chastain, and Igor Panyutin for discussions that resulted in a better understanding of different aspects of triplex science. We are especially indebted to Richard R. Sinden, who read the entire manuscript and made numerous suggestions for changes, to Jan Klysik, who read several chapters and made valuable suggestions, and to Adam Jaworski, who offered insightful criticisms of Chapter 6.

Contents

Introduction

Interest in triple-helical nucleic acids has been stimulated by the recognition of their potential biological roles and genetic applications. DNA triplexes can be formed in natural homopurine–homopyrimidine (PuPy) sequences, which represent up to 1% of eukaryotic genomes. Although direct evidence of participation of triplexes in biological processes has not yet been obtained, a growing body of data suggests that triplexes can be involved in regulation of DNA replication, transcription, recombination, and development. Interest in DNA triplexes has been further enhanced by the first findings that triplex-like structures can exist in vivo. Appropriately designed third-strand oligonucleotides that hybridize to targeted duplex domains can be used to control gene expression, serve as artificial endonucleases in genome mapping strategies, extract and purify the specific duplex DNA, and so forth. Triplex regulation of DNA functions seems very promising because of the demonstrated ability of oligonucleotides to penetrate cell walls via liposome- and receptor-mediated endocytosis, or to be taken up by cells directly. It is important to emphasize that full-length oligomers may persist in the cell for at least several hours after being taken up.

In the course of transcription, a local wave of supercoiling that could promote triplex formation may develop behind the moving RNA polymerase complex. The PuPy tracts necessary for transcription and capable of forming triplexes have been found in the 5′ flanking regions of some eukaryotic genes, which offers hope that the role of triplexes in this particular process will be elucidated soon. It has been suggested that triplex-like secondary structures in DNA are involved in the ordered transcriptional switch from γ-globin synthesis in fetal erythroid cells to β-globin synthesis just before birth. The triplex structure causes specific termination of DNA polymerization in vitro and may participate in the regulation of DNA replication in vivo. A triple-stranded structure is also presumed to be an intermediate in DNA recombination.

Triple-strand formation also represents the basis for numerous site-specific manipulations with duplex DNA. Appropriately designed third-strand oligonucleotides that hybridize to targeted duplex domains might be used to control gene expression, serve as artificial endonucleases in genome mapping strategies, modulate the sequence specificity of DNA-binding drugs, selectively alter the sites of protein activity, provide non-enzymatic ligation of double-helical DNA by alternate-strand triple helix formation, quantitate polymerase chain reaction products, and provide physical genome mapping by electron microscopy. Triplex regulation of DNA expression and its interactions with a range of molecules seems very promising.

All of these findings have occurred within the last five years, and have opened a new and rapidly growing field of research. At the same time, it would be naive to see only successes in the understanding of triplexes and overlook numerous difficulties in fulfilling some very high initial expectations. Not all of the early hopes have been fulfilled, but expectations in the field of triplexes are very high. Many laboratories are involved in research on the nature of triplexes and their potential biological role, as well as in identifying applications of triplexes to biotechnology.

Different aspects of triple-stranded structures have been discussed in several reviews (Wells et al., 1988; Frank-Kamenetskii, 1990, 1992; Hélène, 1991, 1992; Palecek, 1991; Yagil, 1991; Cheng and Pettitt, 1992; Strobel and Dervan, 1992; Sun and Hélène, 1993; Thuong and Hélène, 1993; Mirkin and Frank-Kamenetskii, 1994; Radhakrishnan and Patel, 1994d). However, the amount of new material in the field is growing fast. In this book we systematically describe the properties of triplexes, the methods of investigation, possible triplex-stabilizing interactions, triplex recognition schemes, potential biological roles of triplexes, and genetic and pharmacological applications of triplex methodology. We discuss in more detail the issues not reviewed previously, and briefly outline the better known material.

1
The Discovery of Triple-Stranded Nucleic Acids

Basic Physicochemical Properties of Nucleic Acids

This section will briefly outline some of the basic features of nucleic acids that are essential to understanding the properties of triple-stranded forms.

Bases, Nucleosides, Nucleotides, Polynucleotides

Nucleic acids are polymer molecules built of the basic blocks, nucleotides. A mononucleotide consists of a specific base, a five-membered pentose sugar, ribose or deoxyribose (which lacks the hydroxyl at the C2′ position) type, and a phosphate group (Figure 1.1). Adenine and guanine are derivatives of purine and are called *purine bases*. Cytosine, thymine, and uracil are derivatives of pyrimidine and are called *pyrimidine bases*. Each of the bases forms a nucleoside when attached to a ribose via a glycoside bond between a specific nitrogen atom of the base (N9 of purines and N1 of pyrimidines) and the C1′ carbon atom of the sugar. To produce a mononucleotide, the phosphate group can be attached to several positions of the sugar (usually, 5′ or 3′ carbons). The same phosphate group can be simultaneously attached to the 5′ carbon of one nucleoside and the 3′ carbon of another nucleoside, making a dinucleoside–monophosphate. The bidirectional reactivity of the phosphate group makes possible a polymerization of nucleotides into a long molecule (polyribonucleotide) (Figure 1.2). All nucleotides in the polymer are linked by similar 3′,5′-phosphodiester bonds. The polynucleotide sequence is written from the 5′ to the 3′ end using a one-letter code for the bases and sometimes indicating available phosphate groups.

Nature uses a number of specific enzymes to produce polynucleotides. Using ribonucleotide–triphosphates, an RNA polymerase synthesizes polyribonucleotides according to the sequence of a specific DNA template. Polydeoxyribonucleotides can be synthesized starting from deoxyribonucleotide triphosphates using a DNA polymerase and the sequence information of DNA molecule, or reverse transcriptase and the sequence information of RNA. Shorter RNA and DNA oligonucleotides (up to

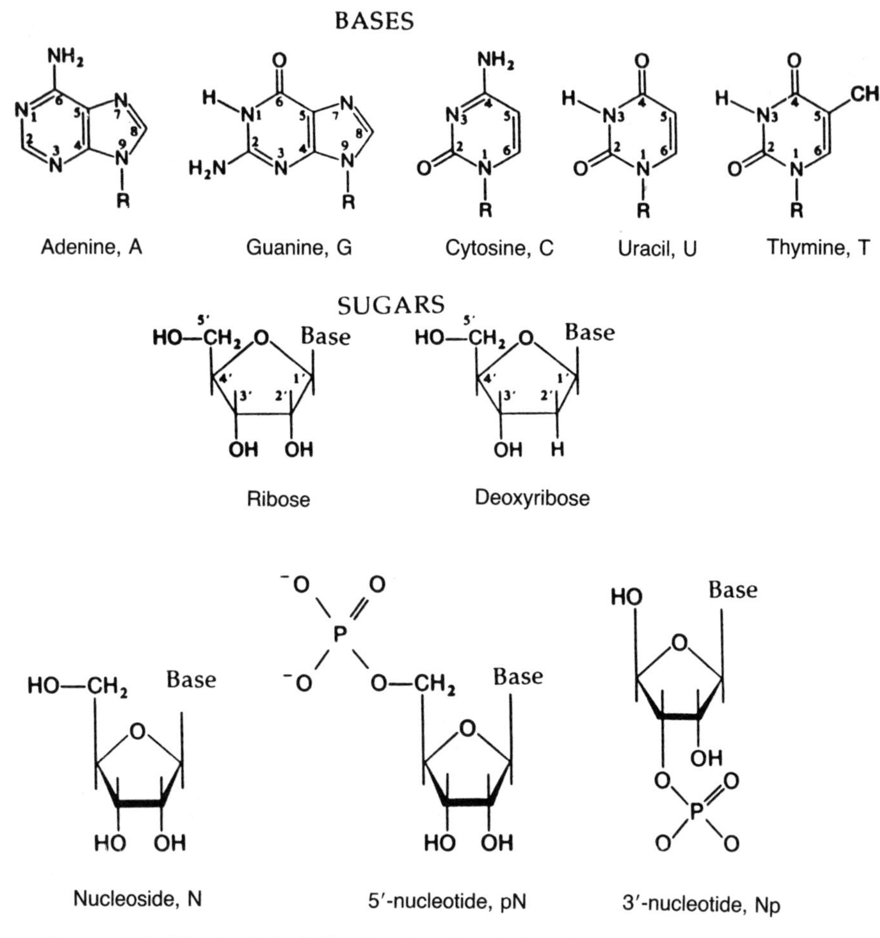

FIGURE 1.1. The basic building blocks of nucleic acids. The figures of the bases in the top row indicate the atom numbering in heterocycles and the glycosidic bonds (N–R) connecting the bases to either ribose (R) or deoxyribose type sugars, making up the nucleosides (left in the bottom row). The middle row shows the two sugar types found in nucleic acids. The difference between them is the presence (in ribose) or absence (in deoxyribose) of the 2′ hydroxyl group. Of importance are the 5′ and 3′ hydroxyl groups, which are the primary sites of phosphate group addition, resulting in 5′ and 3′ nucleotides, respectively (center and right in the bottom row). The bidirectional reactivity of the phosphate group allows it to react with the 5′ and 3′ hydroxyls simultaneously, providing the basis for the formation of long polynucleotide molecules.

several dozen nucleotides) can be obtained via chemical synthesis, which is now highly automated.

A number of important characteristics of the bases, nucleosides, and nucleotides determine their ability to interact with each other and form

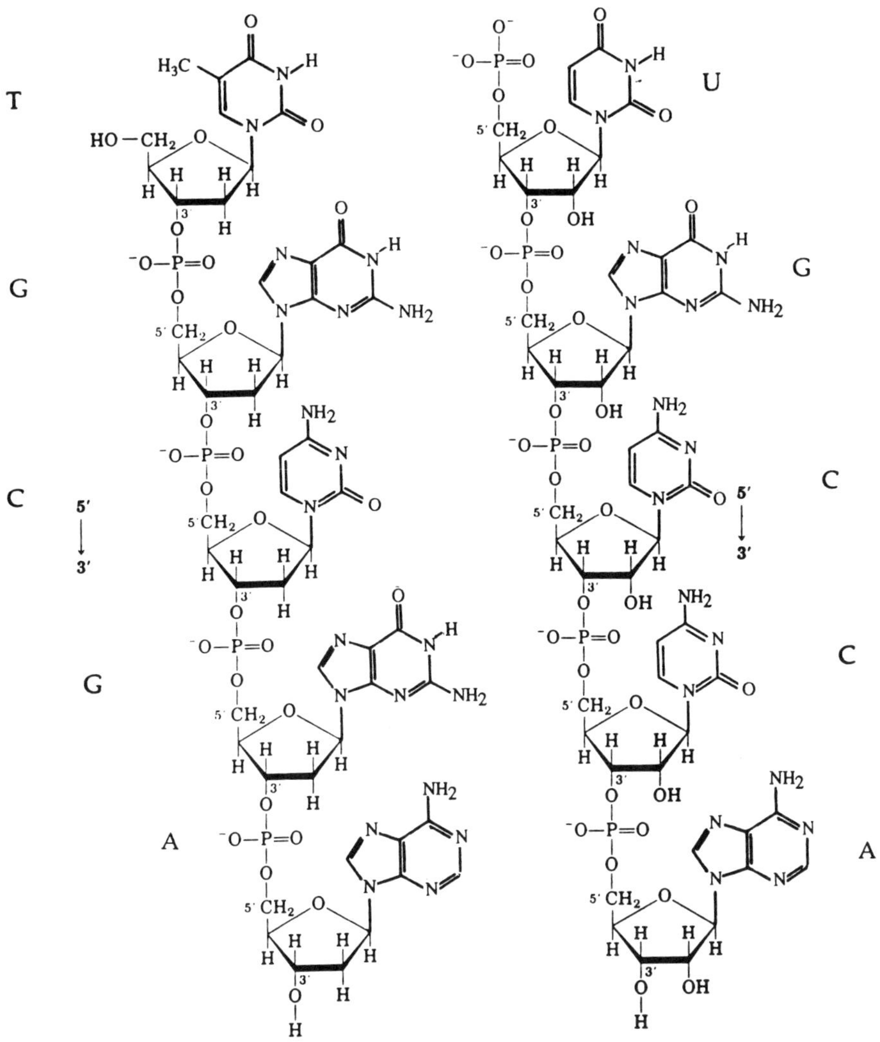

FIGURE 1.2. Examples of polyribonucleotide (right) and polydeoxyribonucleotide (left) chains. The sequence is written form the 5′ to the 3′ end: dTGCGA and rUGCCA. Where important, the available phosphate groups are indicated: d(TpGpCpGpA). Note that one oxygen on each phosphate group is ionized.

ordered structures. The bases have several nitrogen atoms that can accept or release protons following a change in pH. Lowering the pH favors addition of the protons to specific nitrogens (protonation). On the contrary, at elevated pH, nitrogens tend to lose protons (deprotonation). Figure 1.3 shows the relevant nitrogen atoms alongside the values of pK (i.e., the value of pH at which the corresponding nitrogen is half-

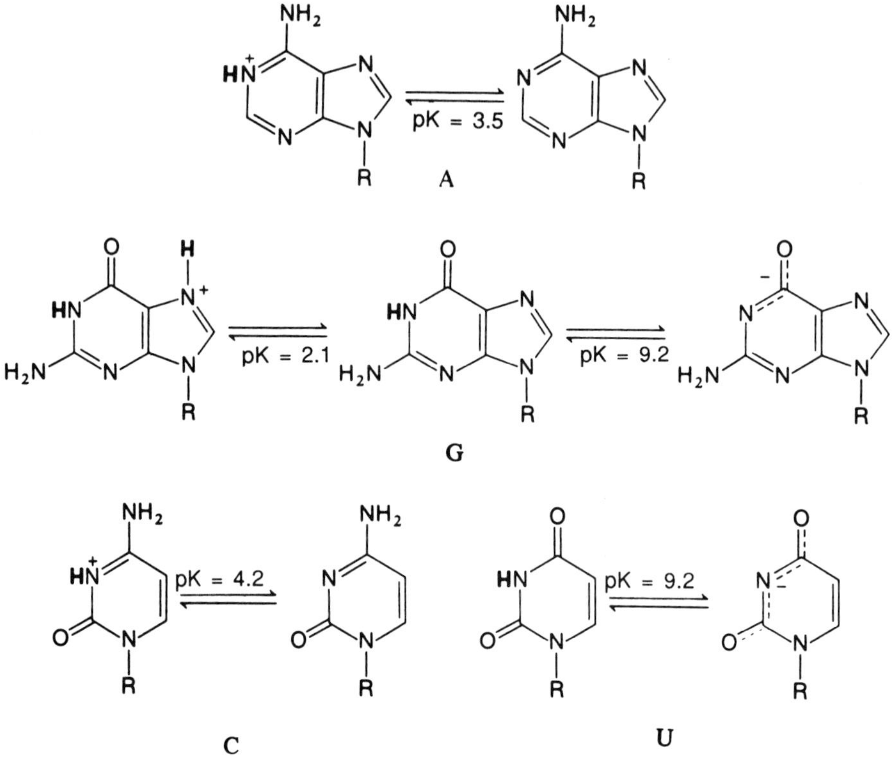

FIGURE 1.3. Ionization equilibriums in the bases. The N1 of adenine is able to be protonated at low pH (i.e., to attract the proton from a medium with a high proton concentration). The pH value at which half of the adenine molecules are protonated and positively charged (p*K*) is close to 3.8. Therefore, when pH < p*K* the adenines are mostly protonated, and when pH > p*K* they are mostly neutral. Guanine has two p*K* values corresponding to proton addition to the N7 below pH 2.1 and proton loss from the N1 above pH 9.2. Guanine is neutral within a broad pH range around the values corresponding to the physiological medium of the cell. The pK for the N3 of cytosine is ca. 4.3. Uracil is able to lose the proton from the N3 and become negatively charged above pH 9.2. The N3 of thymine has a pK value close to 10. The exact pK values of ionizable groups are slightly changed when the bases are included in nucleosides and nucleotides, and they may be changed further when polynucleotides form ordered structures. From: *Biophysical Chemistry* Vol I by Cantor and Schimmel. Copyright © 1980 by W.H. Freeman and Company. Used with permission.

protonated). Common bases are not charged at neutral pH, and their structures correspond to the forms in Figure 1.1. As discussed later, the presence of protons determines the ability of specific nitrogens to form the hydrogen bonds with corresponding groups of other bases, resulting in multiple base associates.

The sugars (either ribose or deoxyribose) have no protonation sites. Their characteristic property is a nonplanar configuration, so that the 2′ or 3′ carbons can significantly deviate from the plane where the rest of the sugar ring atoms are. Corresponding sugar conformations are shown in Figure 1.4. The different specific sugar conformations have been found in polynucleotides which form helically ordered structures as discussed below.

The base–pentose orientation is also structurally important. Figure 1.5 shows the *syn* and *anti* conformations of the guanine base relative to pentose. The *anti* conformation is slightly more energetically preferable in single-stranded polypurines and is strongly preferable in polypyrimidines (Saenger, 1984).

A phosphomonoester group (present in isolated mononucleotides or at the ends of polynucleotides) contains two sites capable of releasing a proton (Figure 1.6). Corresponding pK values are close to 1 and 6. In a phosphodiester, the only protonated site has a pK value close to 1; therefore, at neutral pH, polynucleotides have charged backbones with one negative charge for each nucleobase in the chain. When the polynucleotide chain forms, the bases retain their ability to be protonated or deprotonated, although the precise pK values can vary significantly. For example, pK for the N3 of cytosine varies from 4.3 (free base) to 7.3–7.5 (in polycytidylic acid) (Inman, 1964a,b; Gray et al., 1988).

Generally, an isolated polynucleotide chain does not have a completely disordered structure (random coil). One of the best-known examples of the ordered polynucleotide is polyadenylic acid. Interactions of nonpolar adenine bases with water and dissolved salts are not favorable, and

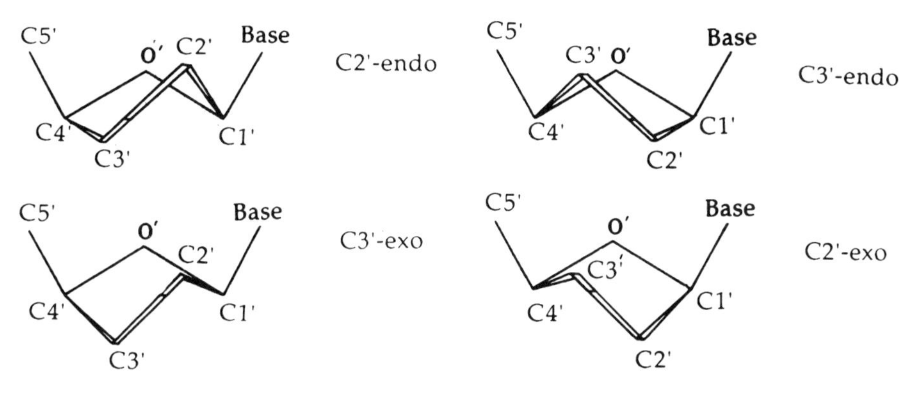

FIGURE 1.4. Four major conformations of ribose and deoxyribose rings. Four atoms (C1′, C4′, O′, and either C2′ or C3′) are in almost the same plane, whereas the C3′ or C2′, respectively, significantly deviate from the in-plane positions. These conformations are determined by the interactions of the sugars with the attached bases and, in the case of polynucleotides, with other closely surrounding groups.

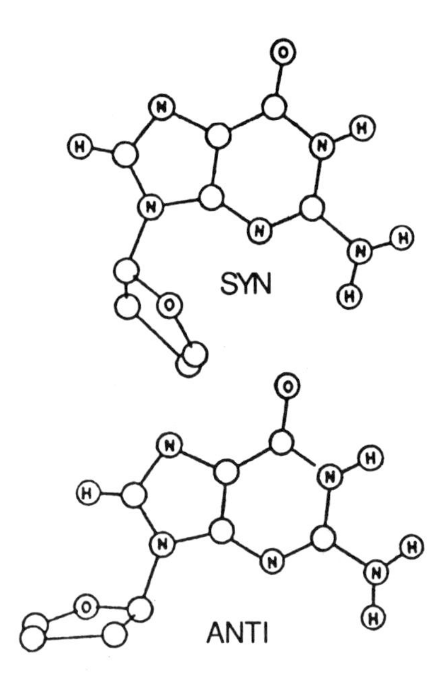

FIGURE 1.5. Guanine with an attached pentose ring in the *syn* (top) and *anti* (bottom) conformations. The interactions between the sugar and the base in these two cases include different atoms and should be energetically inequivalent. The *anti* conformation is preferable in most polynucleotides.

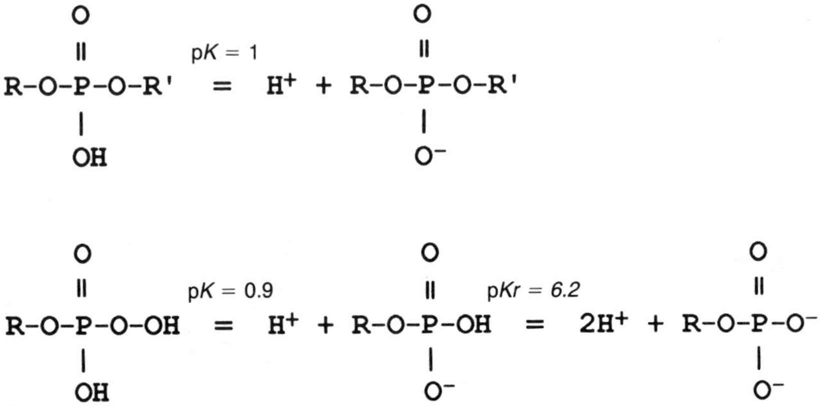

$$\begin{array}{c} \mathrm{O} \\ \| \\ \mathrm{R{-}O{-}P{-}O{-}R'} \\ | \\ \mathrm{OH} \end{array} \quad \overset{pK\,=\,1}{=} \quad \mathrm{H^{+}} + \begin{array}{c} \mathrm{O} \\ \| \\ \mathrm{R{-}O{-}P{-}O{-}R'} \\ | \\ \mathrm{O^{-}} \end{array}$$

$$\begin{array}{c} \mathrm{O} \\ \| \\ \mathrm{R{-}O{-}P{-}O{-}OH} \\ | \\ \mathrm{OH} \end{array} \quad \overset{pK\,=\,0.9}{=} \quad \mathrm{H^{+}} + \begin{array}{c} \mathrm{O} \\ \| \\ \mathrm{R{-}O{-}P{-}OH} \\ | \\ \mathrm{O^{-}} \end{array} \quad \overset{pKr\,=\,6.2}{=} \quad \mathrm{2H^{+}} + \begin{array}{c} \mathrm{O} \\ \| \\ \mathrm{R{-}O{-}P{-}O^{-}} \\ | \\ \mathrm{O^{-}} \end{array}$$

FIGURE 1.6. Protonation of a phosphate group when it participates in the formation of the phosphodiester bond (top) and when it is at the polynucleotide end (bottom). At the physiologically important pH of around 7, the presence of one negative charge per nucleotide creates the repulsive forces between different polynucleotide molecules.

therefore adenine bases associate to minimize base exposure in a polar medium. The stacked bases form an ordered structure whose stability depends on the base protonation state, which makes the exposure of bases in the medium more or less favorable depending on whether the base is in a charged or a neutral state (Steiner and Beers, 1959).

Base Pairs and Double-Stranded Nucleic Acids

Information transfer from DNA to protein requires several kinds of RNA molecules. The process starts with the reading of the DNA sequence by RNA polymerase, which synthesizes a single-stranded messenger RNA (mRNA). The mRNA molecule is then used as a template for protein synthesis by a ribonucleoprotein complex, called the ribosome, which also contains single-stranded ribosomal RNA (rRNA). Another class of essential molecules for this step is the transfer RNAs (tRNAs), which deliver amino acids to the ribosome in order to elongate the protein chain. The molecules of tRNAs are not single-stranded. To minimize the free energy of the molecule, the polynucleotide chain folds into the form shown in Figure 1.7. Some distant parts of the tRNA molecule are brought into close proximity, and the resulting structure is stabilized by the hydrogen bonding and base stacking interaction described below. Each tRNA molecule is responsible for the transport of one specific amino acid to the ribosome. There are three major classes of tRNAs according to the structure of the loops and the presence of some specific modified bases, yet all of them have a similar spatial structure — the "cloverleaf structure." These spatial structures are formed so as to maximize the number of stabilizing pairing and stacking interactions between bases. Therefore, tRNA molecules contain many double-stranded regions. Many viruses (for example, retroviruses) carry genetic information (i.e., information about the structure of viral proteins) in their RNA molecules. These RNAs are single-stranded but can be packed in viral capsids forming double-stranded regions. During propagation of these viruses in the host cells, their single-stranded RNA is duplicated and forms the so-called replicated form (RF), which is double stranded.

In nature, double-stranded nucleic acids (both double-stranded DNAs and RNAs, and DNA–RNA hybrids) occur more often than single-stranded nucleic acids. The formation of double-stranded structures is based on the presence of hydrogen bond donors and acceptors in nucleobases (Figure 1.8). Table 1.1 lists some biologically important hydrogen bonds (Watson et al., 1989). The stronger (and shorter) bonds have greater charge differences between the donor and acceptor atoms. In the absence of surrounding water molecules, the hydrogen bond energies are 3 to 7 kcal/mol. Figure 1.9 shows two complementary base pairs of DNA (similar pairs are formed in the double-stranded RNA, except that thymine is replaced by uracil). These base pairs were first

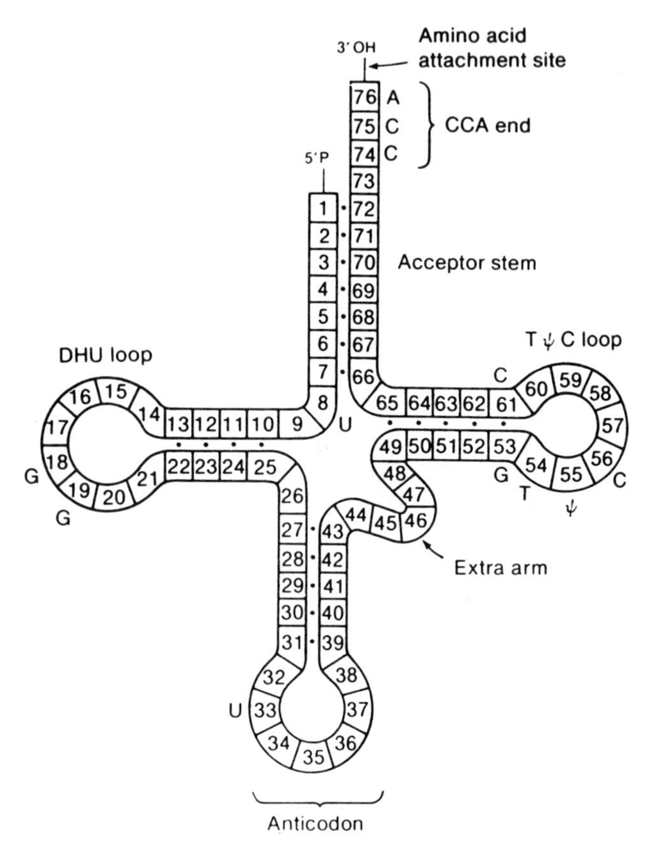

FIGURE 1.7. The generalized cloverleaf structure of tRNA. Several loops are connected to one another by double-stranded stems formed via base pairs (indicated by dots). The bases present in almost all different tRNAs are explicitly indicated. tRNA contains some unusual bases, of which two, dihydrouracil (DHU) and pseudouridine (ψ), are shown. The interactions of the acceptor stem and the TψC loop, and the DHU and anticodon loop result in a more complex L-shaped three-dimensional configuration of tRNA. Reproduced from Freifelder: Molecular Biology. © 1987 Boston: Jones and Bartlett Publishers. Reprinted with permission.

proposed by Watson and Crick (1953) to explain the available data on DNA structure in terms of a double-helical structure which is now known as the B form.

The stability of double-stranded nucleic acids depends on their nucleotide composition and sequence. Note that the GC base pair contains three hydrogen bonds, whereas the AT base pair has only two. It is intuitively clear and has been confirmed experimentally (see below) that it is more difficult to dissociate the base pair with three bonds than the

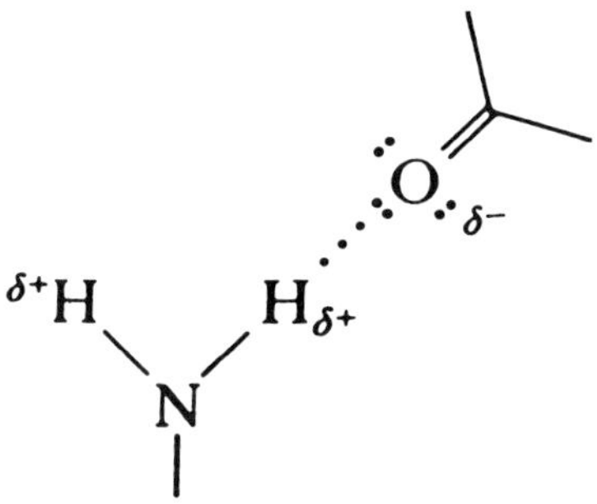

FIGURE 1.8. Schematic of the hydrogen bonds in which two electronegative atoms (N and O) partly share a proton of the NH_2 group. In this scheme the nitrogen atom is a donor and the oxygen atom is an acceptor of the hydrogen bond.

pair with two bonds. Interactions of the bases stacked over one another also contribute to the overall stability of the double-stranded structure. The Watson–Crick base pairs are isomorphous, that is, both of them have the same C1′–C1′ distance and the same N9–C1′–C1′ and N1–C1′–C1′ angles. This allows the formation of a smooth regular double helix without local distortions. The glycoside bonds connecting the bases to the sugar–phosphate backbones are in the *cis* configuration relative to each other. This results in fewer functional groups on one side of the base pairs (bottom in Figure 1.9) than on the other (top) side.

The double-stranded structure of nucleic acid contains two helical polynucleotide strands wound around each other (double helix) and kept together by hydrogen bonds between the bases of opposite strands. The two strands in the double helix are antiparallel: their 5′,3′-phosphodiester bonds run in opposite directions. If inverted 180°, the double helix looks the same because of the strand complementarity. Figure 1.10 depicts the model of the right-handed double helix in the B conformation, where the sugar–phosphate backbones are shown as two ribbons and the base pairs as horizontal lines connecting these ribbons. Two grooves of unequal width are clearly seen. The top sides of the base pairs, shown in Figure 1.9, are exposed in the major groove, whereas the lower parts are exposed

TABLE 1.1. Biologically important hydrogen bonds.

Bond	Approximate length (Å)
O–H ... O	2.70 ± 0.10
O–H ... O^-	2.63 ± 0.10
O–H ... N	2.88 ± 0.13
N–H ... O	3.04 ± 0.13
N^+–H ... O	2.93 ± 0.10
N–H ... N	3.10 ± 0.13

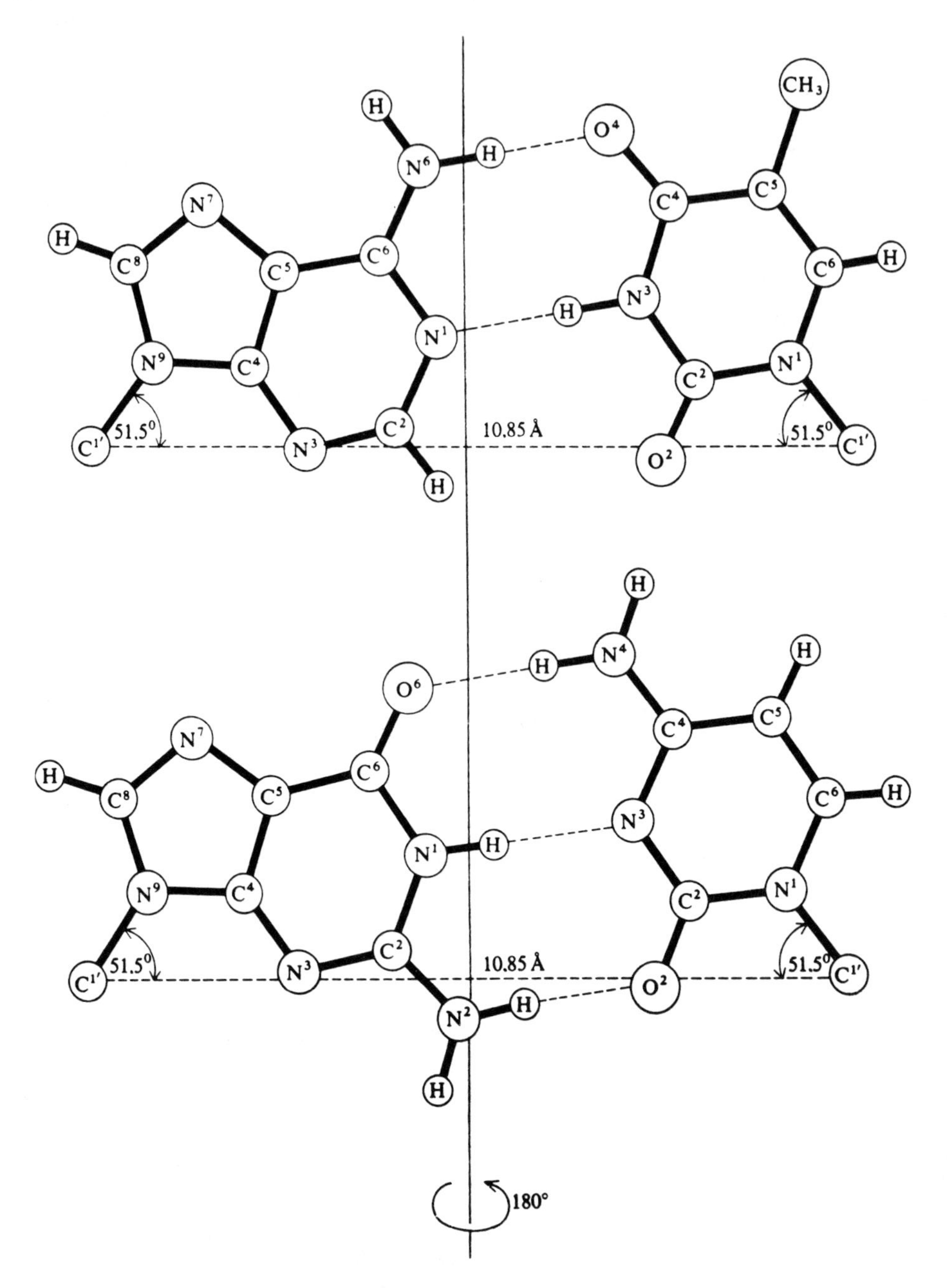

FIGURE 1.9. The Watson–Crick base pairs AT (top) and GC (bottom). The pseudo symmetry axis C2 is shown as a vertical line. The *cis* configuration of the glycoside bonds of the bases in the pair results in the formation of two grooves of unequal width when the polynucleotide strands wind around each other to form a double helix, as shown in Figure 1.10. A larger amount of the functional groups is exposed in the major groove than in the minor groove. From: *Biophysical Chemistry* Vol I by Cantor and Schimmel. Copyright © 1980 by W.H. Freeman and Company. Used with permission.

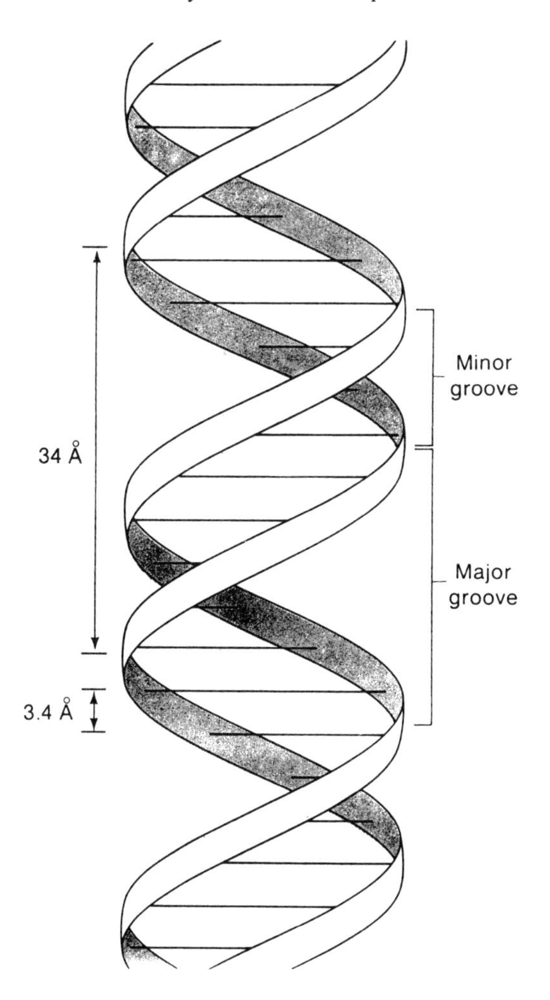

FIGURE 1.10. The ribbon model of B form DNA. The two ribbons represent the sugar–phosphate backbones and the horizontal lines the base pairs. The helix is right-handed and, according to X-ray data from DNA fibers, has 10 base pairs per helix turn and ca. 3.4 Å rise per base pair. The diameter of the helix is ca. 20 Å.

in the minor groove. The phosphate groups are readily accessible to interactions with water and dissolved ions, the sugar rings are less accessible, and the bases are least exposed to the medium. Yet, certain groups in the bases are accessible enough to provide the basis for the interactions with small molecules, proteins, and other polynucleotide chains. Such interaction is crucial for triple-stranded structures.

Negatively charged oxygens of the phosphate groups attract positively charged cations from the surrounding medium, which diminishes the repulsion of the two similarly charged polynucleotide backbones and, therefore, stabilizes the entire double helix. The cationic stabilization of double-stranded polynucleotides originates from two major contributions. First, the cations form positively charged clouds around negatively charged phosphates. As a result, an effective Debye–Hückel screening of one phosphate group from another occurs. This phenomenon manifests itself in an increase of the melting temperature (i.e., the temperature necessary to dissociate two strands of the duplex from one another). The effect depends on salt concentration (Figure 1.11) (Schildkraut and Lifson, 1965; Frank-Kamenetskii, 1971). Second, some cations bind directly to the phosphates. This phenomenon was revealed in the experiments on the determination of the molecular weight of specific DNA molecules in NaCl and CsCl solutions. The molecular weight of DNA in CsCl solution is higher than in NaCl solution, in good quantitative agreement with a suggestion that every phosphate group carries one tightly bound light (Na^+) or heavy (Cs^+) ion (Freifelder, 1987).

Another helix-stabilizing contribution comes from the stacking interactions of the adjacent bases in polynucleotide chains. The melting temperature of the duplex decreases in the presence of the agents that

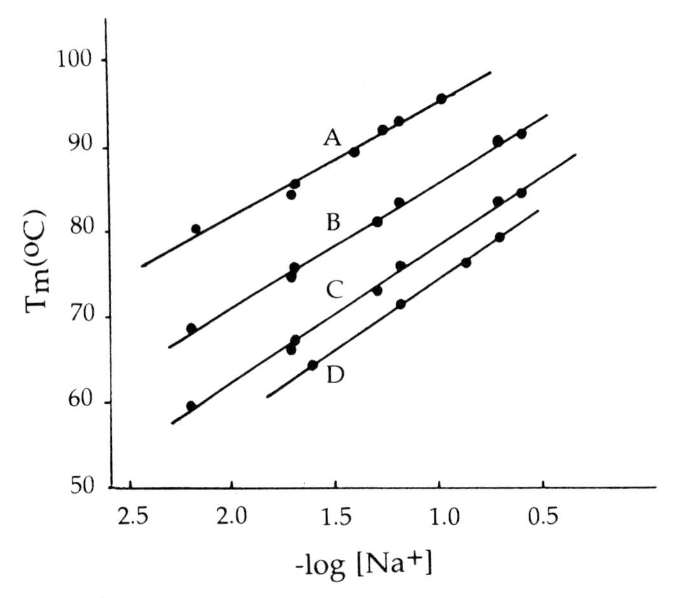

FIGURE 1.11. Dependence of DNA melting temperature (T_m) on the concentration of sodium ions in the medium. In accordance with the higher number of hydrogen bonds in the GC pair than in the AT pair, the four lines also show the dependence of T_m on the percentage of GC pairs: (A) %GC = 72; (B) %GC = 50; (C) %GC = 33; (D) %GC = 24. Adapted from Frank-Kamenetskii (1971) *Biopolymers* 10, 2623–2624. Copyright © 1971 John Wiley and Sons, Inc. Reprinted with permission.

reduce the hydrophobic interactions of the bases. For example, methanol increases the solubility of the bases (both free and involved in DNA helices) by enhancing the interaction of weakly soluble adenines and guanines with the surrounding medium (Freifelder, 1987).

The thermal dissociation of two DNA strands depends on the proportion of GC pairs, as shown in Figure 1.11. A reason for this is the stabilization of the GC pair by three hydrogen bonds compared with two hydrogen bonds in the AT pair. Under otherwise identical conditions, there is a linear dependence of the melting temperature T_m on the percentage of GC pairs in DNA molecules.

There are several structural families of double-stranded nucleic acids (see Dickerson, 1992, for review). The key structural features of double-stranded nucleic acids are listed in Table 1.2 (the structural parameters are defined in Figures 1.4, 1.5, 1.10, and 1.12). The B form is believed to be the predominant state of DNA in living organisms. Upon dehydration, the DNA structure can undergo several transitions, depending on the ambient conditions. When Na^+ is used as a counterion, DNA at low humidity in the fiber or at high concentrations of ethanol or specific salts in solution adopts an A conformation which is underwound relative to the B form (Brahms and Mommaerts, 1964; Fuller et al., 1965; Ivanov et al., 1974; Ivanov and Krylov, 1992; Xu et al., 1993). This B to A form transition of DNA is also accompanied by a change in the sugar conformation from C2′-endo (C3′-exo) to C3′-endo. When Li^+ is used as a counterion, DNA at low humidity becomes more tightly wound (relative to the B form), resulting in the C form (Langridge et al., 1961). This form is also believed to occur in DNA solutions containing up to 6 *M* concentrations of monovalent cations or high concentrations of methanol (Ivanov et al., 1973). A structure different from those of the A and C forms was observed in poly (dA) · poly(dT) fibers (Alexeev et al., 1987; Nelson et al., 1987). This structure (called the B′ form) is more rigid than

TABLE 1.2. Comparison of polynucleotide double helices (X-ray data).

	RNA-11	A DNA	B DNA	B′ DNA	C DNA[a]	Z DNA[b]
Helix sense	Right-handed	Right-handed	Right-handed	Right-handed	Right-handed	Left-handed
Base pairs per turn	11	11	10	10	9.3	12
Helix twist (degrees)	32.7	32.7	36.0	34.1; 36.8	38.7	−10; −50
Rise per base pair (Å)	2.64	2.90	3.36	3.5; 3.3	3.32	3.70
Base pair inclination (degrees)	—	13	0	0	6	−7
Glycosidic orientation	*anti*	*anti*	*anti*	*anti*	*anti*	*anti, syn*
Sugar conformation	C3′-endo	C3′-endo	C2′-endo	C2′-endo	C2′-endo	C2′-endo, C3′-endo

[a] B′ DNA values are for conformational states I and II.
[b] Two values for Z DNA correspond to CpG and GpC steps.

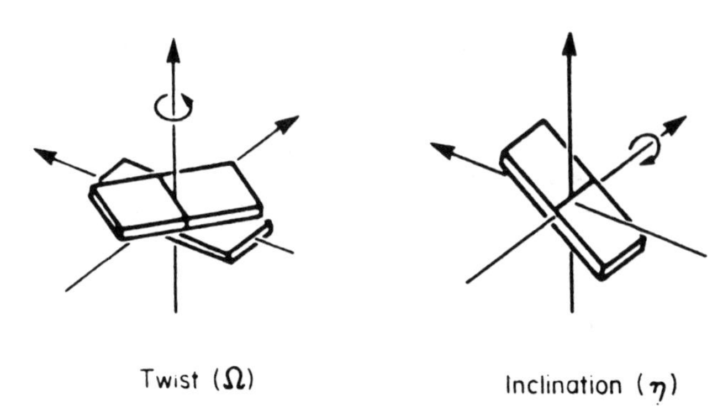

FIGURE 1.12. Definitions of some local rotational helix parameters. The twist angle is the angle of rotation of the base pair relative to the previous base pair in a winding double helix. The inclination angle is the angle between the plane of the base pair and the double helix axis. From Dickerson et al. (1992). Reprinted with permission.

the B form and has an intrinsic curvature of the helix axis (see Crothers et al., 1990 and Hagerman, 1990, for reviews).

Contrary to right-handed polynucleotide and DNA helices, poly-[d(GC)] · poly[d(GC)] can form a left-handed helix (Z form), which is favored by dehydration, polynucleotide-bound multivalent cations, or the presence of some modified nucleobases (e.g., methylated cytosines or brominated guanines) in polynucleotides (Pohl and Jovin, 1972; Wang et al., 1979; Drew et al., 1980; Rich et al., 1984; Harder and Johnson, 1990; Loprete and Hartman, 1993). In left-handed Z DNA, the orientations of glycosidic bonds alternate between *syn* (for purines) and *anti* (for pyrimidines), and corresponding sugar conformations also alternate between C3′-endo and C2′-endo. A comparison of the B and Z DNA (Figure 1.13) reveals major differences between these forms, including the zigzag shape of the sugar-phosphate backbone in the Z form. After the initial discovery of the Z form in alternating poly[d(GC)] sequences, this structure was found also in alternating poly[d(GT)] · poly[d(CA)] sequences and even in nonalternating sequences of purines and pyrimidines (Feigon et al., 1985; Wang et al., 1985).

In contrast to the other members of the family of double-stranded DNA structures (A, B, Z families), all of which have an antiparallel strand orientation, a novel, parallel-stranded (ps) DNA has been recently discovered (Pattabiraman, 1986; Ramsing and Jovin, 1988; van de Sande et al., 1988; Shchyolkina et al., 1989; van Genderen et al., 1990; Rippe and Jovin, 1992). The ps double helices can be formed with the AT base pairs via reverse Watson–Crick hydrogen bonding (see Jovin et al., 1990, for review). They are also consistent with the symmetrical non-Watson–Crick (C^+C, GG, AA, and TT) homo base pairs (van Genderen et al.,

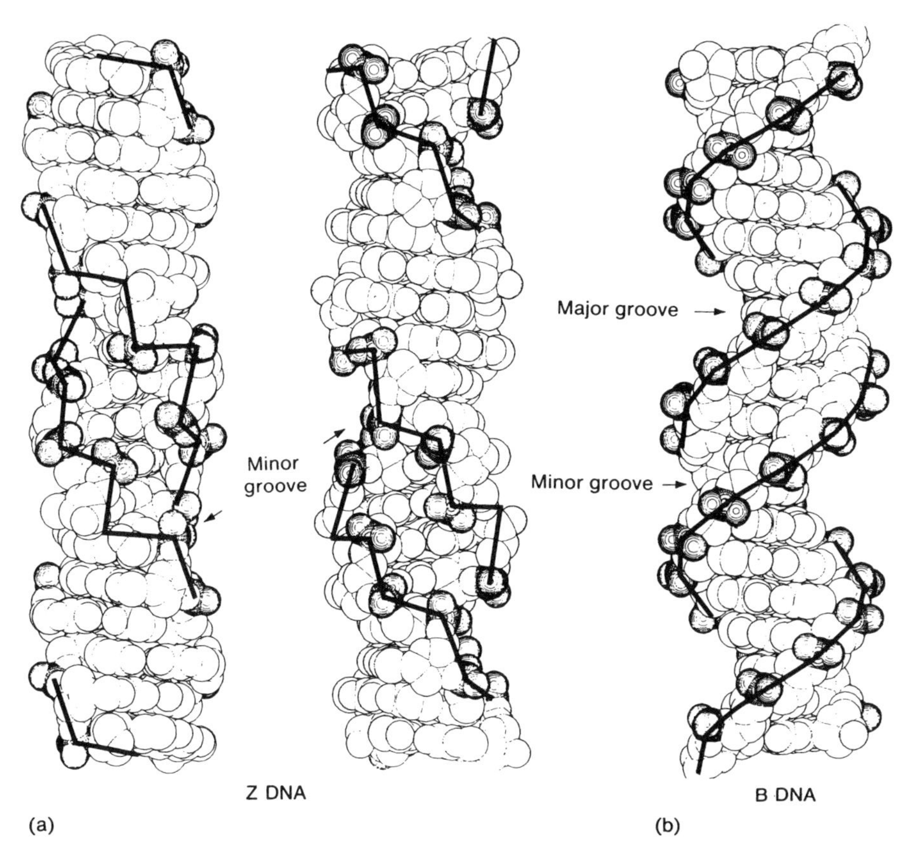

FIGURE 1.13. Comparison of (a) left-handed Z DNA and (b) right-handed B DNA. The heavy lines follow the sugar-phosphate backbone, which has a zigzag shape in Z DNA and is smooth in B DNA. The minor groove of Z DNA deeply penetrates the helix axis. In B DNA the grooves are shallow. The diameter of the Z helix is ca. 18 Å and that of the B helix is ca. 20 Å. Reprinted with permission from *Nature*, Wang et al. (1979). © 1979 Macmillan Magazines Limited.

1990; Rippe et al., 1992; Robinson et al., 1992) and Hoogsteen-type base pairs (Liu et al., 1993). Figure 1.14 shows possible base-pairing in the ps nucleic acids. The exact biological sense of ps nucleic acids is not known, however, such structures could arise during genetic DNA recombination or in highly folded single-stranded RNA molecules (Ramsing and Jovin, 1988).

The structural features of double-stranded nucleic acids listed in Table 1.2 are derived from X-ray analysis of polynucleotide fibers. To some extent, the interactions of separate duplexes in large associates (fibers) may influence the exact values of structural parameters. For example, according to X-ray data, there are 10 base pairs per turn of B DNA.

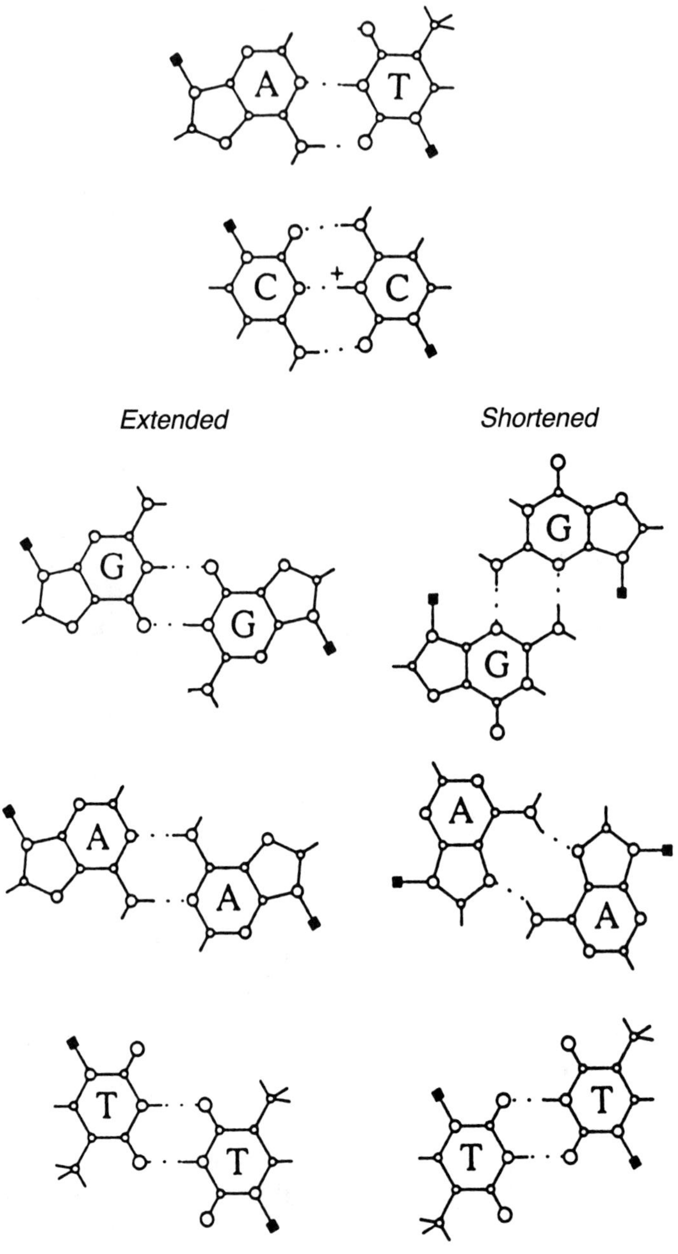

Extended

Shortened

FIGURE 1.14. Possible base-pairings in parallel-stranded nucleic acids. The top base pair is a reverse Watson–Crick one. The CC^+ base pair is hemiprotonated with the proton shared by the two N3 atoms of the cytosines. Two possible pairing schemes for each of the other bases (GG, AA, and TT) are shown. They differ in the C1′–C1′ distances, which are longer for base pairs on the left side than in those on the right side. The *trans* configuration of the glycosidic bonds results in two grooves of equal width. Reprinted with permission from Robinson et al. (1992). © 1992 American Chemical Society.

However, determination of the helical periodicity of DNA in solution using enzymatic cutting from only one side of the double helix gave 10.5 base pairs per turn (see, e.g., Rhodes and Klug, 1980).

X-ray studies of single crystals of synthetic oligonucleotides revealed a variation of local helix parameters from the mean values obtained from the fiber diffraction data (see Dickerson, 1992, for the latest review). For example, the mean value of the helical twist angle between base pairs is 36.1°, and the range extends from 24° to 51°. The mean rise per base pair is 3.36 Å, and the range is 2.5 to 4.4 Å. The heterogeneity of the local duplex conformation holds in solution. Thus, from the nuclear magnetic resonance (NMR) data, the purine and pyrimidine bases in B DNA have been determined in different conformations within the general C2′-endo type (Chuprina et al., 1993).

An important feature of the double-stranded nucleic acid is its dynamic structure in which the bases have significant mobility. For example, the equilibrium positions of adjacent pyrimidines in the B DNA correspond to their orientations 36° relative to each other (Table 1.2). However, they spend some time oriented so that their C5=C6 bonds are paralled, allowing the formation of cyclobutane pyrimidine photodimers (Kochetkov and Budovskii, 1972; Wang, 1976; Cadet and Vigny, 1990). The fluctuating nature of the double helix is also manifested in a transient exposure of groups participating in the interstrand hydrogen bonds. This explains the hydrogen exchange between the hydrogen-bonded protons of the bases and the tritium ions of tritiated water, or the reactivity of formaldehyde with the hydrogen-bonded amino groups of bases (Printz and von Hippel, 1965; Lukashin et al., 1976).

Some Methods of Investigating Nucleic Acids

A large number of physical, chemical, and biochemical methods have been developed for the study and genetic manipulation of nucleic acids. Many physical methods employ the electronic properties of the bases and their changes during the formation of ordered polynucleotides (Cantor and Schimmel, 1980a). Figure 1.15 shows the molar absorption curves for the major nucleoside-monophosphates and a typical UV-absorption spectrum of double-stranded DNA. The measurement of UV absorption is the simplest and most convenient method to determine concentrations of nucleic acids, whereas the temperature dependence of UV absorption allows the determination of the melting temperature of multistranded nucleic acids (see Chapter 2, for more details). Some other techniques, namely, circular dichroism and infrared and NMR spectroscopy, provide information on the type of helical structure, the groups involved in hydrogen bonding, or even detailed structures of short nucleic acid fragments. X-ray studies of polynucleotide fibers have provided information on a variety of helical forms and their structural characteristics. X-ray

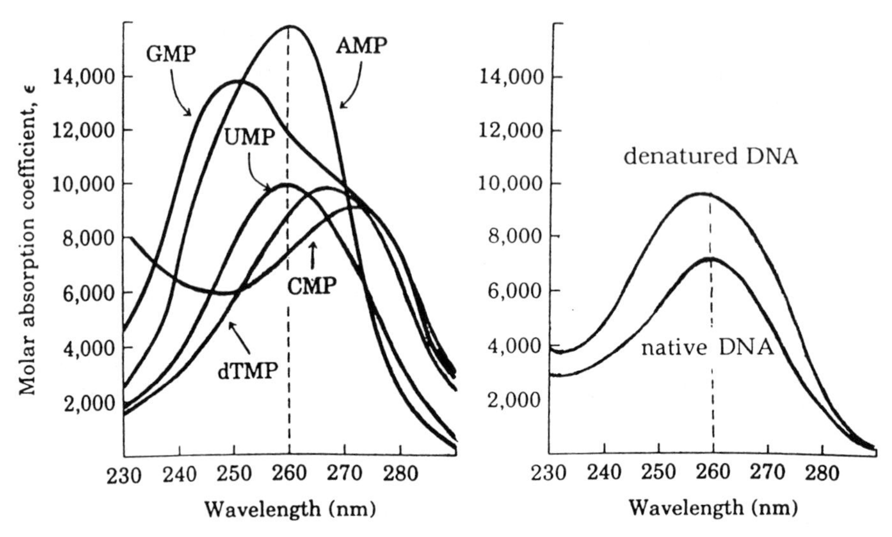

FIGURE 1.15. Left: The molar absorption curves for the major mononucleotides. The absorption maxima of nucleotides are between 248 nm (GMP) and 276 nm (CMP). For nucleic acids with approximately the same amount of AT and GC pairs (50% GC), the absorption maximum is close to 260 nm (vertical dashed line). Right: In the double-stranded form of DNA, the molar absorption coefficient ε at the maximum wavelength is ca. 6,600 L/cm mol nucleotide. If the strands are separated (denatured DNA), the DNA absorption rises ca. 1.4 times.

studies of the crystalline samples of oligonucleotide duplexes have allowed the determination of exact helical parameters and even structural heterogeneities that depend on the specific nucleotide sequence of DNA.

Many chemical methods employ the reactivities of different groups in nucleic acids to specific agents. The best-known examples are reactions with hydrazine, dimethylsulfate, and piperidine, which are used in the Maxam–Gilbert sequencing technique (Maxam and Gilbert, 1980). Some other chemicals differ in reactivity to various structural forms of nucleic acids and have been used to determine locally denatured structures, stretches of left-handed DNA inside largely right-handed molecules, etc. (see Lilley, 1992, for a short review; see Chapter 2, for more details).

Many nucleic acid-specific enzymes are appropriate for DNA structural studies. Some of them are listed in Table 1.3. DNA polymerases are used in the Sanger dideoxy-sequencing technique (Sanger et al., 1977), in a selective enrichment of nucleotide sequences via the polymerase chain reaction (Mullis et al., 1986). The restriction endonucleases are widely used to produce a site-specific fragmentation of DNA, whereas the ligases can covalently link such fragments (Sambrook et al., 1989). Single-strand-specific nucleases recognize unpaired regions in DNA and are very useful

TABLE 1.3. Some specific enzymes used in nucleic acid studies.

Enzyme	Activity	Type of application
DNA polymerases (Klenow fragment, Taq, Sequenase, etc.)	5′→3′ DNA synthesis	3′ end DNA labeling Determination of modified template sites via primer extension, sequencing, etc.
RNA polymerase	RNA synthesis on DNA template	Gene expression as a function of template structure, ambient conditions, etc.
Topoisomerase	Isomerization of one topological version of DNA into another	Preparation of circular DNA with various superhelical densities
Deoxyribonuclease (DNase) etc.	Digestion of double-stranded DNA	Periodicity of duplex DNA in solution, DNA sites protected by bound ligand
Ribonucleases	Digestion of double- and single-stranded RNA	RNA complexes with proteins and other ligands
S1, P1, mung bean nucleases	Digestion of single-stranded DNA	Determination of locally unwound helix
Restriction endonuclease	Single-point cut at a specific sequence	Preparation of predetermined DNA fragments for sequencing or cloning
DNA ligase	Joining of pieces of phosphodiester backbone	Cloning of DNA fragments into vector plasmid, etc.
Polynucleotide kinase	5′ DNA phosphorylation	5′ end DNA labeling

in the determination of sequence-specific locally unwound structures (Vogt, 1980; Laskowski, 1980, for reviews). Deoxyribonuclease I (DNase I) digests double-stranded DNA, providing a probe of structure and a probe of accessible sites in duplex DNA and those protected by the bound ligand (Rhodes and Klug, 1980; Gross and Garrard, 1988; Tullius, 1989).

Circular and Superhelical DNA

Although the basic structure of DNA is the B form, it may be altered as a function of the local base sequence, ionic strength, temperature, pH, and topology. The first structurally different form of DNA was described by R. Sinsheimer, when he found that ϕX 174 bacteriophage DNA exists as a closed circular form (Fiers and Sinsheimer, 1962). Later, J. Vinograd detected DNA supercoiling in circular viral DNAs (Vinograd et al., 1965). Vinograd and colleagues came to the insight that most natural DNA species potentially exist as circular molecules (Vinograd et al., 1965). The relatively small plasmid DNA molecules are the simplest examples of circular DNA molecules. They are covalently closed circles consisting of unbroken complementary single strands (Figure 1.16, top).

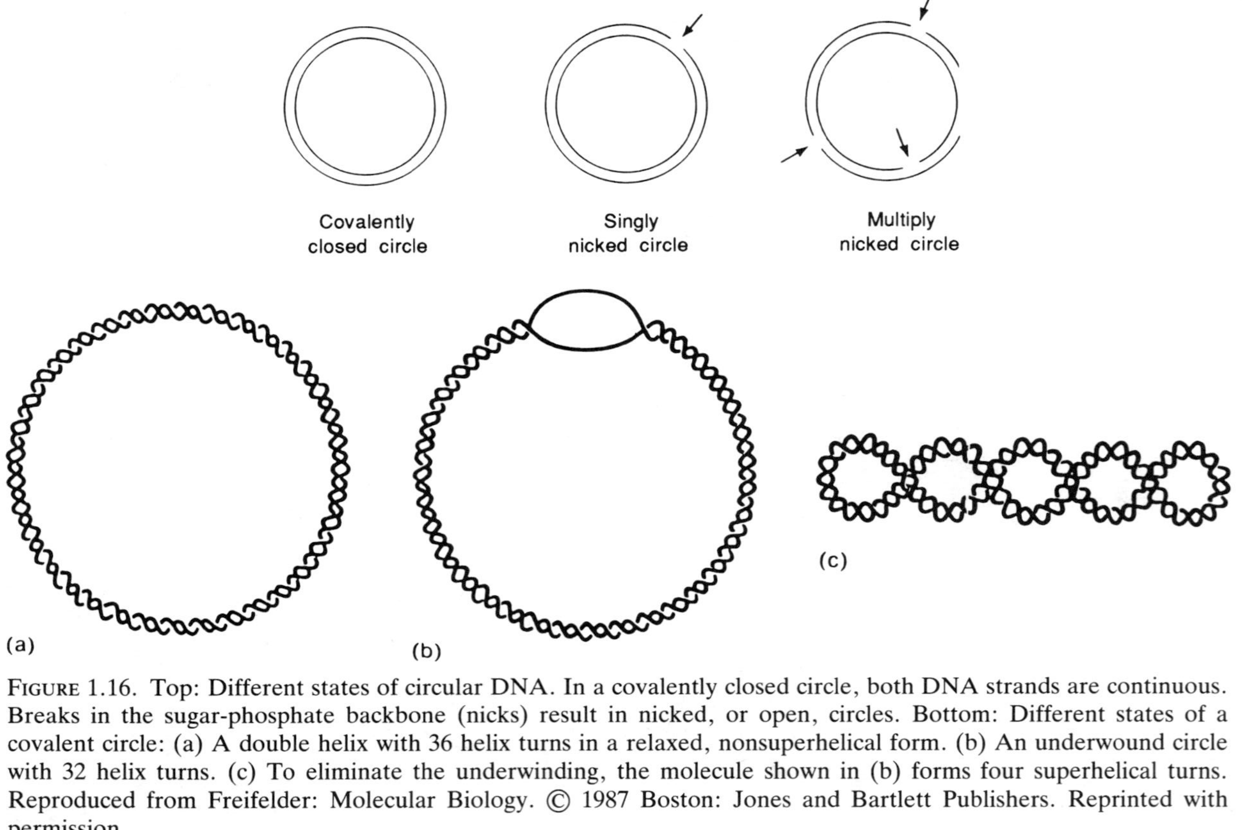

FIGURE 1.16. Top: Different states of circular DNA. In a covalently closed circle, both DNA strands are continuous. Breaks in the sugar-phosphate backbone (nicks) result in nicked, or open, circles. Bottom: Different states of a covalent circle: (a) A double helix with 36 helix turns in a relaxed, nonsuperhelical form. (b) An underwound circle with 32 helix turns. (c) To eliminate the underwinding, the molecule shown in (b) forms four superhelical turns. Reproduced from Freifelder: Molecular Biology. © 1987 Boston: Jones and Bartlett Publishers. Reprinted with permission.

If the circular molecule has one or more strand breaks, it is a nicked circle. Usually, the covalently closed circles are twisted (Figure 1.16, bottom) into a form called supercoiled DNA. The supercoiled and nicked DNA molecules differ in compactness (package density) and can be separated by centrifugation in a density gradient, or by electrophoresis in an agarose gel. Much longer chromosomal DNA is folded into compact circular structures by proteins and is thereby segregated into independent topological domains which can functionally act as separate circular molecules (Stonington and Pettijohn, 1971; Worcel and Burgi, 1972; Pettijohn and Hecht, 1973; Sinden and Pettijohn, 1981).

The most fundamental property of supercoiled DNA is its constant linking number (Lk), which is the number of times the strands wind about each other. This value can be expressed as the sum of two geometrical characteristics: Tw (twist), the number of times one strand rotates around the DNA duplex axis, and Wr (writhe), a measure of global deformation of the duplex axis (see Cozzarelli et al., 1990 and Vologodskii, 1992, for reviews):

$$\mathrm{Lk} = \mathrm{Tw} + \mathrm{Wr}$$

It is important that Lk cannot be changed without breaking one strand of the circular DNA molecule, rotating one strand around the other, and rejoining the strand at the breakpoint. Since DNA in solution tends to retain its B form with 10.5 base pairs per helical turn, any changes in Tw should be compensated for by changes in the tertiary structure. For example, these changes can be realized in winding of the duplex axis (formation of superhelical turns, Figure 1.16, bottom). The closed circular DNA molecules are characterized by the linking difference, often referred to as the number of superhelical turns,

$$\Delta\mathrm{Lk} = \mathrm{Lk} - \mathrm{Lk}_o$$

where Lk_o is the linking number of a relaxed circular DNA. For the overwhelming majority of natural DNA molecules, $\mathrm{Lk} < \mathrm{Lk}_o$. In a relaxed form, such DNA molecules would be underwound relative to 10.5 base pairs per helical turn. By convention, such DNAs are called negatively supercoiled, and their superhelices are right handed. The length-independent measure of superhelicity, superhelical density, $\sigma = \Delta\mathrm{Lk}/\mathrm{Lk}_o$, for isolated plasmids and viral DNAs is typically about -0.05. Because the population of DNA molecules (topoisomers) is distributed around the most probable topoisomer, this is an average value. When the absolute value of the superhelical density increases, the DNA molecule becomes more structurally strained. Experimentally, sets of topoisomers with different average superhelical densities are generated by incubating DNA with a special enzyme, topoisomerase (see Gellert, 1981 and Wang, 1985, for reviews) in the presence of various concentrations of ethidium bromide.

Negative supercoiling can be generally subdivided into two forms, restrained and unrestrained (Sinden, 1994). In an isolated protein-free circular DNA, or in DNA which has a toroidal coil that is not stably wrapped around a protein, the supercoil is unrestrained. In this case, the torsional stress propagates over the entire DNA molecule. Upon introduction of a nick, the negative supercoils are relaxed by the rotation of one strand relative to the other. In a DNA molecule, where the stable coiling is created, for example, by wrapping around the protein, the supercoil is said to be restrained. No unwinding results from the introduction of the nick into this restrained supercoil. Both restrained and unrestrained forms contribute to the supercoiling of long natural DNA.

Torsional tension can be partially relaxed if some kind of locally unwound conformation is produced (Wells and Harvey, 1988). Therefore, DNA supercoiling can promote a number of unusual DNA structures that are sequence- and condition-specific. For eukaryotic DNA molecules, the probability of having relatively long sequences appropriate for one or another nonstandard conformation is sufficiently high. Gel electrophoresis is the simplest means of detecting such structures, since a local change in twist is accompanied by a compensatory change in writhe. The altered shape of the molecule results in change in the electrophoretic mobility (see Bowater et al., 1992, for review, and Chapter 2, for more details). Several local unusual DNA structures have been classified that can be promoted by a superhelical stress or by a specific sequence (Palecek, 1991; Yagil, 1991.) For the latest detailed discussion of non-B DNA structures and their possible biological significance, see Sinden (1994).

Unwound DNA

Separation of the DNA into single strands (Figure 1.17a) requires a large input of free energy, which can come from supercoiling, from interaction with specific proteins (e.g., single-strand-specific proteins) or RNA, and so forth. Experimental data allow the conclusion that only a very small part of an arbitrary sequence could be found in the single-stranded form, the most appropriate sequence being an AT-rich one.

DNA Slippage

The slippage structure (Figure 1.17b) has been implied to form in the DNA regions containing direct sequence repeats. The experimental data

→

FIGURE 1.17. Unusual DNA structures that can form locally inside a mostly B-form molecule. The symmetry and sequence motif requirements are indicated. Reprinted with permission from Yagil (1991), © CRC Press, Boca Raton, Florida, and reprinted with permission from *Nature*, Lyamichev et al. (1989a), and Sen and Gilbert (1989) © 1989 Macmillan Magazines Limited.

a. SINGLE STRANDS

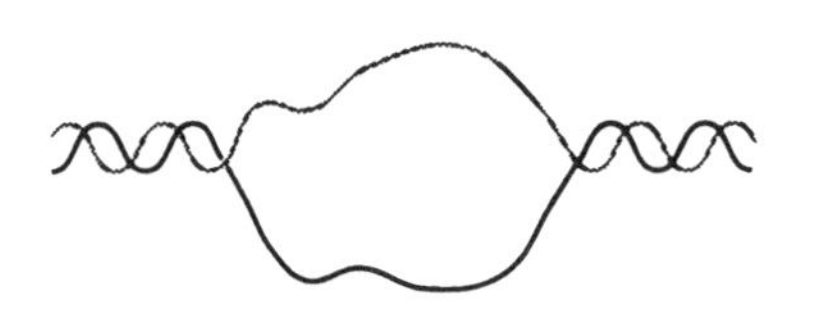

Symmetry: None Motif: A.T rich

b. SLIPPAGE STUCTURE

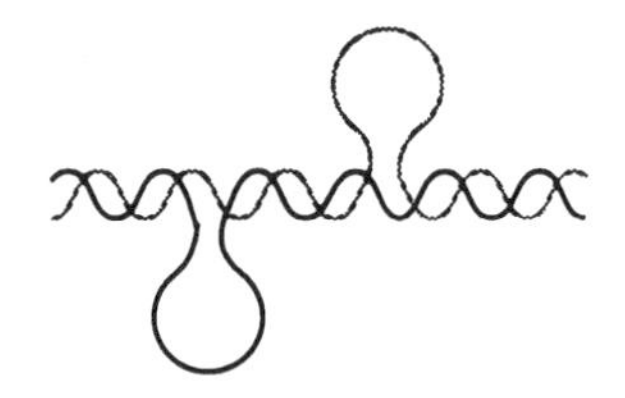

Symmetry: Repeat Motif: R.Y

c. CRUCIFORM

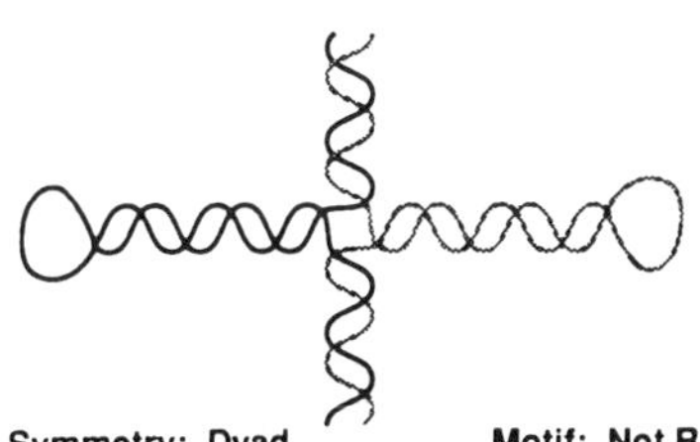

Symmetry: Dyad Motif: Not R.Y

d. PARANEMIC DUPLEX

Symmetry: None Motif: Any

e. ALTERNATING RIGHT- LEFT HELIX

-B- Junction -Z-

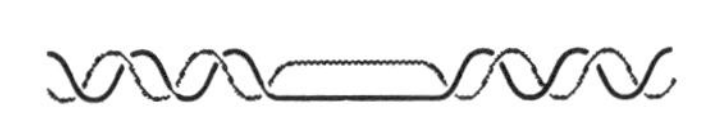

Symmetry: None Motif: G-C

f. FORM V

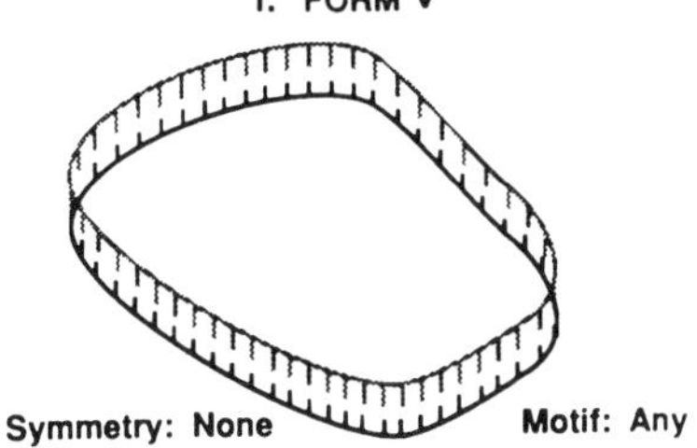

Symmetry: None Motif: Any

g. D - LOOP

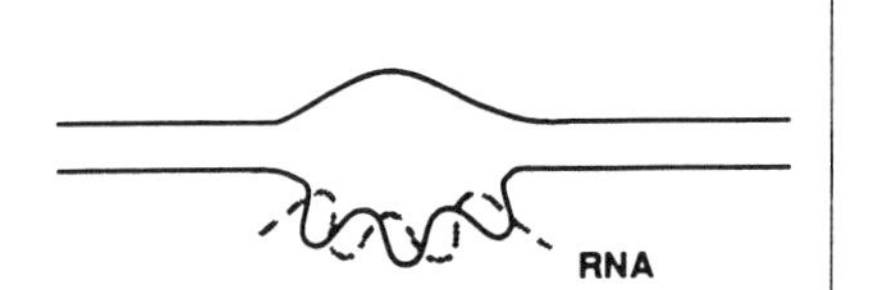

Symmetry: None Motif: A.T rich

h. C,A HAIRPIN

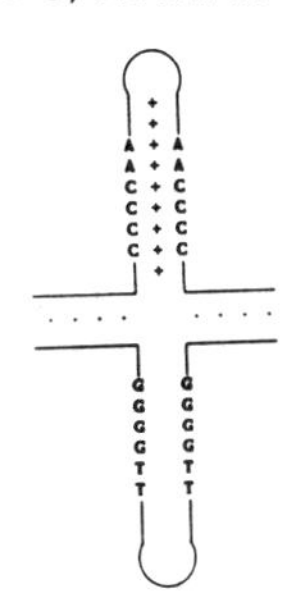

i. QUADRUPLEX

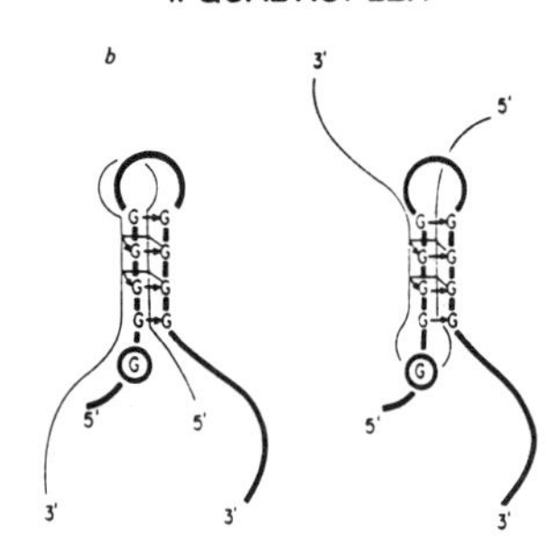

j. TRIPLEX

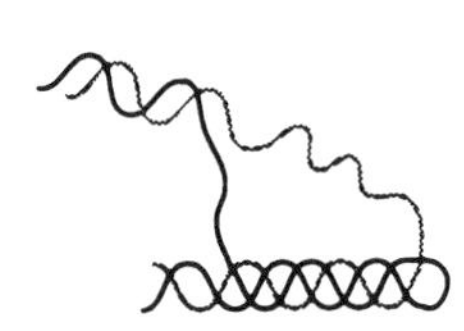

Symmetry: Mirror Motif: R.Y

on single-strand-specific nuclease sensitivity and chemical probing (see Chapter 2) of regular homopurine-homopyrimidine sequences were interpreted in the framework of the slippage model; however, these data were not convincing. The formation of DNA structures with slipped loops in both strands of short direct repeats was recently shown by NMR experiments (Ulyanov et al., 1994). The slippage structures may be involved in the development of some genetic diseases resulting from an expansion of simple trinucleotide repeats (Sinden and Wells, 1992).

Cruciforms and Four-Way Junctions

The cruciform (Figure 1.17c) can be formed in the palindrome (inverted repeat) DNA sequences of the general form

A B C D E E′D′C′B′A′
A′B′C′D′E′E D C B A

in which A and A′ are the complementary bases capable of forming a pair. The two inverted sequences may be separated by a spacer. A cruciform structure is extruded when both polynucleotide strands fold in the middle of an inverted repeat and form two base-paired stems and two single-stranded loops. Thus, excessive supercoil tension in the covalently closed DNA molecule can be partially relieved by the formation of two hairpins protruding more or less perpendicular to the axis of the main helix (see Lilley, 1980; Panayotatos and Wells, 1981; Murchie et al., 1992, for review). The free energy of the cruciform is lower than that of single strands or slippage structures, because it has unpaired bases only at the hairpin loop and at the hairpin–regular duplex junction. These regions are sensitive to the single-strand-specific nucleases. A related structure, the four-way junction, can be considered an equivalent of the Holliday junction, an important intermediate in genetic recombination. It is normally completely base-paired and X-shaped. The structure of the fourway junction is elucidated using four long oligonucleotides (see Duckett et al., 1990, for review).

Paranemic Duplexes

In some cases the data on chemical probing and susceptibility of DNA regions to the single-strand-specific nucleases can be explained by the presence of paranemic duplexes (Figure 1.17d) which are completely unwound but retain the hydrogen bonds and base stacking (Yagil, 1991).

Z-Forms

After the discovery of the left-handed Z form in the crystals of short alternating oligonucleotides $d(GC)_n$, such fragments were cloned in plasmid DNA to test the possibility that an excessive superhelical stress would force a local Z form (Figure 1.17e). These attempts were indeed successful (Klysik et al., 1981; Peck et al., 1982). Moreover, it was found that some other sequences can be driven into a left-handed duplex conformation. The Z form has no true single-stranded regions. However, the junctions between the B and the Z structures can be probed by the chemical agents and nucleases specific to single strands (Johnston, 1992a).

Form V DNA

Form V DNA was discovered during the hybridization of two covalently closed complementary single strands of the bacteriophage ϕX174 (Stettler et al., 1979). The resulting structure seen in the electron microscope was more than 90% double-stranded (Figure 1.17f). It was suggested that form V DNA consists of alternating left- and right-handed segments, and, as a consequence, has no net winding and has a zero linking number (Brahms et al., 1983).

D Loops and R Loops

The locally unwound state of DNA can be stabilized by the interaction of the single-stranded region with the complementary DNA or RNA strand (Champoux and McConaughy, 1975; Richardson, 1975; Wiegand et al., 1977). During the formation of these so-called displacement loops (Figure 1.17g) the number of single strands does not change. Relatively little energy is required to form the displacement loop–regular helix junctions. The available data confirm the existence of displacement loops in a variety of experimental systems.

CA Hairpins

The hypersensitivity of the telomeric sequence $d(C_4A_2)_n \cdot d(T_2G_4)_n$ cloned in plasmid DNA to the single-strand-specific nucleases (Budarf and Blackburn, 1987) was explained by the formation of CA hairpins stabilized by CC^+ and AA^+ base pairs (Figure 1.17h) (Lyamichev et al., 1989).

Quadruplexes

The G-rich sequences of several telomeric sequences were shown to form self-associated four-stranded structures (quadruplexes) (Henderson et al., 1987). Two models have been suggested in which all guanine stretches run in either parallel orientation or alternating parallel-antiparallel orientation (Figure 1.17i) (Sen and Gilbert, 1990). The C-rich strands of telomeric sequences are also capable of forming four-stranded structures where two parallel-stranded CC^+ base-paired helices are "zipped together" in an antiparallel orientation (Gehring et al., 1993; Ahmed et al., 1994).

Triplexes

H DNA (Figure 1.17j), the most recently discovered structure, is promoted by a high level of supercoiling in a circular DNA. It is formed in the homopurine–homopyrimidine mirror repeats under specific conditions. A three-stranded helix may also be formed between the homopurine–homopyrimidine tract in duplex DNA and an external single strand of appropriate sequence. The properties of these three-stranded structures are the major subject of this book.

Physicochemical Studies of Model Triple-Stranded Structures

Although many of the recent publications on intra- and intermolecular DNA triplexes only occasionally cite earlier studies of triple-stranded structures, the fundamental basis for many modern developments with mixed sequence triplexes has originated from those investigations on simple polynucleotide systems.

The Discovery of Nucleic Acid Triplexes

From the historical point of view, the first triple-stranded model for nucleic acid was proposed by Pauling and Corey in 1953 on the basis of X-ray data. In this model, designed to describe a predominant state of DNA, three polynucleotide strands made a helix with a seven-nucleotide pitch, presumably the uncharged phosphates were localized close to the helix axis, while the bases were oriented outside of the molecule (Pauling and Corey, 1953). This model has not become popular because of the more realistic double-stranded model of Watson and Crick (1953).

The existence of triple-stranded (triplex) nucleic acid structures, as opposed to the recognized double-stranded form, was first shown in 1957 when Felsenfeld, Davis, and Rich (Felsenfeld et al., 1957) published the UV-mixing curves that unequivocally indicated the formation of a 1:2 complex of poly(A) and poly(U). This study and several immediately following studies (Felsenfeld and Rich, 1957; Rich, 1958a,b) laid a solid basis for an understanding of the general properties of the triple-stranded nucleic acids.

First, it was suggested that in the triple-stranded polynucleotide structure, two hydrogen bonds are formed between the A and U bases according to the Watson–Crick scheme, and the third base (U) "can make two strong hydrogen bonds, namely, by bonding uracil O6 and N1 to adenine N10 and N7" (Felsenfeld et al., 1957). In modern notation, this means exactly the same type of Hoogsteen hydrogen bond that is adopted in the description of base triads (Figure 1.18).

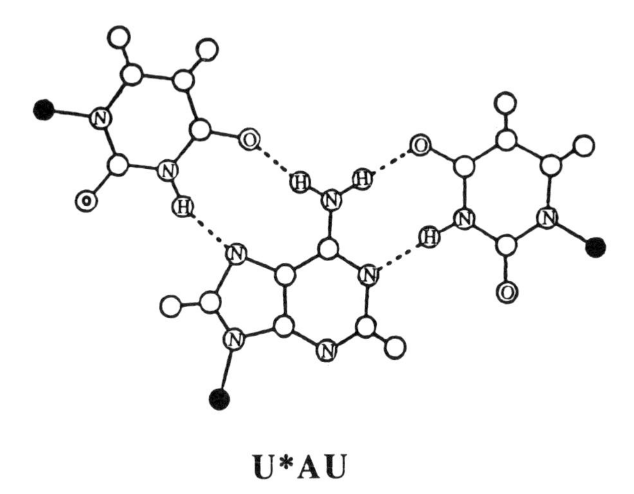

FIGURE 1.18. A proposed structure for poly (rU) · poly(rA) · poly(rU). One U base makes two hydrogen bonds with the A according to the Watson–Crick scheme, and another U base makes two hydrogen bonds with the A. Reproduced from Davis (1967).

Second, kinetic studies showed that the formation of triple-stranded structures was slower than that of double-stranded structures. This process is accelerated by increasing the ionic strength when the presence of the larger number of available cations leads to a decrease in repulsion of the negatively charged polynucleotide backbones. Divalent cations are much more effective in this respect than monovalent cations.

Third, the neutralization of negatively charged phosphate groups by metal cations (more effective upon specific binding than upon unspecific Debye–Hückel screening) was proposed.

Fourth, the triple-stranded structure was not a unique feature of the poly(A) + poly(U) system. UV-mixing curves and X-ray data showed the triple-stranded complexes poly(I) + poly(A) + poly(I) (Rich, 1958a). The base-pairing pattern was suggested for this complex (Figure 1.19). Based on the X-ray data (Rich, 1958b), polyinosinic acid was also suggested to form a triple-stranded complex (this complex was later identified as a four-stranded helix).

To describe triplexes in this book, we placed the pyrimidine strand of the duplex in the first place, the purine strand of the duplex in the second place, and the pyrimidine or purine strand in the third place. Similarly, the first letter in the base triads corresponds to the base in a pyrimidine (or pyrimidine-rich) strand of the duplex, the second letter corresponds to the base in the purine (or purine-rich) strand of the duplex, and the third letter (often after the * sign) corresponds to the base in the third strand.

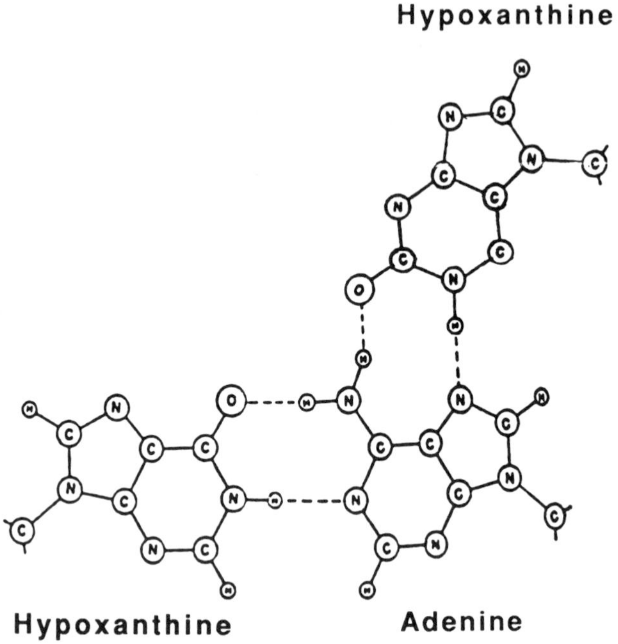

FIGURE 1.19. A proposed structure for the base triad in poly(I) · poly(A) · poly(I) where all strands are of the purine type. Adapted from Rich (1958a).

Physicochemical Studies of Model Triplexes

The hydrogen-bonding pattern in triple-stranded polynucleotides, initially suggested on the basis of the presence of the hydrogen bond donors and acceptors, was confirmed experimentally. First, the X-ray data of Hoogsteen (1959, 1963), who observed a crystalline 9-methyladenine: 1-methylthymine complex with a pairing scheme different from that of Watson–Crick pairing, justified the stereochemical suggestion of Felsenfeld et al. (1957). The infrared data of Miles (1964) confirmed the involvement of the N6 and N7 atoms of adenine in hydrogen bonding with uracil of the third strand; however, the exact pattern was suggested to be reverse Hoogsteen. It is now clear that both possibilities can be realized, depending on the proportion of purine and pyrimidine strands in the triple helix.

Extensive investigations of the poly(A) + poly(U) system, as a function of Na^+ concentration and temperature, allowed the construction of a phase diagram that indicated several regions where single-, double-, and triple-stranded structures, or mixtures thereof, stably exist (Fresco, 1963; Massoulie et al., 1964a; Stevens and Felsenfeld, 1964; Krakauer and Sturtevant, 1968). These studies were expanded further to acidic pH, where adenine becomes protonated, and to alkaline pH, where uracil loses a proton (Michelson et al., 1967). Phase diagrams for the poly(G) + poly(C) system at various pH values were also reported (Thiele and Guschlbauer, 1969).

Attempts to find the triplex-forming polynucleotides other than poly(A) and poly(U) led to investigation of mixtures of poly(C) and poly(G). Lipsett (1963, 1964) examined the complexes of poly(C) with short oligomers GG and GGG. Figure 1.20 shows two possible triads for such ternary complexes (Lipsett, 1964). At neutral pH, a complex with the stoichiometry 1G:1C is rapidly formed. Lowering the pH below 6 resulted in a complex with a ratio 1G:2C that contained protonated cytosine residues. This complex was ionic-strength-dependent, and its stability was lowered by increasing Na^+ concentration. Slow formation of a complex with the stoichiometry 2G:1C at neutral pH was also found. The interaction of poly(G), as opposed to oligo(G), and poly(C) resulted in a 1:1 complex only (Pochon and Michelson, 1965). The triple-stranded complex 2G:1C was found only after reducing the degree of polymerization of either poly(G) or poly(C) down to ~15 by alkaline treatment. It was suggested that a long poly(G) third strand does not perfectly fit a relatively rigid double helix poly(C) · poly(G) and, therefore, the formation of a triplex structure is unlikely. However, the strand breaks in the duplex

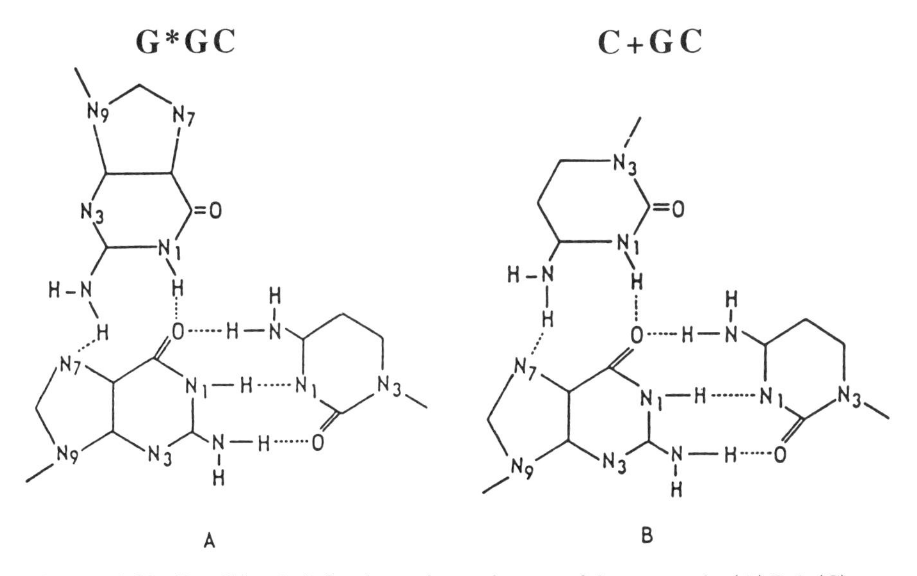

FIGURE 1.20. Possible triads for the various mixtures of three strands. (A) Poly(C) + oligo(G) + oligo(G); (B) poly(C) + oligo(G) + poly(C). Note that in the latter case, the N1 (N3 in modern notation) of the third-strand cytosine is protonated. Reproduced from Lipsett (1964).

and/or the third strand facilitate a conformational adjustment of the three strands.

These experiments demonstrated that the formation of triple-stranded complexes in the poly(G) + poly(C) system is more complex than in the poly(A) + poly(U) system, and it is necessary to take into account some other stabilizing factors aside from the cationic neutralization of phosphates. The complexity of the poly(G) + poly(C) system and the irreproducibility of the 2G:1C complex formation prompted some authors to consider this complex an artifact (Haas and Guschlbauer, 1976). It was only a careful examination of various conditions influencing the formation of the 2G:1C complex that provided the evidence that the triple-stranded structure with CG*G triads is not a metastable artifact structure (Marck and Thiele, 1978).

The early studies of triple-stranded nucleic acids dealt with the homopolymer samples (see Felsenfeld and Miles, 1967, for review). Mainly, these complexes had one polypurine and two polypyrimidine strands. When Rich (1960) and Riley et al. (1966) described the poly(dT)·poly(dA)·poly(U) triplex, it became clear that all common RNA and DNA bases are capable of participating in triad formation. In addition, three-stranded complexes formed by one polypyrimidine and two polypurine strands and by three polypurine strands were reported. Figure 1.21 shows some possible triads for the ternary complexes of inosinic acid with polycytidylic acids (Miles, as cited by Chamberlin, 1965; Thiele and Guschlbauer, 1969).

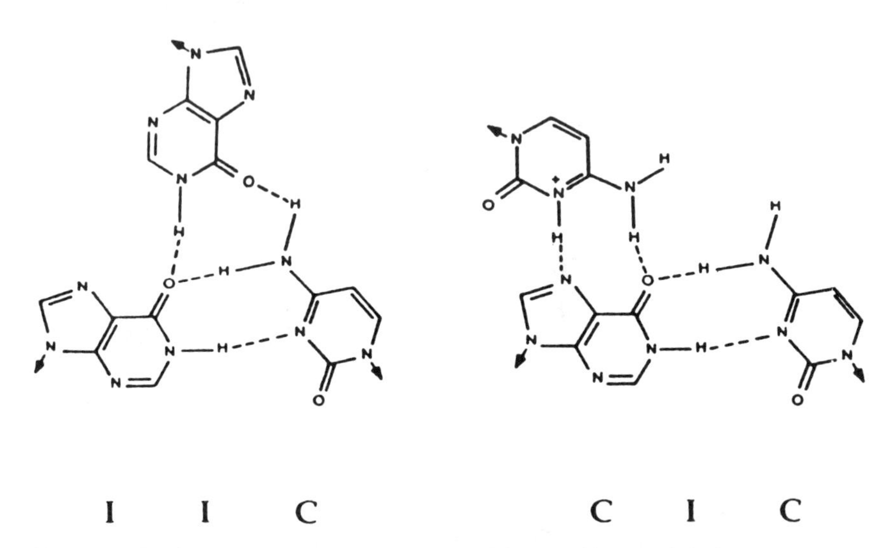

FIGURE 1.21. Proposed base triads for poly(I) · poly(I) · poly(C) (Miles, as cited by Chamberlin, 1965) and poly(C) · poly(I) · poly(C). Modified from Thiele and Guschlbauer (1969).

The next major step was made by Morgan and Wells, who systematically studied the triple-strand-forming properties of various polynucleotides, including simple repeating sequences and random copolymers (Morgan and Wells, 1968). They found that not only homopolymers, but also homopurine and homopyrimidine polymers of appropriate mixed sequences, can form triple-stranded complexes. Namely, poly(dA) · poly(dT) · poly(U) and poly[d(TC)] · poly[d(GA)] · poly[(UC$^+$)] triplexes were characterized by optical and hydrodynamic methods. No appropriate condition was found for triple-helical complexes between poly[d(TG)] · poly[d(CA)] or poly[d(AT)] · poly[d(AT)] and any polyribonucleotide.

In parallel, a number of polynucleotide complexes containing the purine and pyrimidine analogs have been examined (see Michelson et al., 1967, for review of early studies). When the Hoogsteen hydrogen-bonding positions in the bases are blocked, complex formation with an appropriate counterpart does not occur. Alkylation of the 6-amino group in adenine allows at best the formation of a 1:1 complex of poly(A) with poly (U) (Griffin et al., 1964; Ikeda et al., 1970; Hattori et al., 1974). Methylation of the N7 of purine nucleosides (guanosine, inosine) reduces the stability of double-stranded structures, due to base protonation, and also eliminates the ability to form Hoogsteen hydrogen bonds necessary to bind the third strand (Michelson and Pochon, 1966; Pochon and Michelson, 1967). Polynucleotides containing 7-deazaadenosine (tubercidin), where the Hoogsteen hydrogen-bonding N7 is replaced by C, are able to participate in triple-stranded complexes only when the proportion of tubercidin to adenosine is low, and triads with one Hoogsteen bond occur only occasionally (Torrence and Witkop, 1975; Seela et al., 1982).

A wide variety of pyrimidine analogs were tested for duplex and triplex formation. Polyuridylate analogs included 5-fluoro-, 5-chloro-, 5-bromo-, 5-iodo-, 5-ribosyl-, 5-methyl-, 5-amino-, 5-methoxy-, and 5-hydroxy-derivatives (Michelson et al., 1962; Massoulie et al., 1966; Hillen and Gassen, 1975, 1979). The triple-stranded complexes with poly(A) showed a marked increase in stability in the cases of poly-5-bromo-, 5-iodo-, and 5-chloro-uridylates (Figure 1.22), whereas 5-fluoro- and 5-hydroxy-derivatives resulted in decreased stability of ternary complexes (Massoulie et al., 1966). The 5-bromo substitution in cytosine also had a stabilizing effect, although it was demonstrated only for double-stranded complex poly(dI) · poly(dBrC) in comparison with poly(dI) · poly(dC) (Inman and Baldwin, 1964). The reason for such stabilization by halogen substitution was deduced from a supposed shift of the electron density in the corresponding bases and a concomitant increase in the hydrogen bond strength (Inman and Baldwin, 1964; Kyogoku et al., 1967). This hypothesis was in accord with most, but not all, available experimental data, yet no other explanation was presented (Michelson et al., 1967). Methylation and methoxylation at the 5 position of pyrimidines also led to the stabilization

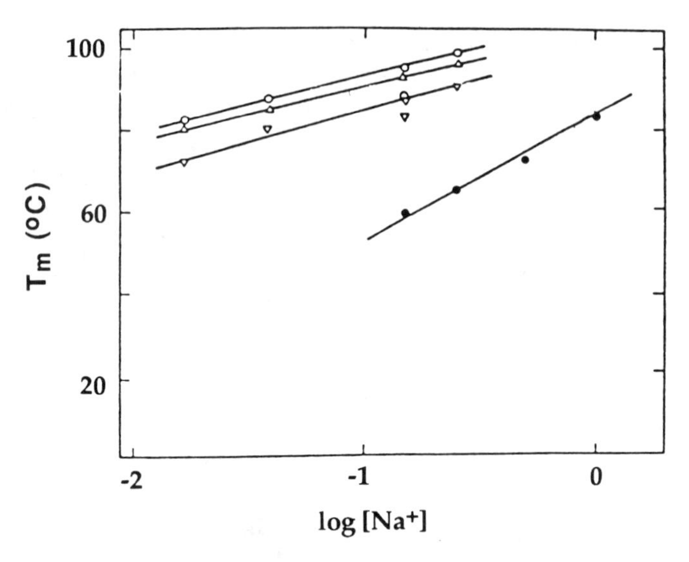

FIGURE 1.22. Variation of T_m according to ionic strength and type of substituent (5-bromine, -chlorine or -iodine-uracil) for some triple-stranded complexes: (●) rA · rU · rU; (▽) rA · rClU · rClU; (△) rA · rIU · rIU; (○) rA · rBrU · rBrU. The halogen substitution shifts the electron density in the corresponding bases and stabilization occurs due to a concomitant increase in the hydrogen bond strength (Inman and Baldwin, 1964; Kyogoku et al., 1967). Adapted from Michelson et al. (1967).

of ternary complexes. Thus, the triple-stranded complexes of poly-5-methyluridylate (polyribothymidylate) and 5-methoxyuridylate with poly(A) had higher melting temperatures than a similar complex including polyuridylate (Massoulie et al., 1966; Hillen and Gassen, 1979). A comparison of the triple-stranded complexes poly[d(TC)] · poly[d(AG)] · poly[d(TC)] and poly[d(Tm^{5}C)] · poly[d(AG)] · poly[d(Tm^{5}C)], in both of which the third-strand cytosines should be protonated, showed that the 5-methylation of cytosine allows triplex formation at neutral, as opposed to acidic, pH for unmodified cytosine (Lee et al., 1984). The triple-stranded complex of poly(A) with poly-5-aminouridylate had a melting temperature comparable to that for a similar complex including poly-uridylate (Hillen and Gassen, 1975).

A number of purine analogs bearing substitutions at positions other than those which are involved in hydrogen bonds were synthesized. Most of the purine analogs were adenine derivatives: 2-amino-adenine ($2NH_2A$) (Howard et al., 1966a, 1976; Howard and Miles, 1984), 2-dimethylamino-adenine (Ishikawa et al., 1973), 2-fluoro-adenine (Broom et al., 1979), 8-bromo-adenine (Howard et al., 1974, 1975), 8-methyl-adenine (Limn et al., 1983), and 2-amino-8-methyl-adenine (Howard et al., 1985; Kanaya et al., 1987). Both 2-amino- and 2-fluoro-substituted polyadenylates form triple-stranded complexes with polyuridylates and polythymidylates (Howard et al., 1976; Broom et al., 1979; Howard and Miles, 1984;). However, the properties of these multistranded complexes are unusual. For example, only triple helix poly(U) · poly(2FA) · poly(U) was formed from single strands, and no conditions were found appropriate for double helix poly(U) · poly(2FA) formation (Broom et al., 1979). In another example, $2NH_2A$ forms three Watson–Crick hydrogen bonds with U (Figure 1.23, top), and the resulting duplex was so stable that the poly(U) · poly($2NH_2A$) · poly(U) triplex always underwent a triplex → duplex + single strand transition (Howard et al., 1976). This was in contrast to the poly(A) + poly(U) system (see phase diagram in Chapter 3), where conditions could be found for both triplex → duplex + single strand and triplex → three single strands transitions. Poly-8-bromo- and 8-methyl-adenylates are not capable of forming the triple-stranded complexes. Moreover, they have poor double helix-forming abilities, presumably because the unfavorable contacts of bulky substituents at the C8 with 2′-OH of ribose result in the *syn* or alternating *syn–anti* conformations of the bases relative to the sugar–phosphate backbone (Howard et al., 1974, 1975; Limn et al., 1983; Kanaya et al., 1987). The stabilizing effect of the 2-amino substituent which provides a third hydrogen bond for duplex formation cannot compete with the destabilizing influence of the 8-methyl substituents, and as a result, the triplex does not form (Howard et al., 1985).

A similar analysis is relevant for analogs of guanine. Polyisoguanylic acid can form multistranded self-complexes. In an acidic medium, the protonated form of poly(isoG) forms a triple-stranded complex with poly(I) (Golas et al., 1976). Another analog, poly-8-amino-guanylic acid ($8NH_2G$), forms a tetra-stranded structure at neutral pH and a double-stranded structure with poly(C) at alkaline pH, where $8NH_2G$ becomes protonated (Hattori et al., 1975). This structure is stabilized by the additional interbase hydrogen bonds, which link two neighbor bases in a stack (vertical hydrogen bonds). As a result, under no conditions does this rigid duplex dismutate into a triple helix between poly($8NH_2G$) and poly(C). However, when the purine strand is composed of monomers, $8NH_2GMP$, the strand adjustment is easier, and because of the presence of three Hoogsteen bonds a very stable triple helix can be formed (Figure 1.23, bottom) (Hattori et al., 1975).

The available data on the triplexes containing modified bases show the complexity of substituent effects. The substituents are capable of changing the hydrophobicity of the bases and the pK values of protonation sites, as well as shifting the conformational equilibrium in nucleosides. In some cases the base-ionization equilibrium is shifted to acidic or alkaline pH, and therefore triple- or even double-stranded structures can be observed only under conditions far from physiological pH. The base substitutions can also lead to somewhat strained conformations of polynucleotides

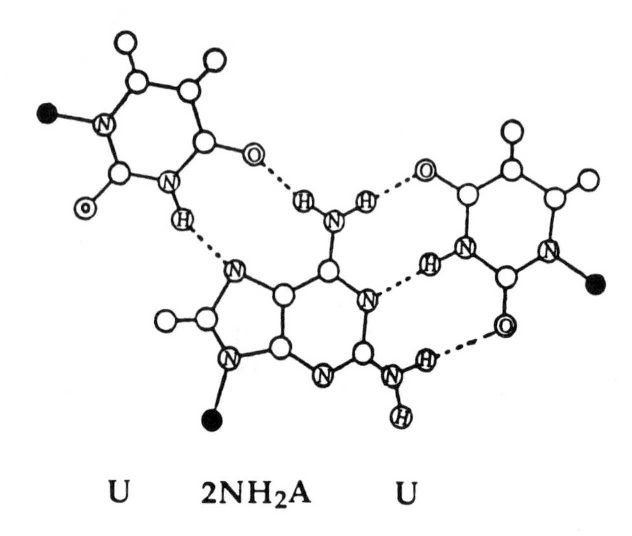

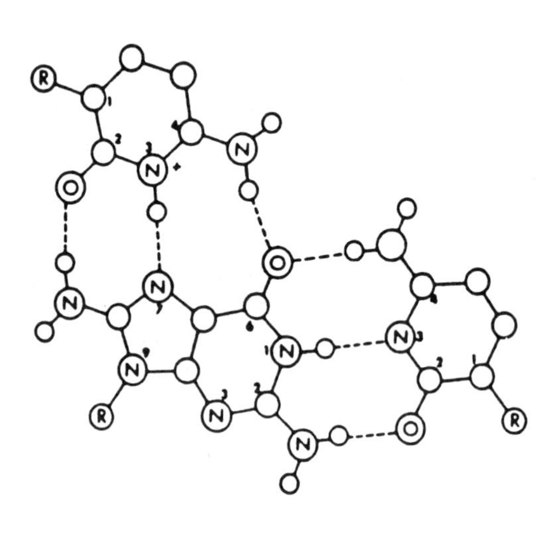

FIGURE 1.23. Additional hydrogen bonds in the triads U · $2NH_2A$*U (top, Howard et al., 1976) and C · $8NH_2G$*C (bottom, Hattori et al., 1976). Because of the high stability of the Watson–Crick base pair U · $2NH_2A$, the formation of a triple-stranded helix in the presence of excess poly ($2NH_2A$) is impossible. The equilibrium is shifted to the double-stranded form. The binding of poly(C) to poly($8NH_2GMP$) by three Hoogsteen bonds results in a highly stable triplex. Reprinted with permission from Howard et al. (1976) © 1976 American Chemical Society, and Hattori et al. (1976) *Biopolymers* 15, 523–531. Copyright © 1976 John Wiley and Sons, Inc.

due to unfavorable interactions of the bases with the sugar–phosphate backbones. This causes the polynucleotide with an inappropriate conformation to be unable to serve as the third strand.

Excessive strain on polynucleotide chains probably occurs in some cases of triplexes with unmodified bases (Pochon and Michelson, 1965; Broitman et al., 1987). Experimentally, such strain is revealed in the length dependence of the stability of triple-stranded complexes. We mentioned earlier that triplex formation in the poly(G) + poly(C) system requires oligomeric ($n < 15$) rather than polymeric strands (Pochon and Michelson, 1965). A similar effect was found in the case of the poly(U) · poly(A) · poly(A) triplex, when the upper limit of the poly(A) length was about 150 nucleotides (Broitman et al., 1987). This can be expected based on an imperfect adjustment in a rigid structure of the "lock-and-key" type. Slightly unfavorable structures of the polymer building block may be accommodated at short triplex lengths. However, structural deformations can accumulate during polynucleotide elongation up to the point where third-strand binding becomes unfavorable (Broitman et al., 1987). The introduction of periodic discontinuities into the duplex or third strand can make both of them more flexible and adjustable to each other. Because of the random distribution of third-strand oligomers on the duplex (Thrierr and Leng, 1972), some breakpoints in the duplex are left unstiffened by the binding of the third strand, providing the basis for easier triplex formation compared with the case of a long intact duplex and a polymer third strand.

If there is a polymer template for triplex formation, the length of the complementary block can be as little as one nucleotide (nucleoside). When the monomer blocks, guanosine, GMP, and $8NH_2GMP$, were hybridized with poly(C), their ability to form hydrogen bonds and stack over one another was sufficient to stabilize the duplexes in which the purine strands were composed of monomers. Other poly(C) molecules were then bound as third strands to these stacked monomer purines. Such triple-stranded complexes were dynamic and required an excessive amount (~100-fold) of the monomer component (Howard et al., 1966b; Huang and Ts'o, 1966; Sarocchi et al., 1970; Hattori et al., 1975).

In view of the two naturally available (ribo- and deoxyribo-) types of strands in triple-stranded complexes, the question of the relative stability or even potential competition between different associates is of interest. The variability in these complexes may be in the chemical differences between RNA and DNA (the 2′OH and thymine methyl groups, respectively) or in the structural differences (the number of bases per helix turn, charge density, width and depth of the major and minor groove, etc.). Figure 1.24 shows that for adenine-, uracil-, and thymine-containing complexes, a completely polyribonucleotide structure is generally more stable than those containing polydeoxyribonucleotide strands. Note also that the relative stability of triple-stranded complexes is ionic strength-

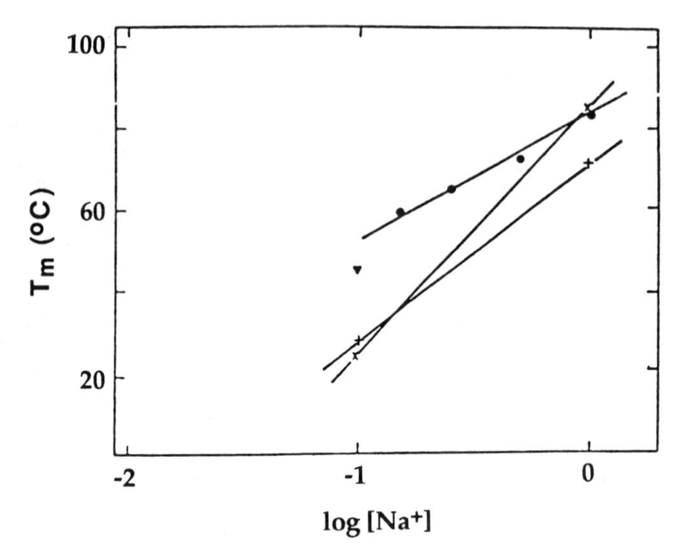

FIGURE 1.24. Variation of T_m according to ionic strength and type of strands (ribo- and deoxyribo-) for some triple-stranded complexes. (●) rA · rU · rU; (▼) dA · rU · rU; (×) dA · dT · dT; (+) dA · rU · dT. Adapted from Michelson et al. (1967).

dependent. Although the displacement reactions in the double- and triple-stranded polymeric systems including polynucleotides with modified bases were studied in some detail (Sigler et al., 1962; Inman, 1964; Michelson et al., 1967; De Clercq et al., 1976), there have been no consistent data on the possibility of displacement of the deoxyribo third strand by its ribo analog. Morgan and Wells (1968) were unable to observe the displacement of poly(dT) or poly[d(TC)] from triple-stranded complexes by poly(U) or poly[(UC)]. On the contrary, the formation of a triple-stranded complex between poly(dG) · poly(dC) and poly(C) might involve the displacement phenomena (Haas et al., 1976). Clearly, there have not been enough data to arrive at a definite conclusion.

By the mid-1980s, the general understanding of details of triplex structure was based on the data obtained on polymer complexes. The accumulated data showed that triplex formation requires either one homopurine and two homopyrimidine strands or one homopyrimidine and two homopurine strands (Felsenfeld et al., 1957; Inman, 1964; Lipsett, 1964; Miles, 1964; Riley et al., 1966; Felsenfeld and Miles, 1967; Morgan and Wells, 1968; Thiele and Guschlbauer, 1969; Marck and Thiele, 1978). Two duplex strands form the ordinary Watson–Crick hydrogen bonds and the bases of the third strand form the Hoogsteen (Hoogsteen, 1959, 1963) or reverse Hoogsteen hydrogen bonds with purines of the duplex. For the cytosine to be involved in the base triad it must be hemiprotonated, which requires a low pH value (pH < 5). The triplex formation is sequence-specific, since T, U, and C of the third strand bind to the AT, AU, and GC pairs of the duplex to form the TA*T, UA*U, and CG*C^+ triads of the PyPuPy triplex, and the third-strand A and G bind to the AU and GC pairs to form the CG*G and UA*A triads of the PyPuPu triplex (Lipsett, 1964; Marck and Thiele, 1978; Broitman et al., 1987). The formation of UA*U, CG*C^+ and CG*G triads was found in polyribonucleotides and that of TA*T, CG*C^+ and CG*G triads in polydeoxyribonucleotides (Inman and Baldwin, 1964; Lipsett, 1964; Riley et al., 1966; Felsenfeld and Miles, 1967; Michelson et al., 1967; Morgan and Wells, 1968; Thiele and Guschlbauer, 1969; Marck and Thiele, 1978; Broitman et al., 1987). The geometry of base triads implies that the third strand is positioned in the major groove of the duplex. The *anti* conformation of the bases relative to the sugar–phosphate backbone requires the two pyrimidine strands in the triplex to be antiparallel (Morgan and Wells, 1968; Thiele and Guschlbauer, 1969; Arnott and Selsing, 1974). X-ray data for the triple-stranded polyribo- and polydeoxyribonucleotide complexes implied the A form geometry for all three strands (Arnott and Bond, 1973; Arnott and Selsing, 1974; Arnott et al., 1976).

Early Hypotheses About the Biological Roles of Triple-Stranded Nucleic Acids

Early studies of polymer triplexes revealed many structural details of the triple-stranded complexes, which have long been considered a peculiar feature of some synthetic polynucleotides with a relatively obscure biological significance.

In 1966 Miller and Sobell published the first hypothesis on the role of triple-stranded nucleic acids in gene repression. A few years earlier, Jacob and Monod (1961) had proposed a model of transcription. In this model, the repressor factor, which keeps DNA nontranscribed most of the time, was suggested to be a polynucleotide. Before the nature of the repressor was clearly determined, Miller and Sobell proposed that it was a ribonucleoprotein whose oligoribonucleotide sequence was complementarily bound to a target DNA through Hoogsteen pairing. This hypothesis was not correct concerning the nature of the repressor (which later was shown to be a protein). However, the interesting and intriguing aspects of their hypothesis were the possibility of oligoribonucleotide recognition of a DNA sequence in the major groove at the operator locus without strand separation and a consequent inhibitory effect on RNA synthesis.

The involvement of triplexes in gene expression was confirmed by the following experimental data: the formation of polynucleotide triple-stranded complexes prior to the addition of RNA polymerase inhibited

RNA synthesis on the normally transcribed double-stranded template (Morgan and Wells, 1968; Murray and Morgan, 1973). This finding was important because it was also shown that polypyrimidine-rich regions in bacteriophage DNA can be sites for RNA polymerase initiation (Szybalski et al., 1966). The replication of double-stranded polynucleotides was also inhibited by third-strand binding (Murray and Morgan, 1973). The inhibition complex may exist for a rather long time, because the degradation of the RNA strand in the DNA–RNA triplex is slower than that of the free RNA molecule (Murray and Morgan, 1973). However, inhibition of RNA and DNA polymerases by triplex formation in vitro was not considered biologically important, since the occurrence of the homopurine-homopyrimidine stretches in DNA was expected to be statistically low, not significantly higher than at random base distribution in a given sequence (Morgan and Wells, 1968). Later, a hybridization analysis of DNA from various species showed that in eukaryotes the homopurine–homopyrimidine stretches occur more often than expected on a statistical basis (Birnboim et al., 1979). This finding renewed interest in the role of triple-stranded complexes in gene expression. Minton (1985) presented a detailed description of the hypothesis in which the newly transcribed mRNA can be involved in a triple helix, thereby leading to a premature termination of transcription at the homopurine–homopyrimidine stretches in DNA. It was shown that a fraction of the RNA product can remain complexed with DNA (Champoux and McConaunghy, 1975) and lead to transcriptional inhibition (Mizuno et al., 1984). It was also suggested that regulatory oligoribonucleotide binding upstream of the genes takes part in the repression of transcriptional initiation (Minton, 1985).

The homopurine–homopyrimidine stretches could also have some functional significance in the structural organization of chromosomal DNA. First, it was hypothesized that RNA binding to the duplex DNA according to the triplex mechanism can stabilize segregated circular domains in the chromosome (Pettijohn and Hecht, 1973). Second, as suggested by Lee and Morgan (1982), the triple-stranded structure may be one of the possible nucleic acid complexes that provide a tight packaging of the chromosomal DNA (Figure 1.25). However, the major renewal of the interest in triplexes is mainly associated with the studies of the nuclease-sensitive sites in chromatin.

Studies of Nuclease S1 Susceptibility of Specific DNA Sequences

The DNA in both prokaryotic and eukaryotic cells is kept in a highly condensed form so that it fits within the dimensions of the cell. In eukaryotes, the interaction of DNA with some specific proteins (histones) results in a tightly packaged nucleoprotein complex, chromatin. However, a significant part of the genomic DNA is important in transcription

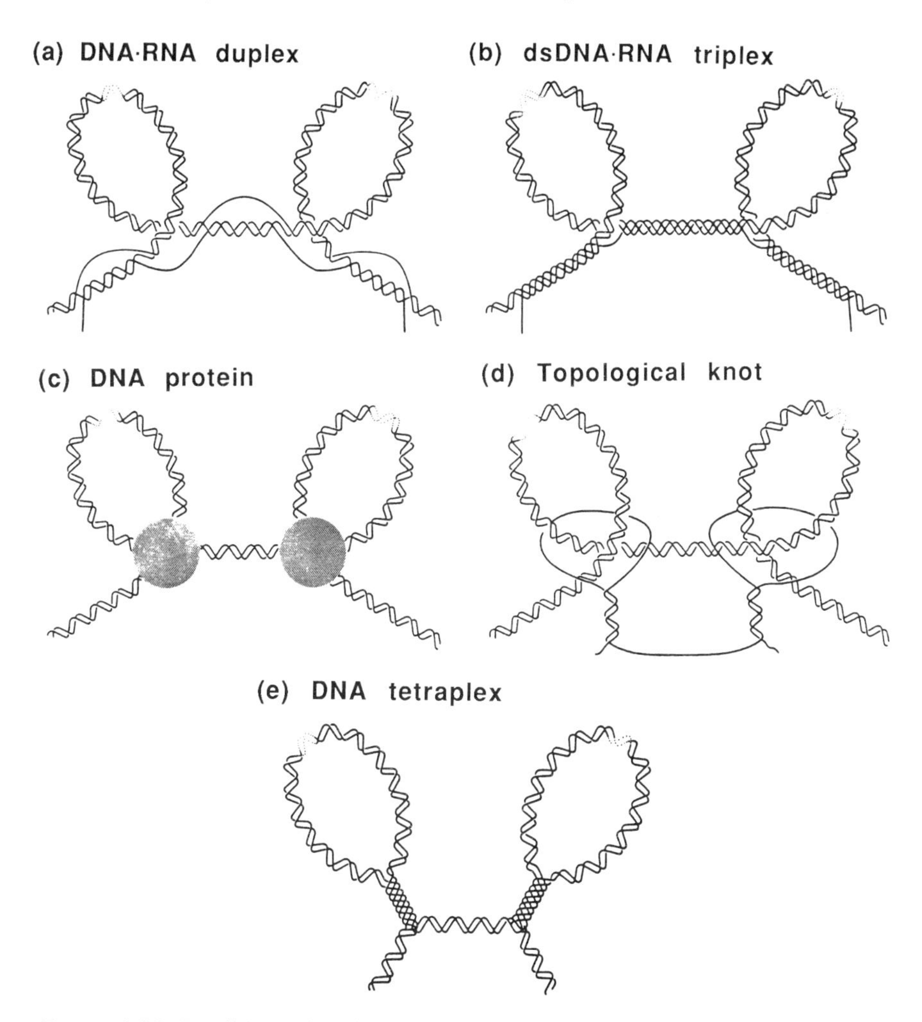

FIGURE 1.25. Possible models for the constraints that may stabilize the chromosomal organization of *E. coli* DNA. The topologically closed DNA loops are constrained by the formation of (a) a DNA · RNA duplex; (b) a double-stranded DNA · RNA triplex; (c) a DNA-protein complex; (d) a topological knot; or (e) a DNA tetraplex. Protein molecules are shown by hatched circles. Reproduced with modifications from Lee and Morgan (1982).

or regulatory activities, during which it interacts with a number of nonchromatin proteins. Protein recognition of specific DNA sequences requires that they be accessible. Thus, in active chromatin, there are unfolded regions that are, typically, nuclease-hypersensitive (see Gross and Garrard, 1988; Freeman and Garrard, 1992, for review). Particular attention was attracted by the sensitivity of some DNA sequences to S1, P1, and mung bean nucleases specific to single strands. Such a susceptibility meant that those sequences had unusual structures different from the double-stranded right-handed B form that is believed to be the normal state in living cells. Such unusual structures may have some role in the gene activation processes. Indeed, the RNA polymerase locally unwinds DNA upon binding, and this alteration of the DNA helix was suggested to be linked to transcribing the DNA template by the RNA polymerase (Saucier and Wang, 1972). For example, DNA unwinding was shown to be a rate-limiting step in the formation of transcription complexes (Mangel and Chamberlin, 1974). RNA polymerase could bind rapidly to supercoiled DNAs containing sites of a single-stranded type as shown by reactions with formaldehyde (Dean and Lebowitz, 1971) and S1 nuclease (Beard et al., 1973; Germond et al., 1974).

Certain homopyrimidine–homopurine (PyPu) sequences (where pyrimidine or purine bases are present in different strands), which often map to the promoter sites of eukaryotic genes have been identified as sites of anomalous sensitivity to the nuclease S1 (Hentschel, 1982; Larsen and Weintraub, 1982; Dybvig et al., 1983; Goding and Russel, 1983; Schon et al., 1983; Shen, 1983; Evans et al., 1984; Htun et al., 1984; McKeon et al., 1984; Ruiz-Carillo, 1984; Cristophe et al., 1985; Margot and Hardison, 1985; Financsek et al., 1986; Fowler and Skinner, 1986; Hoffman-Liebermann et al., 1986; Yavachev et al., 1986; Boles and Hogan, 1987; see Wells et al., 1988; Yagil, 1991, for reviews). The single-stranded character of accessible sites was confirmed by specific DNase I and chemical probing (Kohwi-Shigematsu et al., 1983; Pulleyblank et al., 1985; Evans and Efstratiadis, 1986).

The occurrence of long PyPu stretches in eukaryotic genes including 5′ flanking regions, the capability of the PyPu polymers to form a triplex, and a facilitated initiation of transcription on a supercoiled DNA template were combined in a model for the stimulation of gene expression by the intramolecular triplex/single strand structure (Figure 1.26, top) (Lee et al., 1984). DNA unwinding as a rate-limiting step in the intitiation of transcription could be skipped, since a looped single-stranded part left after the formation of the triple-stranded structure could serve as the RNA polymerase entry point. In the other model, the single-strand-binding protein could bind and stabilize the locally unwound site in superhelical DNA (Lee et al., 1984). Therefore, gene expression would be inhibited because the promoter site is not accessible to RNA polymerase (Figure 1.26, bottom). Yet, the available data were insufficient to draw a definite conclusion whether the triplex structure can be formed in natural DNAs.

Experimental Evidence That the S1-Sensitive PyPu Tracts in Supercoiled DNA Form Intramolecular Triplexes

The experimental data on the nuclease S1 susceptibility of the PyPu tracts in a supercoiled DNA led to several suggestions about the possible

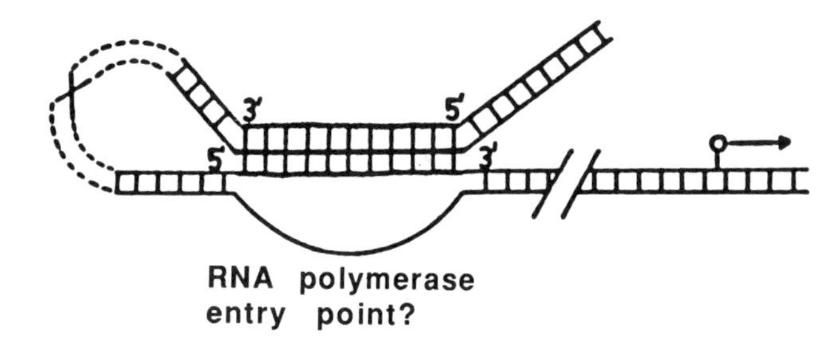

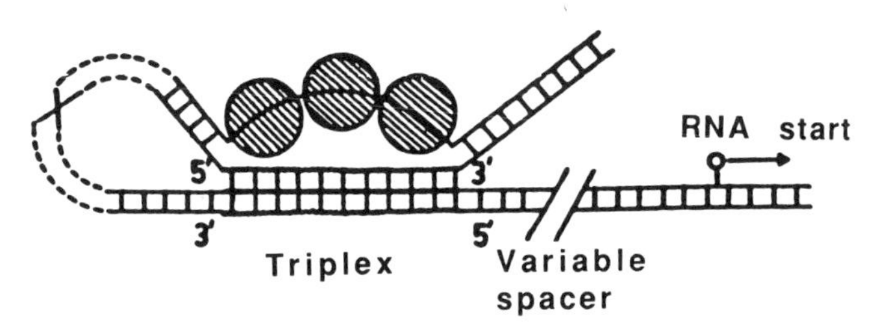

FIGURE 1.26. Possible roles for intermolecular triple-stranded structure in DNA transcription. Top: Stimulatory effect. The preformed unwound structure serves as the RNA polymerase entry point, thereby facilitating the initiation of transcription. Bottom: Inhibitory effect due to binding of protein molecules to the single strand left after triplex formation. RNA polymerase binding is inhibited due either to the occupancy of its entry point by the bound protein or to the unfavorable DNA structure near the promoter created by the triplex plus single-strand complex. Reproduced from Lee et al. (1984) *Nucleic Acids Res* by permission of Oxford University Press.

structures involved. First, it was hypothesized that a slipped structure (Figure 1.17a) can be formed (Hentschel, 1982; Glikin et al., 1983; Mace et al., 1983). Based on their DNase I footprinting and chemical probing data, Evans and Efstratiadis (1986) rejected the slippage model and proposed a heteronomous DNA with a dinucleotide repeat unit. Also on the basis of the chemical probing data, Pulleyblank et al. (1985) proposed a structure in which the AT base pairs of the Watson–Crick type alternate with the GC^+ pairs of the Hoogsteen type. Using data on the nuclease S1 sensitivity and pH dependence of supercoil relaxation in a model plasmid obtained from 2D gel-electrophoresis analysis, Lyamichev et al. (1985) put forward their first model of an H form (stabilized by H^+ ions), which contained the purine single strand and the pyrimidine hairpin stabilized by the protonated CC^+ pairs (Figure 1.27).

However, a modern model of intramolecular triplexes seemed ready to be born. The important step that was required to develop such a model was the construction of an energetically plausible structure. The nuclease

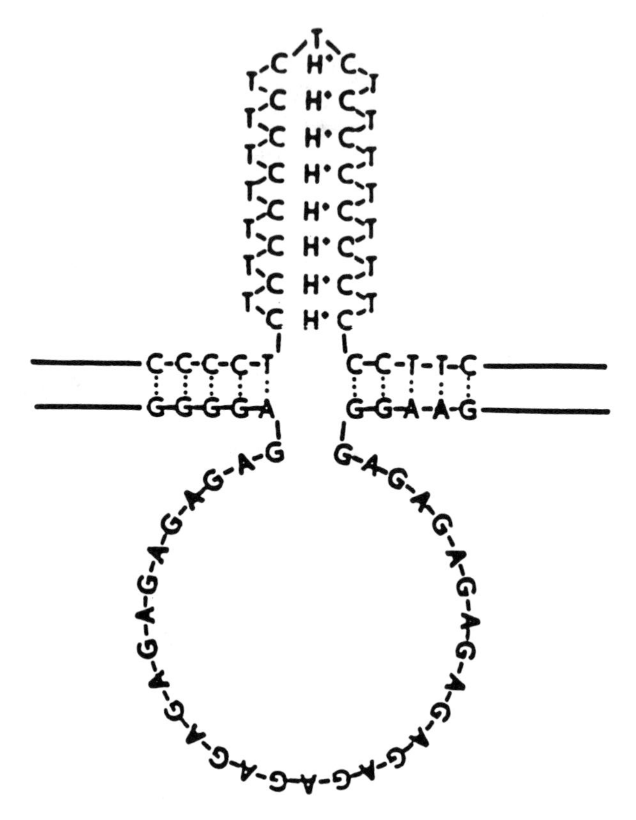

FIGURE 1.27. The first model of the intermolecular H form in which a homopyrimidine strand forms stacked protonated CC^{+} pairs with thymines looped out. The homopurine sequence is completely single stranded. From Lyamichev et al. (1985). Reprinted with permission.

S1 digestion pattern for long PyPu tracts in the 5′ flanking region of the human thyroglobulin gene inserted into a supercoiled DNA led to a suggestion (Christophe et al., 1985) about the formation of the "hairpin-triplex" structure (Figure 1.28) similar to that proposed by Lee et al. (1984). In this structure the regions that took part in the triplex were separated by about 100 base pairs along the sequence, which would require the formation of a double-stranded hairpin. Since the persistence length of DNA is longer, the formation of such a short double-stranded hairpin seems unlikely. Fowler and Skinner (1986) also suggested the formation of an unspecified triple-helical structure to explain altered conformation in the PyPu tract cloned in plasmid DNA.

Having reexamined the accumulated experimental data and models, Frank-Kamenetskii et al. proposed their second explicit model of H form DNA (Lyamichev et al., 1986; Mirkin et al., 1987). This was the most

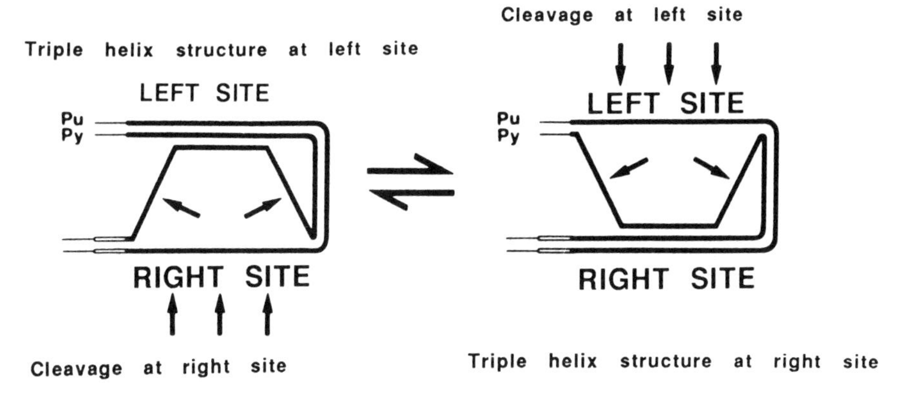

FIGURE 1.28. The "hairpin-triplex" structure. The duplex DNA bends so as to provide contact between one homopurine–homopyrimidine part of the molecule and a single strand formed in another homopurine–homopyrimidine part of the molecule. One S1 nuclease-sensitive single-stranded tract is assumed in the purine strand and two shorter (i.e., a few nucleotides long) single-stranded tracts are formed in the pyrimidine strand. Reproduced from Christophe et al. (1985) *Nucleic Acids Res* by permission of Oxford University Press.

plausible model of the intramolecular triplex, which is now widely accepted and is even used in popular textbooks (see, for example, Lehninger et al., 1994, p. 338). This model does not require a regular polynucleotide sequence, as was the case in previously studied model polymers. In H form DNA, the PyPu tracts are mirror repeated and separated by several base pairs, and upon triplex formation a single-stranded fold forms (Figure 1.29). This structure forms more easily at high torsional stress (the H DNA formation is driven by high supercoiling energy) and lower pH (the condition for the protonation of cytosine, which then forms the CG^*C^+ triad). The extrusion of the H form leads to only two duplex ends; therefore, its initiation is associated with a relatively modest energy loss.

The data of Christophe et al. (1985) could be explained by the formation of two separate H structures, because two mirror-repeated PyPu tracts in the sequence are available (Christophe et al., 1985; Frank-Kamenetskii, 1988).

Structure-dependent chemical modification experiments are more sensitive to various local structures, because they can detect even a small portion of DNA molecules in an altered conformation, contrary to two-dimensional electrophoresis (Bowater et al., 1992), which detects structural transitions only if they occur in a considerable fraction of DNA molecules. Chemical probing of the accessibility of various groups in nucleobases or their protection in the triplex structure confirmed the validity of the proposed H form model for intramolecular PyPuPy type triplexes consisting of two pyrimidine strands and one purine strand

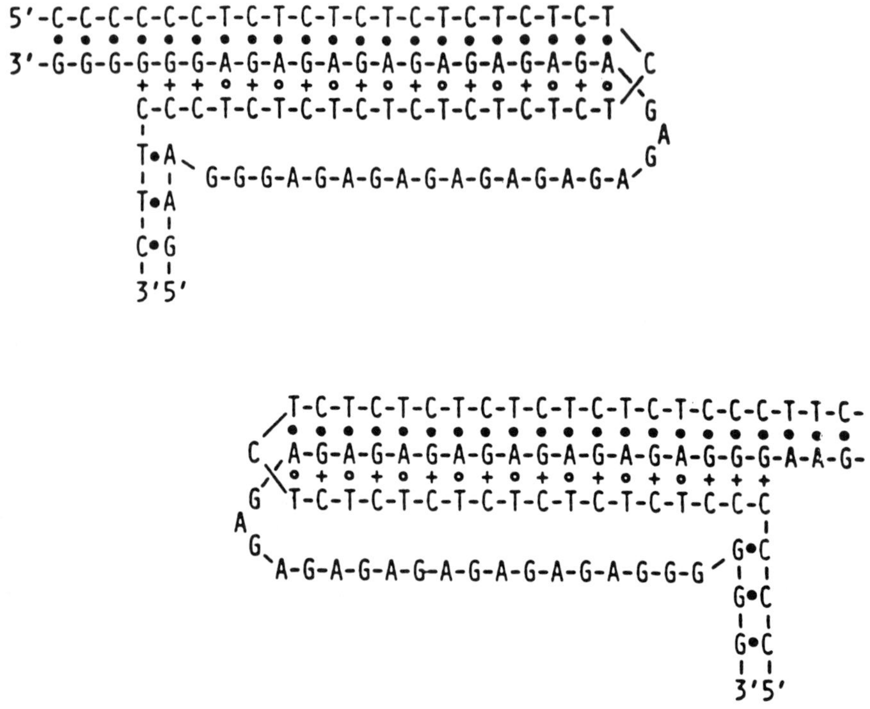

```
5'-C-C-C-C-C-C-T-C-T-C-T-C-T-C-T-C-T-C-T-C-T
   • • • • • • • • • • • • • • • • • • • • • \
3'-G-G-G-G-G-G-A-G-A-G-A-G-A-G-A-G-A-G-A-G-A\ C
         + + + o + o + o + o + o + o + o + o X
         C-C-C-T-C-T-C-T-C-T-C-T-C-T-C-T-C-T/ G
         |                                    A
         T•A\                                G
         | |  G-G-G-A-G-A-G-A-G-A-G-A-G-A-G-A/
         T•A
         | |
         C•G
         | |
         3'5'


        /T-C-T-C-T-C-T-C-T-C-T-C-T-C-T-C-T-C-C-C-T-T-C-
       / • • • • • • • • • • • • • • • • • • • • • • •
      C /A-G-A-G-A-G-A-G-A-G-A-G-A-G-A-G-G-G-A-A-G-
       X o + o + o + o + o + o + o + o + + +
      G \T-C-T-C-T-C-T-C-T-C-T-C-T-C-T-C-C-C
     A                                    |
      G\                                 G•C
        A-G-A-G-A-G-A-G-A-G-A-G-A-G-G-G/ | |
                                         G•C
                                         | |
                                         G•C
                                         | |
                                         3'5'
```

FIGURE 1.29. The recognized model of the H form of DNA, which may have two isomers due to the mirror symmetry of the homopurine–homopyrimidine tract. The major elements of the structure are the triple helix, in which the Watson–Crick duplex binds the homopyrimidine strand by Hoogsteen base pairing, and the unpaired half of the homopurine strand. To make the Hoogsteen hydrogen bonds, the third-strand cytosines are protonated. From Lyamichev et al. (1986). Reprinted with permission.

(Vojtiskova and Palecek, 1987; Hanvey et al., 1988; Htun and Dahlberg, 1988; Johnston, 1988; Vojtiskova et al., 1988; Voloshin et al., 1988) and PyPuPu type triplexes consisting of one pyrimidine strand and two purine strands (Kohwi and Kohwi-Shigematsu, 1988).

Intermolecular Triplexes Between DNA and Oligonucleotides

Triplexes can also form in natural DNA sequences (given an appropriate third strand). Thus, triplex structures are not restricted to only regular nucleotide sequences but are possible in any sufficiently long PyPu tract. Specifically designed pyrimidine oligonucleotides were shown to form Hoogsteen bonds with the purine strands of the PyPu tracts, resulting in local triple-stranded PyPuPy structures (Figure 1.30) (Le Doan et al, 1987; Moser and Dervan, 1987). The PyPuPu triplex containing a purine-rich oligonucleotide third strand was also designed (Cooney et al., 1988). The conditions necessary to form the intermolecular triplexes are generally the same as for H DNA, except for the absence of superhelical tension.

Current Fields of Interest in Investigation of Triplexes

After it was recognized that triplexes can form in the mixed sequence PyPu tracts available in natural DNA, several important directions of triplex studies developed. The methods employed in these studies are briefly outlined in Chapter 2. Chapter 3 describes the general physicochemical properties of triple-stranded nucleic acids.

The need to confirm the biological role of triplexes and to extend triplex-based methodologies on DNAs with different sequence types

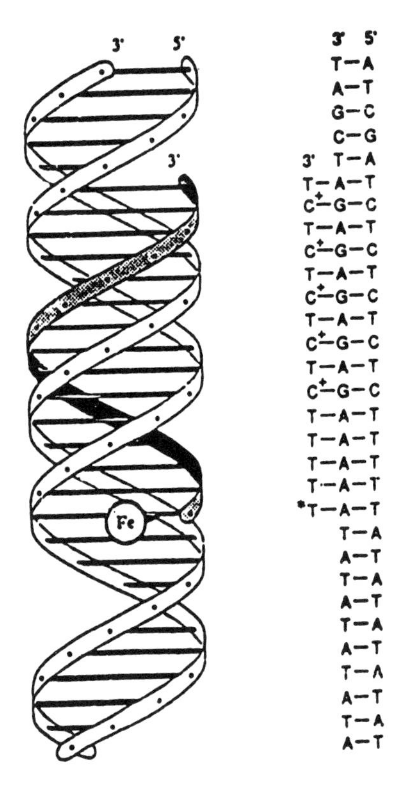

FIGURE 1.30. Intermolecular DNA triplex formed on the duplex with the irregular homopurine–homopyrimidine sequence. To confirm triplex formation, duplex cleavage by Fe-EDTA attached to the third strand is shown. Reprinted with permission from Moser and Dervan (1987) *Science* 238, 645–650. © 1987 AAAS.

prompted a search for new recognition schemes. This topic will be discussed in Chapter 4. Chapter 5 will describe some approaches to the elucidation of interactions that stabilize triplexes. Such knowledge is important for understanding possible modes of triplex formation in the cell and the role of factors that participate in the stabilization of triplexes in vivo.

Since it is widely accepted that the majority of natural DNA molecules are in the form of closed circles or are segregated into supercoiled topological domains, the possibility of H form extrusion in vivo seems plausible. A number of studies have been devoted to the search for triplexes in living cells and their possible roles in DNA function. This subject is discussed at length in Chapter 6. Intermolecular triplexes have provided the basis for the development of several genetic and biotechnological methodologies, which are discussed in Chapter 7.

2
Methods of Triplex Study

Physical Methods for Triplex Study

A number of physicochemical techniques have been used to study the properties of triple-stranded nucleic acids. Specific features of triplexes and relevant methods for their elucidation are compiled in Table 2.1. They will be briefly outlined below.

Spectral Methods

In spectroscopic methods, a sample is subjected to electromagnetic irradiation of a fixed wavelength or a certain wavelength range. Certain parameters of the radiation emerging from the sample are analyzed. Among such parameters are the fraction of absorbed radiation [typically measured by ultraviolet (UV), infrared (IR), and electron spin resonance (ESR) spectroscopies, and some methods of nuclear magnetic resonance (NMR) spectroscopy]; intensity and changes in the wavelengths of emitted, relative to incident, radiation (typically measured by fluorescence and Raman spectroscopy); and the difference between refraction or absorption of polarized radiation [optical rotatory dispersion (ORD) or circular dichroism (CD)]. Each of these spectroscopic methods that has been applied to triplex studies corresponds to a specific frequency (wavelength) range. Different spectral ranges relate to the properties of different structural groups in nucleic acids.

UV absorbance and CD spectroscopies deal with transitions between states that have different electronic structures of the bases. These methods are very sensitive to interactions of nearby bases which are vertically stacked in the strands. Stacking interactions depend on the conformational details of nucleic acid structure. The characteristic wavelength range for UV and CD spectra is 170 to 350 nm.

Nucleic acid components usually do not possess significant intrinsic fluorescence. Absorption of light by the nucleic acid–bound dye molecule, followed by partial vibrational relaxation, results in the emission of light of a different (usually longer) wavelength. The fluorescent properties of the dye molecule depend on its electronic and vibrational structures and their interactions with the nearby electronic structures. The strength of

TABLE 2.1. Physicochemical methods of triple helix characterization.

Method	Characteristic feature	Relevant information
UV spectroscopy	Mixing curve hypochromism	Complex stoichiometry
	Melting curve hyperchromism	Thermodynamic parameters
Circular dichroism	Changes in spectra beyond the sum for the duplex and the third strand	Presence of triplex
		A or B form family
Fluorescence spectroscopy	Inhibition of EtBr fluorescence	Condition-dependent oligomer triplex formation
Infrared spectroscopy	Sugar pucker marker bands	Overall conformation type
	Frequencies relevant to H bonds	Hydrogen bond pattern
Raman spectroscopy	Sugar marker bands	Overall conformation type, sugar–base orientation
NMR	Characteristic chemical shifts	Molecular structure, in particular sugar conformation
	Interproton distances	
Electron spin resonance	Rotational correlation time	Proportion of various triplexes
Calorimetry	Enthalpy change	Thermodynamic parameters
Equilibrium sedimentation	Sedimentation coefficient	Complex stoichiometry
Electron microscopy	Kinks in double-stranded DNA	Presence of intramolecular triplex
		Intermolecular triplexes between:
	Rosettes of several plasmids	Plasmids and polypyrimidine strand
	Streptavidin–DNA complex	DNA and biotinylated oligonucleotide
	Single-stranded loop	DNA and PNA oligomer
X-ray diffraction	Diffraction pattern	Complex stoichiometry, helix pitch
Native gel electrophoresis	Discontinuity in two-dimensional gel	Condition-dependent intramolecular triplex formation
	Comigration of the duplex and the third strand	Condition-dependent intermolecular triplex formation
Affinity chromatography	Retention of specific sequences	Sequence specificity of triplexes
Filter assay	Filter-binding of the complex	Thermodynamic and kinetic parameters for triplex formation
Drug interaction	Binding	Accessibility of duplex grooves and overall conformation type
Antibody binding	Binding	Presence of triplexes in vitro and in vivo

the dye–nucleobase interactions is determined by the conformation of nucleic acid. The binding of dyes to nucleic acids and interaction with nucleic acid bases results in fluorescent emission predominantly in the visible wavelength range (350–700 nm).

The vibrational absorption bands arise from transitions that are somewhat localized on a molecule. Generally, several vibrations of various parts of the molecule are coupled; however, it is often possible to distinguish the bands due to C═O stretching, C—H stretching vibrations, etc. Transitions between the vibrational levels of the ground electronic state of a molecule are of much lower energy than the electronic transitions. The absorption bands from these vibrational transitions occur in the IR range of 2,000–50,000 nm or, more commonly, 200–5,000 cm^{-1}. The vibrational properties of some identified molecular groups depend on their participation in hydrogen bonding, the deviation of cyclic molecules from planarity, etc. Thus, the conformation-dependent spectral band shifts and splittings arise. IR spectroscopy has a somewhat lower sensitivity than UV spectroscopy, and has the drawback of overlapping some IR bands of nucleic acids and the intense IR band of water.

The Raman spectra arise due to the frequency change of radiation after interacting with vibrating groups in the molecule. For large asymmetric molecules, Raman and IR effects produce the same sets of bands. However, unlike the case with the IR spectrum, water has only weak Raman bands that may only slightly distort the spectrum of the system under study.

NMR techniques monitor the absorption of energy associated with transitions of nuclei between the adjacent magnetic energy levels. The NMR spectrum represents energy absorption in the sample as a function of an externally applied magnetic field or the radiofrequency (10–800 MHz) supplied by an external oscillator. Unlike the overlap of bands for similar chromophores in optical absorption techniques, individual nuclei of a given kind within the same molecule usually absorb energy at distinct spectral positions which are often discernable. This is possible because the exact value of the magnetic field strength or radiofrequency corresponding to the energy absorption by the nucleus depends on interactions of its magnetic moment with the magnetic moments of numerous surrounding nuclei. Rapidly developing NMR techniques can identify separate chemical groups in complex molecules and provide information on their interactions with nearby partners and appropriate groups which are remote along the polymer chain but are brought into close spatial proximity by a conformational chain bend.

The magnetic moment of an unpaired electron is 10^3 times larger than that of common nuclei. Therefore, ESR, which deals with transitions between electron magnetic energy levels, requires frequencies in the gigahertz range (i.e., three orders of magnitude higher than in NMR). Since electrons in most molecules are paired, relatively few molecules produce an ESR signal. In the case of nucleic acids, some small molecules with unpaired electrons (spin labels) are attached to specific positions, and the ESR characteristics of the interaction between spin labels and nearby nuclear magnetic moments are analyzed.

UV Spectrophotometry

When two triplex-forming oligomers are mixed in various proportions, they can form 1:1 and 1:2 or 2:1 complexes, depending on the conditions in the medium and the ability of one of the two oligomers to bind as the third strand. It is well known that the interaction of stacked bases in complexes results in a reduction of absorbance relative to the absorbance of the mixture containing single-stranded oligomers in which the bases are less ordered (Cantor and Schimmel, 1980a). By varying the oligomer proportions, one gets a UV mixing curve composed of straight lines throughout the whole graph, except for characteristic points (Cantor and Schimmel, 1980b). Provided that the complexes which are formed are stable (the equilibrium in the mixture is shifted to the complex formation), the hypochromic reduction in absorbance of double- and triple-stranded complexes results in inflection points at 0.5 and 0.33 or 0.67 mole fraction of one strand. These points correspond to the complete involvement of strands in complexes. Every linear part of the graph corresponds to the conditions under which the reaction mixture contains the complex and an excess of one strand. Mixing curves are used to confirm the formation and stoichiometry of complexes and to monitor their dependence on strand composition and ambient conditions (Felsenfeld et al., 1957; Lipsett, 1964; Massoulie et al., 1964a; Stevens and Felsenfeld, 1964; Chamberlin, 1965; Howard et al., 1966a; Pilch et al., 1990b; Plum et al., 1990; Johnson et al., 1991; Kan et al., 1991). Figure 2.1 shows the experimental mixing curves for equimolar solutions of poly(A) and poly(U) (Felsenfeld et al., 1957). In the absence of $MgCl_2$, the point of minimum absorbance corresponds to an approximately 1:1 molar ratio of polymers, and therefore only the duplex poly(A) · poly(U) is formed. In contrast, the point of minimum absorbance at 10 m*M* $MgCl_2$ corresponds to an approximately 1:2 molar ratio of poly(A) to poly(U). Under these conditions, a triple-stranded complex exists.

Heating of the triple-stranded complex usually results in biphasic strand dissociation according to the transitions triplex → duplex + single strand → single strands. Stacking interactions in the liberated strands (in contrast to the bound strands) are weaker, resulting in a hyperchromic increase in UV absorbance. Thus, the process of triplex melting can be monitored using the temperature-dependent changes in UV absorbance (see Felsenfeld and Miles, 1967, for review of early experiments; Manzini et al., 1990; Ohms and Ackerman, 1990; Pilch et al., 1990a,b, 1991; Plum et al., 1990; Xodo et al., 1990, 1991, 1993; Mergny et al., 1991b;

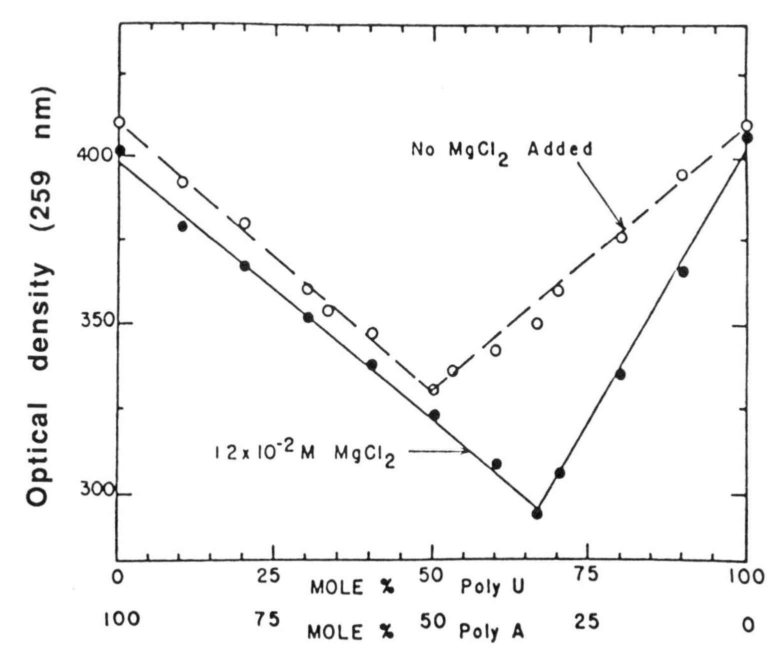

FIGURE 2.1. The optical density of the mixtures of poly(A) and poly(U) under conditions appropriate for the double-stranded form (0.1 *M* NaCl, 0.01 *M* glycylglycine, pH7.4, 25°C) and the triple-stranded form (same medium with $MgCl_2$ added). The total nucleotide concentration is constant, but the poly(A)/poly(U) ratio is varied. Optical densities were measured 2 h after mixing to ensure that complexing has been completed. At the left ordinate, only poly(A) is present in solution; at the right ordinate, only poly(U) is available. When both poly(A) and poly(U) are present, they can form the 1:1 complex, whose absorbance is lower than that of the sum of single strands. The upper left part of the graph corresponds to the presence of the poly(A) · poly(U) duplex and unbound excess poly(A); the right part corresponds to the presence of the duplex and unbound excess poly(U). When poly(A) and poly(U) concentrations are equal, they are completely involved in a double-stranded complex, and the absorbance under these conditions is minimal. In the lower part of the graph, the 1:1 complex also has low absorbance, which can be further diminished when the poly(A) · 2poly(U) complex forms. From Felsenfeld et al. (1957).

Ono et al., 1991a; Roberts and Crothers, 1991, 1992; Egholm et al., 1992a,b; Froehler and Ricca, 1992; Rougée et al., 1992; Jetter and Hobbs, 1993; Miller and Cushman, 1993; Völker et al., 1993). In the case of short helices, the melting curve is gradual, and it is sometimes difficult to determine the temperatures of transitions corresponding to the dissociation of the third strand and melting of the duplex. As will be shown in Chapter 3, under certain conditions the two transitions can be very close in temperature range and it is difficult to resolve them. This difficulty can be overcome using the wavelength differences of the spectra for duplex and triplex. Figure 2.2 shows the difference spectrum for poly(A) · poly(U) duplex and the sum of spectra of single strands, and a similar curve for poly(U) · poly(A) · poly(U) triplex and single strands (Massoulie et al., 1964b). It is clear that the change in absorbance at 280 nm allows selective monitoring of triplex formation or dissociation because zero absorbance change at this wavelength accompanies duplex formation or melting. By contrast, at 283 nm only duplex formation or melting is seen, and triplex formation or dissociation occurs with zero absorbance change.

Figure 2.3 shows the changes in absorbance for the short oligonucleotide duplex and triplex as a function of temperature (Pilch et al., 1990b). The melting temperature for the duplex is determined from the melting curve of the 1:1 complex at 260 nm (panel A). However, it is difficult to

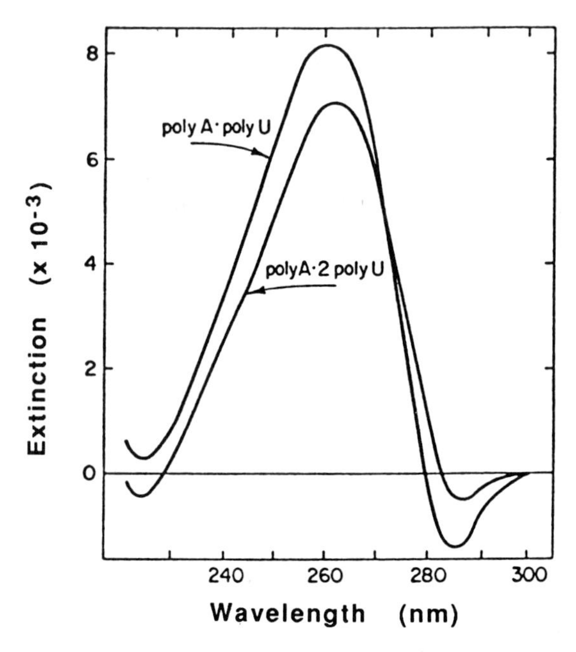

FIGURE 2.2. Molar difference spectra for poly(A) · poly(U) and poly(U) · poly(A) · poly(U). The molar absorbances of the duplex and triplex at 20°C were subtracted from the molar absorbances of 1:1 and 2:1 mixtures of single strands of poly(U) and poly(A) at 85°C. Since duplex formation should be accompanied by a zero absorbance change at 280 nm, the decrease in absorbance for the complex relative to the mixture of single strands indicates triplex formation. The trend is opposite at 283 nm. A zero absorbance change corresponds to triplex formation, whereas an absorbance increase indicates duplex formation. For combinations of other polynucleotides with different absorbance maxima, similar indicatory wavelengths may vary (see Figure 2.3). From Michelson et al. (1967).

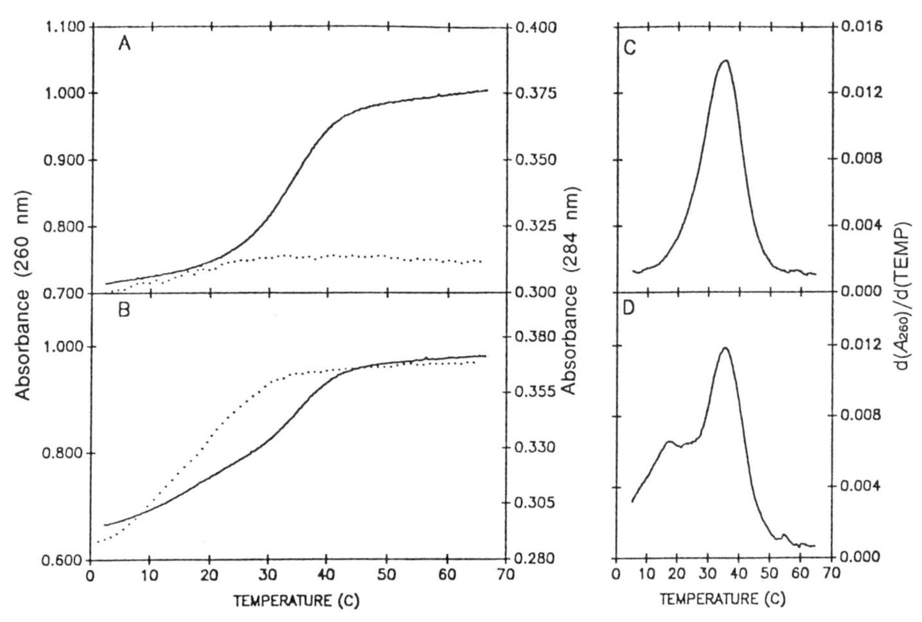

FIGURE 2.3. Helix–coil transitions in solutions containing either 1:1 (A and C) or 1:2 (B and D) molar ratios of $(dA)_{10}$ and $(dT)_{10}$ at 260 nm (–) or 284 nm (···). First-derivative plots of the 260-nm transitions are shown in C and D. Conditions: 10 m*M* Tris · HCl (pH 7.3), 50 m*M* $MgCl_2$. Panel A allows the determination of the duplex melting temperature from the changes in absorbance at 260 nm. At 284 nm only triplex disruption is seen (panel B). The melting temperatures for the duplex and the triplex can be determined with high accuracy from the differential melting curves measured at 260 nm (panels C and D). From Pilch et al. (1990b).

determine precisely the temperature of dissociation of the third strand at this wavelength (panel B). Using the absorbance at 284 nm, which reflects rearrangements in the $(dT)_{10} \cdot (dA)_{10} \cdot (dT)_{10}$ triplex, it is possible to determine the temperature of dissociation of the third strand. Another way to determine dissociation temperature is to plot the first derivative of absorbance, which shows the biphasic nature of the transition more clearly (panel D). The melting temperatures were determined as 17° and 35°C for triplex and duplex melting, respectively.

Whereas the changes in UV absorbance reflect the transitions between the three states (triplex, duplex, and single strands), the theoretical models are mainly applicable to two-state transitions without significantly populated intermediate states (Cheng and Pettitt, 1992a). For transitions in which intermediate states are significantly populated, the integral or differentiated melting curve will be broadened (Marky and Breslauer, 1987), resulting in a disparity between, for example, the van't Hoff enthalpies derived from the optical melting curves and the true enthalpy values determined from the calorimetric experiments (see below). Determining the thermodynamic parameters from optical melting data requires significant care and, necessarily, the development of a suitable model of the transition under study. However, in some cases, even taking into account various probable processes during the triplex melting, a reasonable convergence of the data from the two methods is not necessarily achieved (Manzini et al., 1990; Rougée et al., 1992; Völker et al., 1993).

Optical Rotatory Dispersion and Circular Dichroism

The optical rotatory dispersion (ORD) due to differences in the refractive indices of the medium for left and right circularly polarized light reflects the structural details of such molecular complexes as double- and triple-stranded nucleic acids. ORD was used in the study of triplex formation and dissociation (Sarkar and Yang, 1965a,b; Howard et al., 1966b). However, a drawback of ORD is that it must be measured beyond the molecular absorbance bands that contain important structural information. Therefore, ORD has been largely replaced by circular dichroism (CD), which is based on the difference in absorbance of left and right circularly polarized light.

In many cases, the CD spectra of nucleic acid triplexes differ from the sum of the spectra of constituent double- and single-stranded DNA (Lee et al., 1979; Antao et al. 1988; Gray et al., 1988; Manzini et al., 1990; Pilch et al., 1990b; Xodo et al., 1990; Callahan et al., 1991; Johnson et al., 1991; Miller et al., 1992; Völker et al., 1993). The appearance of an intense negative short-wavelength (210–220 nm) band in the CD spectra indicates the formation of the triple-stranded complex (Figure 2.4) (Manzini et al., 1990). A similar trend was observed in other studies where the target duplexes had moderate AT contents (Lee et al., 1979; Gray et al., 1988; Manzini et al., 1990; Callahan et al., 1991). In the cases of oligomeric or polymeric triplexes containing only TA*T triads and triplexes with a relatively high proportion of TA*T triads, the changes in the CD spectra consist of a reduction in intensity of the positive short- and long-wavelength bands (Durand et al., 1992a,b; Park and Breslauer, 1992). Thus, a difference in the CD spectrum of the complex and the sum of CD spectra of the target duplex and the third strand is used to confirm triplex formation. In one exception, Plum et al. (1990) were unable to detect a significant difference between the CD spectrum of a triplex with a high proportion of TA*T triads and the sum of CD spectra for the constituent duplex and third strand. They suggested that this effect could be due to the occasional compensation of the different sign contributions.

The changes in the CD spectra provide a certain basis for suggesting the overall conformation of strands in the triplex. For example, the appearance of the negative short-wavelength CD band (210–220 nm)

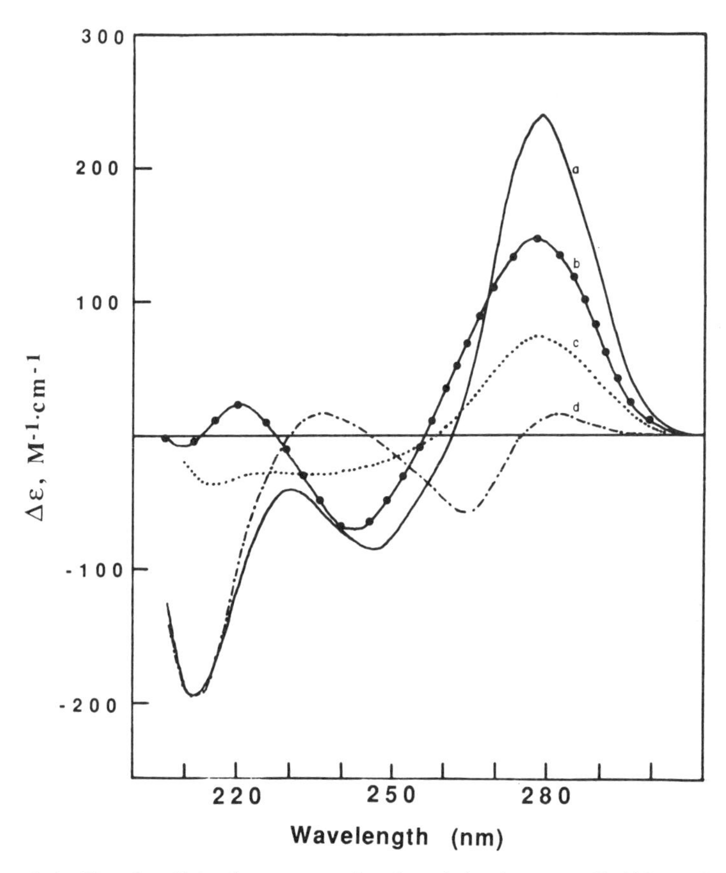

FIGURE 2.4. Circular dichroism spectra for the triplex h + s = 5′-d(GAAGGAG-GAGATTTTTCTCCTCCTTC) + 5″-d(CTTCCTCCTCT) (curve a); harpin duplex h (curve b); single strand s (curve c); and the difference CD spectra for the triplex (h + s) and sum of the noninteracting components h and s (curve d). Conditions: 25°C, 1 m*M* sodium acetate, pH 5, 100 m*M* sodium perchlorate. The ellipticity was calculated in L/cm·mol DNA strands. Triplex formation results in a CD spectrum which differs from those for duplex, single strand, or their sum. Note also the appearance of a significant negative short-wavelength band. From Manzini et al. (1990). Reprinted with permission.

indicates that strands in the triplex may have conformations related to the A form (Johnson et al., 1991, 1992). A positive CD band or shoulder near 190 nm indicates that the purine strands in triplexes made of TA*T or UA*U triads may have conformations related to the B′-form (Johnson et al., 1991).

Spectrofluorometry

To detect triplex formation, ethidium bromide (EtBr) can be used as a probe. Intercalation of EtBr between the adjacent base pairs of double-stranded DNA and concomitant interaction with the nearby stacked nucleobases results in a significant induced fluorescence of EtBr. However, interaction of EtBr with the triple-stranded DNA is weak. Although EtBr was shown to intercalate AT-rich triplex structures (Mergny et al., 1991a; Scaria and Shafer, 1991), it does not bind to CG*C$^+$-containing regions, presumably because of electrostatic repulsion between cationic drug and protonated cytosine (Lee et al., 1979). Another reason could be that a third strand positioned in the major groove makes the DNA stiffer. As a result, local duplex distortion to accommodate an intercalative ligand becomes more difficult (Chen, 1991).

Figure 2.5 shows that EtBr itself has low intrinsic fluorescence (point at the ordinate axis). EtBr is less able to intercalate triplex DNA than duplex DNA. Therefore, EtBr fluorescence increases about 15-fold at pH 8.0 when a double-stranded hairpin is formed, and the increase is weaker at pH 5.0 when the triplex is established. Thus, a relatively low enhance-

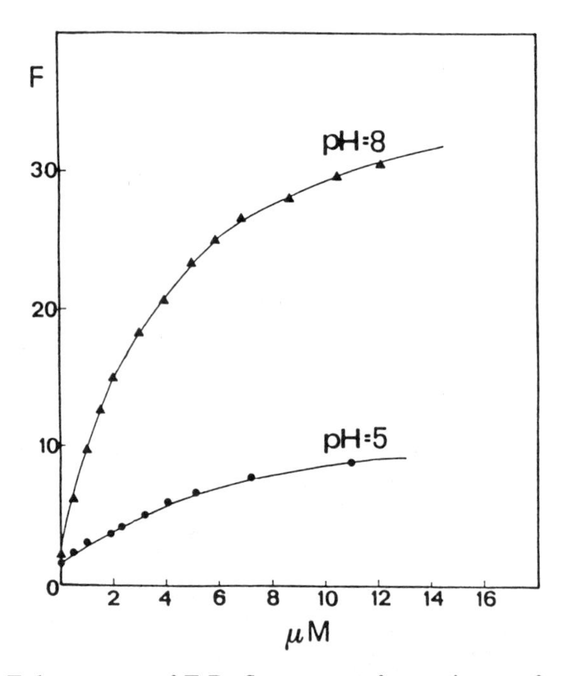

FIGURE 2.5. Enhancement of EtBr fluorescence for a mixture of pyrimidine and purine oligonucleotides (sequences in the legend to Figure 2.4) at pH 5 (stable triplex) and pH 8 (duplex with a dangling single strand). EtBr concentration was 2.5 mM. The equimolar concentrations of oligonucleotides are indicated on the abscissa. Reproduced from Xodo et al. (1990) *Nucleic Acids Res.* by permission of Oxford University Press.

ment of EtBr fluorescence is evidence of complex formation between the duplex and single strand (Lee et al., 1979, 1984; Xodo et al., 1990; Callahan et al., 1991).

Energy transfer between two fluorescent moieties was suggested as a means of studying triplexes (Mergny et al., 1994; Yang et al., 1994). In one of these studies, oligopyrimidines were coupled to an acridine (donor, attached to the 5′ end of one oligomer) or ethidium (acceptor, attached to the 3′ end of the other oligomer) (Mergny et al., 1994). Fluorescent energy transfer from acridine to ethidium was observed when oligomers were bound to adjacent PyPu tracts. A single base pair change in one of the tracts significantly reduced the efficiency of energy transfer. The authors propose this technique for discrimination between fully complementary and singly mismatched sequences. In another experimental design, fluorescent energy transfer was used to study the binding of an acceptor (tetramethylrhodamine)-labeled oligonucleotide to a donor (fluorescein)-labeled DNA target (Yang et al., 1994). A number of thermodynamic and kinetic parameters for the triplexes between homopyrimidine oligonucleotides and their duplex targets have been determined.

Vibrational Spectroscopy

The infrared (IR) spectral region 1,500–1,800 cm^{-1} can be used to discern the formation of several helical nucleic acid structures. For example, when a triplex composed of UA*U triads is formed, a strong 1,657 cm^{-1} band appears in the IR spectrum (Howard et al., 1964; Miles, 1964; Ohms and Ackermann, 1990). This band was assigned to the C4=O group vibration in the third strand U. For the Watson–Crick bonded C4=O group, the same vibration in the double helix is about 1,675 cm^{-1}. Thus, triplex formation is accompanied by the appearance of new characteristic IR bands (Howard et al., 1964; Miles, 1964; Miles and Frazier, 1964a,b; Howard et al., 1966b; Ohms and Ackermann, 1990). The assignment of these bands to the particular groups of nucleobases provides the basis for determining the schemes of base triads (Miles, 1964). The stoichiometry of the complexes can also be determined using computer fitting of IR spectra (Miles and Frazier, 1964b).

The modern Fourier-transformed infrared (FTIR) spectroscopic technique (see Taillandier and Liquier, 1992, for review) in combination with computational simulation allows one to determine molecular details of triplex structures such as sugar conformations (spectral region 750–1,000 cm^{-1}) (Liquier et al., 1991, 1993; Ahkebat et al., 1992; Howard et al., 1992; Ouali et al., 1993a,b). Figure 2.6 shows examples of sugar pucker determination on the basis of FTIR spectra. Panels A and B indicate the reference marker bands for 2′-endo (S) and 3′-endo (N) sugar conformations in double-stranded B and A forms. In experimental polymeric triplexes, both sugar conformations can be present (panels C and D).

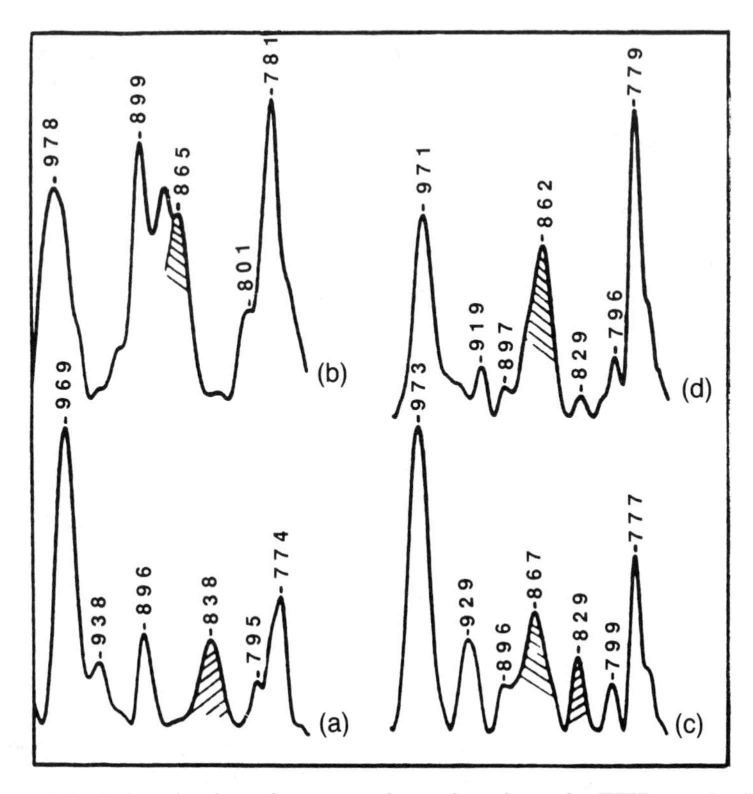

FIGURE 2.6. Determination of sugar conformations from the FTIR spectra in the 1,000–750 cm^{-1} region. To eliminate a strong water signal, spectra were recorded in D_2O. (a) Native DNA in the B form [2′-endo (S) sugar conformation]. (b) Poly(dG) · poly(dC) in the A form [3′-endo (N) sugar conformation]. (c) The poly(dC) · poly(dG) · poly(dC^+) triplex has sugars in both S and N conformations. (d) The poly(dC) · poly(dG) · poly(rC^+) triplex has sugars in predominantly N conformation. Absorptions characteristic of N type sugars (▧) and of S type sugars (▨) are hatched. From Ahkebat et al. (1992). Reprinted with permission.

Two domains of the Raman spectra can be used to characterize the sugar conformations (600–900 cm^{-1}) and their *anti* or *syn* orientation with respect to the bases (1,200–1,400 cm^{-1}) (Ouali et al., 1993a,b). FTIR has been also applied to studies of the hydration-dependent conformation of triple helices (White and Powell, 1995).

Vibrational circular dichroism (VCD) is a relatively new technique. The VCD of a double-stranded poly(rA) · poly(rU) at pH 7 and moderate added salt concentration (0.1 *M*) has been measured in both the base-stretching and the phosphate-stretching regions of the IR range as a function of temperature (Yang and Keiderling, 1993). The nature of the transitions that occur has been identified by comparing the VCD and IR absorption spectra of the initially double-stranded samples with samples of single-stranded poly(rA) and poly(rU) and with samples of triple-stranded poly(rA) · poly(rU) · poly(rU). The large differences in the VCD band shapes allow positive identification of the intermediate and final states. Later VCD was applied to the study of the relative stability of several of the RNA and DNA triplexes: poly(rU) · poly(rA) · poly(rU), poly(dT) · poly(dA) · poly(dT), and poly(rU) · poly(dA) · poly(dT) (Wang, L. et al., 1994).

NMR Spectroscopic Techniques

A number of well-developed nuclear magnetic resonance (NMR) spectroscopic techniques have been applied to confirm the formation and determine the molecular details of short oligonucleotide triplexes (de los Santos et al., 1989; Rajagopal and Feigon, 1989a,b; Mooren et al., 1990; Pilch et al., 1990b, 1991; Sklenar and Feigon, 1990; Umemoto et al., 1990; Kan et al., 1991; Radhakrishnan et al., 1991a,b,c, 1992a,b; Lin and Patel, 1992; Lopez and Lancelot, 1992; Macaya et al., 1992a,b; E. Wang et al., 1992; Radhakrishnan and Patel, 1993). The objects of these studies were intermolecular oligomer complexes (de los Santos et al., 1989; Rajagopal and Feigon, 1989a,b; Mooren et al., 1990; Pilch et al., 1990b, 1991; Umemoto et al., 1990; Kan et al., 1991) and intramolecular triplexes formed in short oligonucleotides folded twice upon themselves (Sklenar and Feigon, 1990; Radhakrishnan et al., 1991a,b,c, 1992a,b; Lin and Patel, 1992; Macaya et al., 1992; E. Wang et al., 1992; Radhakrishnan and Patel, 1993a,b; Bornet et al., 1994). Numerous NMR techniques (see Feigon et al., 1992, for a review of modern NMR strategies) allow the assignment of resonances for the majority of protons or other nuclei (e.g., ^{31}P) and reveal interactions between atoms which are remote along the chemical bonds but are spatially close because of specific conformation. From the NMR data it is possible to determine specific details of the triplex structure: the hydrogen bond patterns, the *syn* or *anti* orientations of bases relative to the sugar–phosphate backbone, the distances between the base planes in the strands, the distances between the bases of the third strands and those of the purine strand of the duplex, etc.

Reconstruction of the molecular structure from the NMR data relies upon the structural model, input data, and computational software (Feigon et al., 1992). For example, the data on short oligonucleotide triplexes were initially interpreted in terms of A′-DNA-like structures (Rajagopal and Feigon, 1989a,b; Mooren et al., 1990; Umemoto et al., 1990); however, it was later shown that the structures are closer to an underwound B DNA than an A DNA helix (Kan et al., 1991; Radhakrishnan et al., 1991a,b, 1992a; Lin and Patel, 1992; Macaya et al., 1992b; Wang, E. et al., 1992; Radhakrishnan and Patel, 1993).

Electron Spin Resonance (ESR)

Electron spin resonance (ESR) experiments using spin-labeled poly(A) and poly(U) allow one to follow the triplex formation between two polynucleotides (Timofeev et al., 1990). An imidazolide spin label was introduced into 2′-oxo-groups of polymer ribose in the proportion of one spin label per 18–20 bases. Formation of a triplex structure which is more rigid than single- and double-stranded structures results in an abrupt increase in the rotational correlation time, which relates to the loss of the spin-label orientation caused by an applied magnetic field. The temperature and pH dependence of rotational correlation time makes it possible to evaluate the proportion of poly(U) · poly(A) · poly(U) and poly(U) · poly(A) · poly(A) complexes in a dynamic equilibrium (Timofeev et al., 1990). Quantitative characteristics of the complexes formed can be obtained by computer simulation of the ESR spectra.

Differential Scanning Calorimetry

The principle of differential scanning calorimetry (DSC) consists of the measurement of heat absorbed by the system at rising temperatures that induce some transitions (e.g., conformational transitions). True heat absorption can be differentiated from incidental temperature effects by using parallel measurement of heat absorption of the buffer itself and the buffer with dissolved molecules under study. Usually, the plot of ΔC_p (excess heat capacity at constant pressure) versus temperature has several peaks corresponding to heat absorption during transitions in limited temperature intervals.

In contrast to optical melting, where the extraction of thermodynamic parameters depends on the chosen model, in DSC this same procedure is model-independent. The transition enthalpy and entropy can be easily determined by integrating the apparent excess heat capacity (ΔC_p) versus temperature (T) and $\Delta C_p/T$ versus T profiles, respectively (Figure 2.7). Differential scanning calorimetry of triple-stranded complexes has been shown to give more consistent results for multistate transitions like triplex melting (Ross and Scruggs, 1965; Krakauer and Sturtevant, 1968; Neumann and Ackermann, 1969; Ohms and Ackermann, 1990; Plum et al., 1990; Xodo et al., 1990, 1991; Park and Breslauer, 1992; Völker et al., 1993).

Equilibrium Sedimentation

When a macromolecular solution is centrifuged at a speed insufficient to precipitate macromolecules, their equilibrium distribution along the centrifuge tube is established after a sufficient time. A component of low molecular weight (e.g., salt) forms a continuous concentration (i.e.,

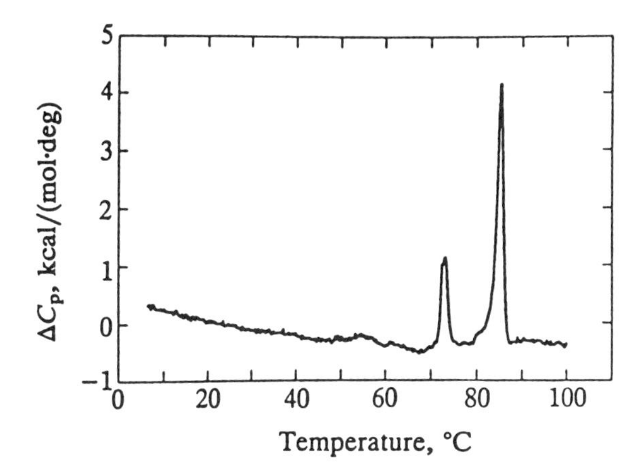

FIGURE 2.7. Calorimetric determination of excess heat capacity (ΔCp) as a function of temperature for the PyPuPy triplex poly(dT) · poly(dA) · poly(dT). Conditions: 10 m*M* phosphate buffer, pH 7.0, 0.1 m*M* EDTA, 0.55 *M* Na^+, 0.4 m*M* DNA. The enthalpies for triplex and duplex melting are determined to be the peak areas at 72°C and 85°C by integrating the ΔC_p versus T profile. The transition entropies can be determined integrating the $\Delta C_p/T$ versus T profile. From Park and Breslauer (1992).

density) gradient in the centrifuge tube, and macromolecules form bands at some positions along the tube according to their buoyant density.

This method of equilibrium sedimentation has often been used in studies of triple-stranded complexes composed of polynucleotides with simple repeating sequences. Due to the differences in buoyant density of single, double-, and triple-stranded complexes, their stoichiometry is easily determined under a variety of conditions (Chamberlin, 1965; Riley et al., 1966; Morgan and Wells, 1968; Murray and Morgan, 1973; Lee et al., 1979). Figure 2.8 shows that the band of a polynucleotide complex is shifted to a new position corresponding to a higher buoyant density when poly[d(TC)] · poly[(GA)] binds poly(UC) to form a triplex (Morgan and Wells, 1968). In such a way, various homopolymers and simple heteropolymers have been tested for their triplex-forming properties.

Electrophoretic Techniques

Two-Dimensional Agarose Electrophoresis

Much work in the study of intramolecular triplexes in plasmid DNA has been done using a two-dimensional agarose electrophoresis technique (reviewed by Bowater et al., 1992). As discussed in Chapter 1, in a covalently closed circular DNA, the number of times the strands wind about each other (the linking number, Lk) is constant. Because the

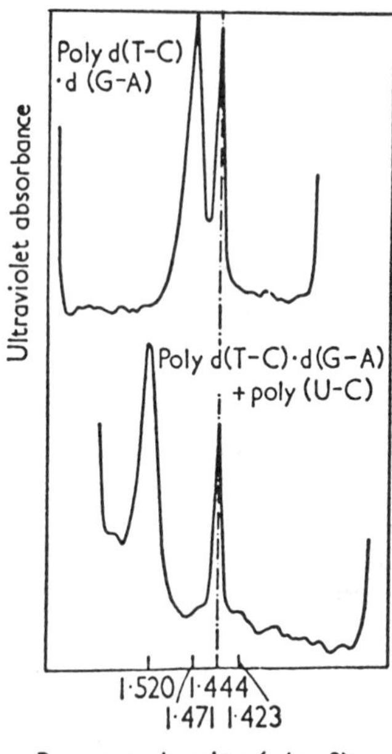

FIGURE 2.8. Determination of buoyant densities of polynucleotide complexes in a cesium sulfate gradient. Shown are the microdensitometer traces along the centrifuge cells. A density marker of T4 DNA ($1.444\,g/cm^3$) was added to each experiment. The duplex poly[d(TC)] · poly[d(GA)] has a buoyant density of $1.471\,g/cm^3$. Upon addition of the third-strand poly(UC), the density of the complex increases to $1.520\,g/cm^3$, indicating the formation of the poly[d(TC)] · poly[d(GA)] · poly(UC^+) triplex. From Morgan and Wells (1968).

population of DNA molecules (topoisomers) is distributed around the most probable topoisomer, this is an average value. More broad sets of topoisomers with different average superhelical densities may be generated by incubating DNA with a special enzyme, topoisomerase, in the presence of various ethidium bromide (EtBr) concentrations. When the superhelical density increases, any change in the linking number is distributed between Tw (twist) and Wr (writhe). The value of Tw always closely corresponds to 10.5 base pairs per helical turn. Therefore, the changes in Lk are realized in a global deformation of the duplex axis, resulting in the formation of superhelical turns. Excessive torsional tension can be partially relaxed due to creation of a locally unwound conformation. Such a structural transition can be revealed by the two-dimensional (2-D) distribution of topoisomer spots, part of which have retarded mobility in the first direction, due to the presence of a locally unwound structure, and normal mobility in the second direction, when the unwound structure is absent. For example, a complete set of plasmid topoisomers is separated first from a single well. Those topoisomers which are partly relaxed due to the presence of local triplex structures have reduced mobilities relative to those without relaxation. Their part of the ladder appears to be shifted backward to topoisomers with fewer superhelical turns. The topoisomer ladder is then run in a second direction, 90° to the first direction, in a gel containing an intercalating agent. This reduces negative supercoiling and eliminates the local triplex. The resulting 2-D distribution of topoisomer spots has a discontinuous "jump" at the topoisomer position beginning from where the relaxation occurs.

Figure 2.9 shows an application of 2-D electrophoresis to the study of H DNA formation in the plasmid pEJ4 containing a regular $(GA)_{16}$ sequence under decreasing pH (Lyamichev et al., 1985). At pH 5.6 (panel A), the H form could extrude at a very high superhelical density. A decrease of pH first to 5.1, then to 4.7, results in a decrease of several superhelical turns required for extrusion of the H form each time; no 2-D discontinuity was observed for the pUC19 plasmid where the H DNA does not form.

This method allows one to study intramolecular triplex extrusion as a function of DNA supercoiling, sequence variation, ambient conditions, etc. (Lyamichev et al., 1985, 1986; Pulleyblank et al., 1985; Mirkin et al., 1987; Htun and Dahlberg, 1989; Belotserkovskii et al., 1990; Collier and Wells, 1990; Klysik, 1992).

Gel Comigration

In the absence of triplex formation, a target duplex and a third strand oligonucleotide move separately in an agarose gel. Their positions can be detected using radioactive labeling of oligonucleotide and EtBr staining of the duplex. Under conditions of triplex formation, oligonucleotide and duplex change their mobilities and move in a single band (comigration) (Lyamichev et al., 1988). Figure 2.10 shows comigration of oligo d(C) with linearized plasmid DNA containing $d(G)_{35} \cdot d(C)_{35}$ insert, and comigration of $d(TC)_5$ with linearized plasmid DNA containing $d(GA)_{45} \cdot d(TC)_{45}$ insert. No cross-reactions of oligomers with other binding sites were found. Very strong triple-stranded complexes were detected, which did not dissociate during 10–20 h of electrophoresis. This technique is appropriate for the study of stable complexes with a lifetime greater than the amount of time necessary for electrophoretic separation.

This comigration of a third strand and a duplex has been used to confirm triplex formation as a function of a number of variables, including pH, temperature, salt type, and concentration, etc. (Lyamichev et al., 1988; Kibler-Herzog et al., 1990; Manzini et al., 1990; Xodo et al., 1990; Pilch et al., 1991; Roberts and Cronthers, 1991, 1992; Jetter and Hobbs, 1993) and to test the stabilities of various nonstandard triads (Yoon et al., 1991).

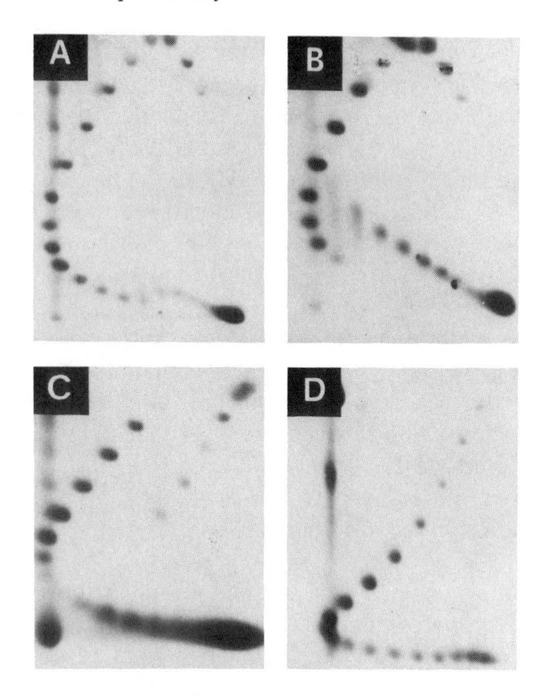

FIGURE 2.9. Two-dimensional agarose gel electrophoresis of the DNA topoisomers. DNA topoisomers were separated in the first direction using sodium citrate buffers of various pHs. For the separation in the second direction, the gels were saturated for 4–5 h by tris-acetate buffer, pH 7.8, containing 2.5 μg/ml chloroquine. This intercalating drug relaxed the negative superhelicity, thereby eliminating the triplex structure. (A) pEJ4 DNA containing a $(GA)_{16}$ insert at pH 5.6. The DNA topoisomers form a smooth spot curve when the triplex is not formed. (B) pEJ4 DNA at pH 5.1. The formation of the triplex in a $(GA)_{16}$ insert results in the spot "jump" between topoisomers at the bottom of the spot curve. (C) pEJ4 DNA at pH 4.7. Lowering the pH results in triplex formation at a low superhelical density (the "jump" at the top of the spot curve). (D) Control pUC19 DNA, which does not contain a triplex-forming insert at pH 4.8. From Lyamichev et al. (1985). Reprinted with permission.

In the case of short duplexes and third strand oligonucleotides (several dozen bp long), the triplex formation results in appearance of a novel band corresponding to a triple-stranded complex. This band migrates slower than an oligonucleotide or duplex DNA. Thus, some of the material detected by EtBr staining or radioactivity labeling is retarded or shifted to a new position. Therefore, this technique is also called band-shift assay, gel-retardation assay, or mobility shift assay.

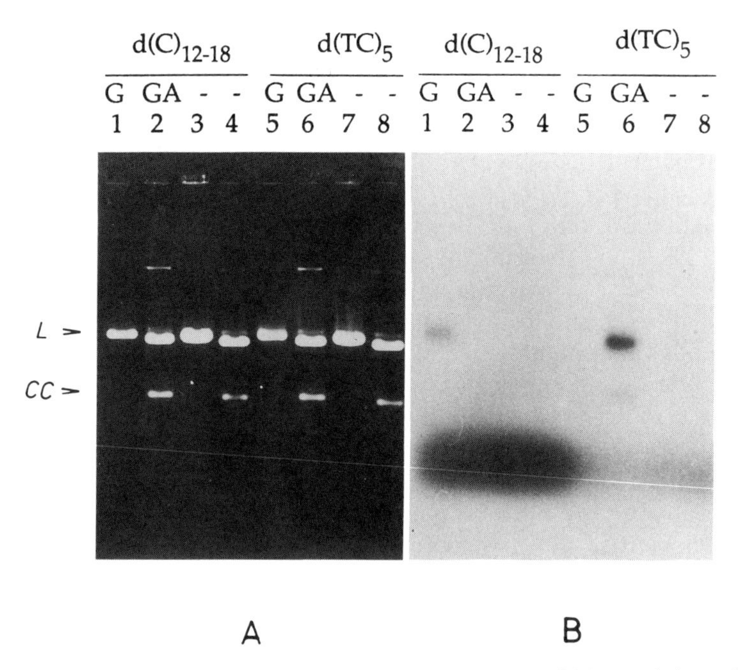

FIGURE 2.10. Gel-electrophoresis of linear plasmid DNA pG35 containing a $(G)_{35} \cdot (C)_{35}$ insert (lanes 1 and 5) and pTC45 containing a $(GA)_{45} \cdot (TC)_{45}$ insert (lanes 2 and 6), and their parental vectors pUC19 (lanes 3 and 7) and pDPL13 (lanes 4 and 8), respectively. When the DNAs were preincubated with either ^{32}p-$(C)_{12-18}$(lanes 1–4) or ^{32}P-$(TC)_{45}$ (lanes 5–8), these single strands formed complexes and comigrated with their own double-stranded targets, as seen by the lowering oligonucleotide mobilities in lanes 1 and 6 (panel B). Conditions: 1.5% agarose gel, 100 m*M* sodium citrate buffer, pH 4.6. Panel A: EtBr-stained gel. Panel B: radioautograph of the same gel. Reproduced from Lyamichev et al. (1988) *Nucleic Acids Res* by permission of Oxford University Press.

Immunological Methods

Antibodies to nucleic acids can serve as reagents for the detection of specific nucleic acid structures and structural transitions (reviewed by Stollar, 1992). Even antibodies to the same structural form of a nucleic acid can recognize different aspects of the structure in question: some of them are structure- and sequence-specific and may interact with the bases in the grooves; others seem to interact predominantly with the sugar–phosphate backbone, whose conformation does not vary greatly with the sequence changes (Stollar, 1992). A major motivation for the preparation of antibodies to unusual nucleic acid structures is their application to the search for such structures in natural nucleic acids within cells.

Early experiments showed that the triple-helical polyribonucleotides in complexes with a positively charged carrier (e.g., methylated bovine serum albumin) were strongly immunogenic (Lacour et al., 1973).

For experiments in vitro, rabbit antibodies were raised to the triple-helical polynucleotides poly(U) · poly(A) · poly(I), poly(U) · poly(A) · poly(U), and poly(I) · poly(A) · poly(I), each recognizing a different conformational feature of the triple-helical immunogen (Rainen and Stollar, 1977). Two of them reacted with double-stranded poly(A) · poly(U) or poly(A) · poly(I) regions of the triple helix, and the third required all three strands for reactivity. It was also shown that structures with poly(A) annealed to any polypyrimidine strand [poly(U), poly(dU), poly(dT)] were immunologically identical; however, they were quite different from an analogous triplex structure built on poly(dA) (Stollar and Raso, 1974). By contrast, antibodies induced by poly(U) · poly(dA) · poly(U) react with poly(U) · poly(dA) · poly(dT) and poly(dT) · poly(dA) · poly(dT), but not with the complexes involving poly(A). Thus, sometimes antibodies can be used not only to distinguish triple-helical structures from double-helical ones, but even to distinguish various triple-helical polynucleotides from each other (Stollar, 1992).

Antibodies specific for poly(A) · poly(U) · poly(U) were used to determine the polynucleotide size and base modifications appropriate for triplex formation (Kitagawa and Okuhara, 1987). The binding sites of rabbit antibodies were as small as 3–4 base triads, that allowed an estimation of minimum sizes for maintaining a stable triple helix as 10 bases for oligo(A) length and 20 bases for oligo(U) length.

Monoclonal antibodies (Jel 318 and Jel 466) were produced by immunizing mice with poly[d(Tm^{5}C)] · poly[d(GA)] · poly[d(m^{5}CT)], which forms a stable triplex at neutral pH (Burkholder et al., 1988; Agazie et al., 1994). To characterize the binding specificities of these antibodies, homopurine–homopyrimidine polymers of repeating sequences were mixed under triplex-forming conditions before returning to neutral pH where immunoblotting or solid-phase radioimmunoassay was accomplished. The patterns of specificities are significantly different. For instance, both antibodies bind to the triplexes derived from poly[dT^{5}mC] · poly [d(GA)] and poly[d(TTC)] · poly[d(GAA)], but Jel 318 binds to poly[d(T)] · poly[d(A)] · poly[d(T)] whereas Jel 466 does not, and Jel 466 binds to poly[d(C)] · poly[d(G)] · poly[d(C$^+$)] whereas Jel 318 does not (Burkholder et al., 1988; Agazie et al., 1994). Jel 318 was shown to bind to fixed mouse metaphase chromosomes and interphase nuclei (Lee et al., 1987; Burkholder et al., 1988); however, some ambiguity in interpretation could arise, as fixation in the initial experiments itself could change the structure under study (Hill and Stollar, 1983). Unfixed, isolated mouse chromosomes also reacted positively with the antibody, particularly when they were gently decondensed by exposure to low ionic conditions at neutral pH (Burkholder et al., 1988). Unlike Z DNA detection (Hill and

Stollar, 1983; Robert-Nicoud et al., 1984), Jel 318 immunofluorescent staining of the polythene chromosomes was independent of the specific fixation methods. A comparison of Jel 318 and Jel 466 binding to chromosomes showed a reciprocal relationship between the pattern for Jel 466, on the one hand, and that for Hoechst 33258 and Jel 318, on the other. Jel 466 did not bind to C and G bands, which are AT-rich, but bound to the R band, which is GC-rich. The opposite was found for Hoechst 33258 and Jel 318. Thus, the staining preferences matched the sequence preferences of two antibodies.

The accumulating data present more evidence that triplex structures can be found in cells using immunological methods coupled with mild chromosome decondensation. However, these data should be interpreted with special caution, since it has not yet been shown whether decondensation simply makes the triplex accessible to antibodies or whether it involves changes that favor triplex formation (Stollar, 1992).

Antibodies against DNA modified with a single-strand selective probe, OsO_4 in complex with 2,2′-bipyridine (see Enzymatic Chemical Probing of Triplex Structure), were raised in rabbits (Kuderova-Krejcova et al., 1991) and mice (Kabakov and Poverenny, 1993). They were used to detect a number of non-B structures (including triplexes) in plasmid DNA in vitro, some modified single-stranded regions in cell nuclei, and in polythene chromosomes from *Drosophila melanogaster* (Kabakov and Poverenny, 1993).

Affinity Methods

Chromatography and filter binding assay methods, which exploit the differences in affinity of the third strands to their duplex PyPu targets as a function of solution conditions or the presence of mismatches in the third strand, may be used to determine some key features of triple-stranded nucleic acids.

Affinity Chromatography

A variant of chromatography in which complementary duplexes can bind to an agarose-linked third-strand polynucleotide has been used to study the sequence-specific formation of synthetic polyribonucleotide triplexes under a variety of ambient conditions (Letai et al., 1988). The on column incubation time for the duplex is typically several minutes, and the complex formation is weakly dependent on a temperature between 4° and 37°C. The method has a significant advantage in that the oligomers, which tend to form self-aggregated structures in free solution [e.g., poly(G) and poly(I)], are too far apart on column to interact with one another. Thus, it was possible to observe the formation of the U/TA*I, CG*I, and CG*G

TABLE 2.2. Binding of nucleic acid duplex ligands by agarose-linked polynucleotides.

Potential ligand	Agarose-linked strand*					
	Poly(A)	Poly(U)	Poly(G)	Poly(C)	Poly(I)	Agarose
Poly(A) · poly(U)	20%	89%	—	—	72%	—
Poly(AU) · poly(AU)	—	—	—	—	—	—
Poly(dA) · poly(dT)	85%	88%	—	—	85%	—
Poly(I) · poly(C)	—	—	57%	**	19%	—
Poly(IC) · poly(IC)	—	—	—	—	—	—
Poly(G) · poly(C)	—	—	43%	29%	32%	—
Poly(dG) · poly(dC)	—	—	80%	23%	35%	—
Poly[d(GC)] . poly[d(GC)]	—	—	—	—	—	—
Poly[d(AG)] · poly[d(CT)]	—	—	—	—	25%	—
Poly[d(AC)] · poly[d(GT)]	—	—	—	—	—	—

* Minus sign indicates the absence of binding.
** Significant binding was observed below pH 5.

triads. Table 2.2 shows the specificity of duplex binding by agarose-linked polynucleotides. Application of the technique was restricted to a limited number of commercially available polynucleotide–agarose conjugates.

However, the development of custom synthesis of oligonucleotides bearing functional end groups suitable for convenient cross-linkng to chromatographic sorbents gives definite hope for more extensive use of affinity chromatography based on the triplex mechanism. Immobilization of a pyrimidine oligonucleotide on a thiol-Sepharose support was accomplished through a disulfide bond at the 3′ oligonucleotide terminus (Pei et al., 1991). After annealing a complementary strand to form a duplex, a large-scale hybridization of a randomized RNA library was performed to select sequences that bind specifically to the target DNA via the triplex mechanism. Virtually any DNA substrate can be linked to a chromatography matrix via avidin–biotin linkage (Fishel et al., 1990). Coupling of an oligonucleotide bearing a primary amino group on the 5′-terminal phosphate to an activated N-hydroxysuccinimidyl silica could be a major step toward high-performance affinity chromatography (Goss et al., 1990). Hopefully, both of the latter methods will be used in the studies and applications based on triplex-mediated chromatography. So far, the triplex-affinity mechanism has been successfully used for analytical purposes and for selective enrichment of natural DNA and RNA sequences (see also Chapter 7).

Filter-Binding Assay

The kinetic and thermodynamic properties of the PyPuPy triplexes as a function of pH and temperature have been studied using the difference in retention of single-, double-, and triple-stranded DNA molecules on a nitrocellulose membrane filter (Sarai et al., 1993; Shindo et al., 1993). The retention efficiency of the filter for single-stranded homopyrimidines is around 45%–65%, an order of magnitude greater than that for double- and triple-stranded DNA molecules. When a previously prepared complex is filtered through the nitrocellulose membrane, the stability of the triplex is inversely proportional to the amount of single strand that can be immobilized on the filter. Figure 2.11 shows the pH dependence of the PyPuPy triplex formation. When pH increases and the protonated triplex becomes less stable, much more double-stranded DNA is required to bind the fixed amount of the radioactive third strand.

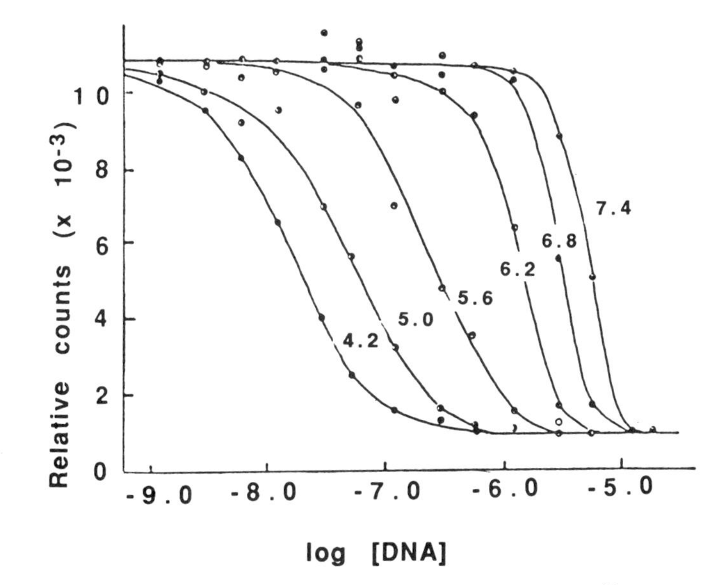

FIGURE 2.11. Representative plots of relative counts in cpm of ^{32}P end-labeled 5′-dTCCTCTTCTTTTCTTTCTT-3′ versus logarithm of concentration of the duplex (5′-dGCAGGAGAAGAAAAGAAAGAACG-3′ + 3′-dCGTCCTCTTCTTTTCTTTCTTGC-5′), assayed at 15°C and different pH values by filter-binding experiments. The single strands are retained on the filter much more efficiently than the duplexes or triplexes. A decrease in filter-bound radioactivity means that less oligonucleotide is present in a single-stranded form; rather it goes through the filter in the duplex-bound form. Under triplex-forming conditions (low pH), the affinity of the Py single strand for its duplex target is high, and therefore less duplex DNA is required to bind the radioactive oligonucleotide and carry it through the filter. Very high duplex concentrations are needed to bind the available fixed amount of the third strand at triplex-destabilizing neutral pHs. Reprinted with permission from Shindo et al. (1993). © 1993 American Chemical Society.

This method allows one to handle many samples simultaneously but requires significant care, because the retention efficiency of the filter is affected by filtering speed, temperature, pH, and, probably, base sequences. Detection of homopyrimidine sequences in complex mixtures of RNA extracted from the cell or generated in vitro can be accomplished by a triplex blotting technique (Noonberg et al., 1994b). The RNA mixture is separated on a denaturing polyacrylamide gel, transferred to a nylon membrane and then hybridized with a radioactive duplex probe. To determine optimal binding conditions, ionic conditions or pH can be changed in the hybridization buffer.

Electron Microscopy

Electron microscopy as a method of direct visualization has been used to demonstrate the formation of triplexes in several variants.

Specimens of supercoiled DNA containing a 60-base-pair-long PyPu tract were prepared for electron microscopy at acidic pH, which promotes PyPuPy triplex formation. A single "stem" or a "denaturation bubble," corresponding to the region of intramolecular foldback, or a looped-out single strand were observed in a large number of DNA molecules (Stokrova et al., 1989). These local DNA distortions largely correspond to sites of significant curvature in supercoiled DNA molecules, which supports the conclusion that the formation of an intramolecular triplex should result in a sharp kink in a DNA molecule (Htun and Dahlberg, 1988).

The formation of intermolecular triplexes was detected by the appearance of "rosettes" when several plasmid DNA molecules carrying the PyPu insert were simultaneously bound to a short single-stranded oligo[d(Tm^{5}C)] with an average length of 170 nucleotides (Lee et al., 1989).

Electron microscopic visualization of the triplex formed between the plasmid DNA and biotinylated oligonucleotides was described (Cherny et al., 1993b). Streptavidin, which binds to the biotin moiety, was used as a marker of binding sites for purine or pyrimidine oligonucleotide targeted to the 17-base-pair-long PyPu sequence of a fragment of human papilloma virus-16 inserted in plasmid DNA. Either the PyPuPu or the PyPuPy triplex can be detected with proper choice of specific oligonucleotide and incubation conditions. More than 80% of DNA molecules bound the streptavidin marker in the correct position of the PuPy tract, and very few cases of nonspecific binding were detected.

The specific triple-stranded structure was found for the DNA–PNA (peptide nucleic acid) complex (Cherny et al., 1993b). A homopyrimidine peptide nucleic acid oligomer forms Watson–Crick and Hoogsteen bonds with the purine DNA strand, thereby displacing the pyrimidine DNA strand. Such DNA looping at the binding site was confirmed by data from electron microscopy (Cherny et al., 1993a; Demidov et al., 1994a).

X-Ray Analysis

X-ray diffraction by nonhomogeneously distributed electron density in molecules is one of the most efficient methods of structure determination. The structural details of crystallized molecules can be determined at 3 Å or better resolution. In general, fiber diffraction patterns of the polynucleotides and nucleic acids contain relatively few diffraction maxima, resulting in lower resolution compared to single crystals. However, the fiber diffraction pattern provides the dimensions of the helix and other information on the possible arrangement of diffracting matter in the fiber. Usually, the structure of nucleic acids in fibers is deduced from comparison of the experimentally observed diffraction pattern with the calculated pattern based upon stereochemically acceptable models.

Some of the polynucleotide triplexes were studied by X-ray diffraction techniques (Rich, 1958a,b; Sasisekharan and Sigler, 1965; reviewed by Davis, 1967). The data obtained, in combination with model building, resulted in sets of parameters describing the helices and schemes of triads for poly(I) · poly(A) · poly(I) (Rich, 1958a), poly(I) · poly(I) · poly(I) (Rich, 1958b), and poly(U) · poly(A) · poly(U) (Sasisekharan and Sigler, 1965; Davis, 1967).

For more than a decade, the model for molecular details of triple-stranded complexes was based on the X-ray studies of Arnott et al. (see Arnott and Bond, 1973; Arnott and Selsing, 1974; Arnott et al., 1976). The data from the analysis of poly(U) · poly(A) · poly(U), poly(I) · poly(A) · poly(I), and poly(dT) · poly(dA) · poly(dT) fibers were interpreted as evidence of a series of similar structures of the A type with 12-fold helix and C3′-endo furanose rings. This conclusion is in apparent contradiction with modern structural data (Liquier et al., 1991; Radhakrishnan et al., 1991b, 1992; Howard et al., 1992; Macaya et al., 1992b; Ouali et al, 1993a,b), which show that the conformations of strands in the triplex instead correspond to the underwound B form. Explanations for this discrepancy could be either the choice of the starting structure for a refinement of the X-ray model or the dependence of triplex conformations on the nucleotide sequence, as well as the structural difference between triplexes in solution and in aggregated fiber form.

More recently, Campos and Subirana (1987, 1991) used an X-ray technique to study the condition-dependent formation of duplex and triplex structures of poly(dG) · poly(dC). Recently, data showing the fiber-type diffraction patterns from single crystals of oligonucleotide triplexes (Liu et al., 1994) and the high-resolution structure of TAA and CG*G triads in model oligonucleotide complexes have been reported (Berger et al., 1995; Meervelt et al., 1995).

A Short Overview of Experimental Physical Methods for Triplex Studies

The methods described above give insights into the various aspects of triplex structures. Some of them have been used to confirm the formation of triple-stranded structures (UV-mixing curves, CD, fluorometry, ESR, sedimentation, gel comigration, electron microscopy). However, more information can be obtained using samples with various sequences and under different ambient conditions.

UV and calorimetric melting have been used for quantitative thermodynamic characterization of intermolecular triplexes. More recently, the filter-binding assay was adjusted to find the thermodynamic parameters for triplexes at temperatures far from melting intervals. For intramolecular H DNA, 2-D electrophoresis has proved to be the method of choice for thermodynamic description of triplex formation.

Sedimentation, UV, NMR and IR spectroscopy, gel comigration, 2-D electrophoresis, and affinity chromatography have been used to determine specific triads and the consequences of imperfect triads for triplex stability. NMR and IR spectroscopy and X-ray analysis can also provide more detailed information on triplex conformation (for example, sugar pucker type or base orientation relative to the backbone). However, the resulting triplex structure seems to be dependent on a starting model of the complex, which is used for molecular modeling and simulation of experimental data.

The physical methods described above differ with respect to their requirements for the quantity of DNA (oligonucleotide) samples. For many of them (UV and CD spectrometry, fluorometry, equilibrium sedimentation, electrophoresis, electron microscopy), a single experiment requires 1 to 10 μg of DNA. Calorimetry, IR spectroscopy, and affinity chromatography require an amount an order of magnitude higher. NMR and X-ray techniques consume milligram quantities of polynucleotides. In the majority of methods, the preliminary incubation to form the triplexes is relatively long compared to the measurement period of only several minutes. In electrophoretic experiments, time requirements for triplex formation and separation are comparable (up to several hours). Equilibrium sedimentation is the longest procedure, requiring about 20 hours.

Generally, to avoid misinterpretation of the data, several methods should be used for the physical characterization of triple-stranded structures. In some cases, the experimental data can be compared with the results of computer simulation, which uses some other experimentally determined values as input parameters.

Theoretical Descriptions of the Triplex Structures

To date, there have been only a few theoretical works concerning triplex formation. The first quantum mechnical calculations were related to the interaction energies in hydrogen-bonded base triads (Pullman et al., 1967). It was shown that the interaction energy is relatively small in IA*I and relatively very great in CG^*C^+ triads. The involvement of a protonated cytosine in hydrogen bonding with guanine (or with a nonprotonated cytosine) greatly increases the energy of the interaction.

The first thermodynamic description of the pH-dependent structural transition was that related to extrusion of the protonated H form in supercoiled DNA (Lyamichev et al., 1985). It predicted a linear pH dependence of the superhelical density at which transition occurs. In the case of the intermolecular protonated triplex, the corresponding theory predicted a very sharp pH dependence of the triplex lifetime in the range $pK_f < pH < pK_t$, where pK_f and pK_t are the pK values for the protonated base in the single- and triple-stranded forms, respectively (Lyamichev et al., 1988). This theory is applicable to any transition which depends on protonation state (Frank-Kamenetskii, 1990, 1992).

The further development of the theory dealt with the number of base pairs in the PuPy tract which can form the H DNA structure and took into account the presence of a homopyrimidine loop inside the structure (Lyamichev et al., 1989b). Using experimental data on pH-dependent H-form extrusion in PyPu tracts of various lengths, the free energy of nucleation of the H form was estimated as 18 kcal/mol and the minimum length of the PuPy tract necessary for extrusion of the H form as 15 base pairs.

Molecular mechanics simulations using the AMBER force field were performed for the PyPuPy and PyPuPu triplexes (Laughton and Neidle, 1991, 1992a,b). Comparison with the Arnott and Selsing triplex model shows that some elements of the A form are retained, especially in the Pu strand (Laughton and Neidle, 1992a). The orientation of the phosphate groups of two Py strands into the major groove creates a channel of increased negative charge density (Laughton and Neidle, 1992a), so that triplex stabilization by polyvalent cations can be explained in terms of charge neutralization (Maher et al., 1990; Hampel et al., 1991; Lyamichev et al., 1991; Singleton and Dervan, 1993). Calculations for the PyPuPu triplex $d(C)_{10} \cdot d(G)_{10} \cdot d(G)_{10}$ show that two likely structures are possible. One has the glycosidic torsion angles of the third strand bases in the *anti* conformation and Hoogsteen hydrogen bonds to the purine strand of the duplex; the other has the third strand purines in the *syn* orientation and uses a reverse-Hoogsteen hydrogen-bonding pattern. The *syn* orientation is slightly preferred in calculations in vacuo; however, if explicit solvent molecules are taken into account, the *anti* orientation is much preferred. When TA*A or TA*T base triads are substituted for the six CG*G triad in the sequence, the *syn* orientation for the third strand base of the TA*A triad in the solvated model is preferable, while the third base in the TA*T triad retains a preference, though reduced, for the *anti* conformation (Laughton and Neidle, 1992b).

The problem of Hoogsteen versus reverse Hoogsteen hydrogen bonds was studied in relation to the natural *c-myc* promoter sequence (Cheng and Pettitt, 1992b). A combination of molecular mechanics, molecular dynamics (CHARMM force field), and Poisson–Boltzmann solvation energy calculations was performed for homogeneous $d(TA^*T)_{27}$ and $d(CG^*G)_{27}$ triplexes. For $d(TA^*T)_{27}$ triplexes, both internal potential energy and solvation contribute to the experimentally known base pairing and strand orientation. Solvation was found to determine the third strand orientation for $d(CG^*G)_{27}$ triplexes, with either Hoogsteen or reversed-Hoogsteen base pairing between the two purine strands being possible.

Recently, the results of a molecular dynamics simulation study of antiparallel reverse-Hoogsteen CG*G heptamer triplex were reported (Mohan et al., 1993b). The simulation explicitly treated 837 water molecules, 21 Na^+ ions, and excess NaCl at a concentration of 1 *M*. A unique hydration spine was found in the groove formed between the cytosine and the third guanine strands of a triplex. A series of water molecules, bound between the NH_2 groups of the third G strand and of the C strand at each base plane, is characteristic of a structure found in a 100-ps simulation of a rigid triplex and in a 500-ps simulation of a flexible triplex structure. This spine of hydration would not be possible in TA*T and TA*A systems, where no favorable disposition of amino groups on pairing bases is available.

Molecular dynamics studies of the poly(dT) · poly(dA) · poly(dT) triplex using the AMBER force field resulted in an equilibrium structure different from both A and B DNA (Sekharudu et al., 1993). The sugar pucker, the major groove width, and the base tilt are analogous to B DNA, whereas the axial base displacement and helical twist resemble that of A DNA.

Molecular modeling studies (Raghunathan et al., 1993) predict a triplex structure of $(dT)_n \cdot (dA)_n \cdot (dT)_n$, with the geometric parameters of the B form in accordance with the infrared spectral data (Howard et al., 1992) and in apparent disagreement with the A-form model of Arnott and Selsing (1974). Energy-minimized model GC-containing triplexes were found to have various combinations of C3′-endo and C2′-endo type sugar puckers, depending on the combinations of polyribo and polydeoxyribo strands in agreement with the FTIR and Raman spectroscopy data (Ouali et al., 1993b).

Ab initio quantum mechanical and statistical mechanical studies on the equilibrium geometries and solution proton affinities of 5-methylcytosine and cytosine have been carried out (Hausheer et al., 1992). The data obtained were used in large-scale numerical simulation models of a DNA triple helical system $d(CT)_{10} \cdot d(GA)_{10} \cdot d(5mC^{+}T)_{10}$. In accordance with an experimental trend, additional enthalpic triplex stabilization by 7.3 kcal/mol per base triad for 5-methylcytosine-substituted third strand relative to the cytosine-substituted one was predicted.

The ability of the methylphosphonate oligonucleotide analog to form a triplex was tested using a molecular dynamics simulation (AMBER force field) (Hausheer et al., 1990). The fully solvated modified triplex formed a hybrid of the A and B conformations and was more stable than the unmodified parent triplex $(dT)_{10} \cdot (dA)_{10} \cdot (dT)_{10}$, presumably because of reduced interphosphate repulsion. This result was in accordance with experimental data from the same group (Ts'o et al., 1992) but contradicted the data of another group, where a complex pattern of stabilization was observed for complexes of oligonucleotides containing methylphosphonate linkages (Kibler-Herzog et al., 1990).

The effect of mismatches in oligonucleotide triplexes has been studied using several computational approaches (Pettitt and Rossky, 1990; van Vlijmen et al., 1990; Mergny et al., 1991b). Local geometric distortions suggested by theoretical considerations manifested themselves in experimentally discovered triplex destabilization (Mergny et al., 1991b). The relative effects of van der Waals and electrostatic interactions as well as hydrogen bonds on energetics of TA*T and CG*G triplexes and similar triplexes with some mismatches were estimated (van Vlijmen et al., 1990).

Ab initio quantum mechanical and molecular mechanics (CHARMM force field) calculations were used to characterize the geometric and thermodynamic consequences of the base-pair reversals (TA $\rightarrow$ AT) in double-helical DNA inside the PyPuPy triplex (Mohan et al., 1993a). Single-base-pair reversals significantly destabilize local structures, but the effect can be alleviated by the use of some nonnatural, xanthine-like bases.

Energy minimization (Sun et al., 1991b; Sun and Lavery, 1992) shows that in α- and β-anomeric oligonucleotides the Hoogsteen or reversed-Hoogsteen types of pairing and corresponding orientation of the third Py strands are dictated by the nucleotide sequence.

Based on a constrained molecular modeling approach, Srinivasan and Olson (1993) generated various DNA, RNA, and hybrid DNA/RNA triple helices and found their thermodynamic parameters to be in agreement with the experimental stability series (Roberts and Crothers, 1991, 1992) (see Chapter 3).

Enzymatic and Chemical Probing of Triplex Structures

Physical methods are suitable for the study of triple-stranded complexes in polymer and oligomer complexes where the triplex occupies the major portion of the structure. However, such examples are unusual in natural

DNA. The formation of intra- and intermolecular triplexes often affects only minor portions of the DNA molecule. For instance, the PyPu sequences in plasmids usually comprise only a few percent of the total DNA length. Therefore, a number of enzymatic and chemical methods appropriate for investigation of local structures in DNA have often been used in triplex studies. Analysis of local DNA structures at the sequence level is accomplished using either Maxam–Gilbert or Sanger sequencing principles.

Analysis of Modifications Using Maxam–Gilbert Sequencing Gel Analysis

Detection of chemically modified bases or bases protected from modification in the triplex structure is often accomplished by end-labeling the DNA fragment and cleaving the backbone at the positions of the adducts by treatment with a suitable agent, such as piperidine, followed by an analysis in a sequencing gel using Maxam and Gilbert reactions as markers. This method has been used in the majority of studies in which chemical, photochemical, and enzymatic modifications have been used to reveal triplex structures.

Primer Extension Analysis

The primer extension analysis, employing modified DNA as a template for an RNA or DNA polymerase that stops at the chemically or photochemically produced adducts, can also be used to analyze structures at the sequence level. The sequencing gel analysis reveals the distribution of newly synthesized radioactive fragments (see Htun and Johnston, 1992, for review). This method is more convenient than the chemical cleavage approach, in that it obviates the troublesome procedures of preparing a singly end-labeled fragment and hot piperidine treatment. Incorporating multiple radioactive labels in the newly synthesized strand and repeated cycles of synthesis increases the sensitivity of analysis. Certain modifications that are not alkali-labile (e.g., methylation of the N1 of adenine) can be detected, as they effectively block elongation of a new chain. Adducts that do not block the progression of polymerases, such as 7-methylguanine, can be revealed by the polymerase approach after a preliminary chemical scission of the modified strand. The drawbacks of this approach are a slight heterogeneity in termination at the modified site and premature termination in a small fraction of newly synthesized strands, both of which increase the radioactive background.

Two variants of this method have been applied to triplex studies. DNA polymerase (Sequenase, Klenow fragment, etc.) which extends the primer cannot overcome chemically (Htun and Dahlberg, 1988; Htun and Johnston, 1992; Kang et al., 1992b) or photochemically modified sites

(Ussery and Sinden, 1993). A comparison between mapping the chemical modification products according to termination of transcription analysis and the hot piperidine technique shows a strong correlation between the results of the primer extension and the chemical cleavage approaches (Htun and Johnston, 1992).

On the other hand, primer extension on the sequence capable of forming an intramolecular triplex was used to show that triplex formation prevents DNA polymerase progression along the template and therefore terminates DNA polymerization (Lapidot et al., 1989; Baran et al., 1991; Dayn et al., 1992b).

Enzymatic Methods

Single-Strand-Specific Nucleases

Mung bean, P1, and S1 single-strand-specific nucleases recognize unpaired parts of the PyPu tract in H form DNA and digest them, producing duplex DNA fragments with blunt ends. Localization of the regions subjected to digestion by single-strand-specific nucleases is accomplished using Maxam–Gilbert sequence analysis, primer extension analysis, and determination of fragment lengths in an agarose gel. Among these enzymes, nuclease S1 from *Aspergillus orizae* has been most frequently used in the study of local DNA structure (see Yagil, 1991, for a list of genes susceptible to single-strand-specific nucleases). This enzyme has an optimum pH between 4 and 5, requires the presence of Zn^{2+} ions, and recognizes structural distortions that result in the more or less single-stranded character of the sugar–phosphate backbone. Nuclease S1 was one of the major tools used in the discovery of unusually structured PyPu tracts in natural DNAs, which were later shown to adopt an intramolecular triplex conformation (see Chapter 1). The drawback of working with nuclease S1 is the need to work in acidic media. Although at acidic pH the protonated triplex forms readily, this fact is not relevant to physiological conditions.

Mung bean and P1 nucleases are nearly as active at neutral pH as at pH values between 4 and 5. They have also been used to reveal single-stranded parts of H DNA (Christophe et al., 1985; Fowler and Skinner, 1986; Blaho et al., 1988; Collier et al., 1988; Hanvey et al., 1988a; Bacolla and Wu, 1991; Klysik, 1992; Beltran et al., 1993; Völker et al., 1993).

DNase I Footprinting

DNase I footprinting is based on the partial cleavage of duplex DNA with a double-strand-specific nuclease and subsequent analysis of susceptible regions using Maxam–Gilbert sequencing. Those sites which are not susceptible to digestion are protected by some bound ligand (see Tullius, 1989, for review). Protection of duplex DNA from DNase I digestion by a third-strand oligonucleotide was used to reveal DNA sequences where a triplex was formed (Durland et al., 1991; Postel et al., 1991; Sun et al., 1991a; Duval-Valentin et al., 1992; Gee et al., 1992; Jayasena and Johnston, 1992b; Krawczyk et al., 1992; Maher, 1992; Nielsen, 1992; Grigoriev et al., 1993a; Kessler et al., 1993; Roy, 1993).

Inhibition of Restriction Nuclease Action

Formation of a triplex can inhibit the action of a restriction/modification enzyme if the recognition site overlaps the PyPu tract (François et al., 1989c; Maher et al., 1989; Hanvey et al., 1990). Such recognition sites include those for *Eco*RI, *Msp*I, *Hae*III, *Taq*I, *Alu*I, *Ava*I, and *Hpa*II. Figure 2.12 shows the scheme of such an experiment with two specially constructed plasmids. When the PyPu tract and the recognition site partially overlap (pRW1704 plasmid), the restriction site is protected against cleavage and methylation. There is no such overlap in plasmid pRW1703, and the *Eco*RI site adjacent to the PuPy tract is therefore normally methylated (Hanvey et al., 1990). The extent of duplex protection by triplex-bound oligonucleotides can be used to study the relative stabilities of triple-stranded complexes formed by oligonucleotides of various sequences. Inhibition of *Ava*I activity by the third strand has been used in a kinetic analysis of triplex formation (Maher et al., 1990).

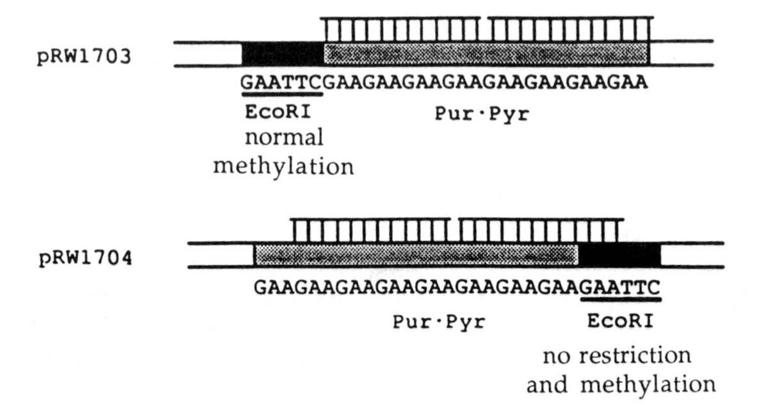

FIGURE 2.12. Protection of the recognition site from restriction and methylation enzymes (*Eco*RI and *MEco*RI) by the intermolecular triplex in the overlapping PuPy tract (lower panel). The recognition site immediately adjacent to the triplex is normally methylated (upper panel). Indicated are the PuPy tract, the *Eco*RI recognition site and two $(CTT)_4$ oligomers forming the triplex. Reproduced from Hanvey et al. (1990) *Nucleic Acids Res.* by permission of Oxford University Press.

Maintenance of a selected restriction site unmethylated due to temporary protection from the triplex-forming oligonucleotide was suggested in a variant of the "Achilles heel" methodology. Rare restriction sites that were not enzymatically methylated due to temporary protection allowed rare chromosome cutting (Dervan, 1992; see also Chapter 7).

Chemical Methods

The application of enzymes to probing DNA structures is limited by the prerequisite conditions that optimize enzymatic activity, the possibility of inducing structural changes in DNA, etc. Therefore, a number of chemical methods are being developed for probing DNA structures. Perhaps the first example of chemical probing in triplex studies was the use of formaldehyde to measure noncomplexed poly(A) during examination of the complexes between poly(A) and poly(U) under various conditions (Massoulie, 1964). Modern variants of chemical probing provide more detailed information at the sequence level. Figure 2.13 shows a general scheme for chemical probing of an intramolecular triplex with Maxam–Gilbert sequence markers.

Unbound Agents

A number of chemical compounds possess different affinities to ordinary B DNA and DNA containing unusual local structures (e.g., triplexes). These chemical agents discriminate between accessible and hidden potential reaction sites. The susceptibility of these sites depends on whether they belong to single-, double-, or triple-stranded structures. The chemicals can react with the functional groups in bases (glycid- and haloacetaldehydes, osmium tetroxide, potassium permanganate, hydroxylamine, diethylpyrocarbonate, and dimethylsulfate) or deoxyribose (uranyl ion, copper-phenanthroline). The specificity of interaction of particular chemicals with nucleic acids inside the triplex is outlined in Table 2.3. The nature of these reactions has recently been reviewed in greater detail (Wells et al., 1988; Palecek, 1991, 1992a; Yagil, 1991; Htun and Johnston, 1992; Johnston, 1992b; Kohwi-Shigematsu and Kohwi, 1992; Sinden and Ussery, 1992).

Aldehydes

Several single-stranded regions exist in the structure of an intermolecular H DNA. Bromo- and chloroacetaldehyde (BAA and CAA, respectively) then react at the base-pairing positions of unpaired adenines and cytosines (Figure 2.14) and, to a lesser extent, of guanines, forming cyclic ethenoderivatives (Kohwi-Shigematsu and Kohwi, 1992). DNA can

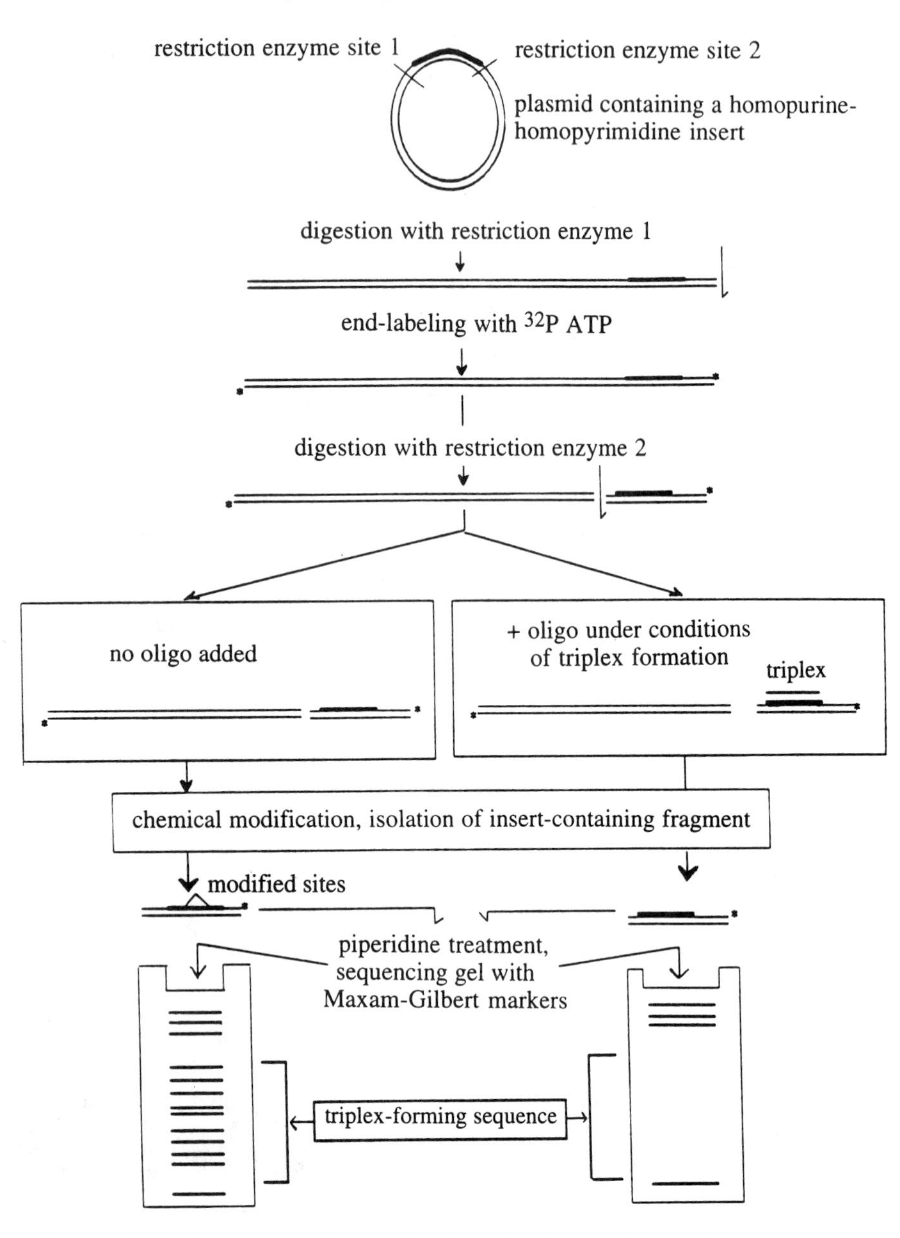

FIGURE 2.13. A general scheme for the chemical probing of intramolecular triplexes using chemical cleavage at modified sites and Maxam–Gilbert sequencing lanes for reference. (According to Soyfer, unpublished.)

TABLE 2.3. Reactions used in studies of triple helix structure at the sequence level.

Agent	Information
Haloacetaldehydes	Detection of unpaired A and C
Glycidaldehyde	Detection of unpaired G and C
Osmium tetroxide	Detection of unpaired T
Potassium permanganate	Detection of unpaired T
Diethyl pyrocarbonate	Detection of unpaired A
Hydroxylamine and methoxylamine	Detection of unpaired C
Dimethyl sulfate	Protection of G in duplex inside triplex
4,5′,8-Trimethylpsoralen	AT and TA dinucleotides in duplex or in single-strand loops
UV light	Protection of pyrimidines inside triplex from photomodification
Restriction enzyme	Inhibition of cleavage in triplex zone
Nuclease S1	Detection of single-stranded regions
Mung bean nuclease	Detection of single-stranded loops
Nuclease P1	Detection of single-stranded loops

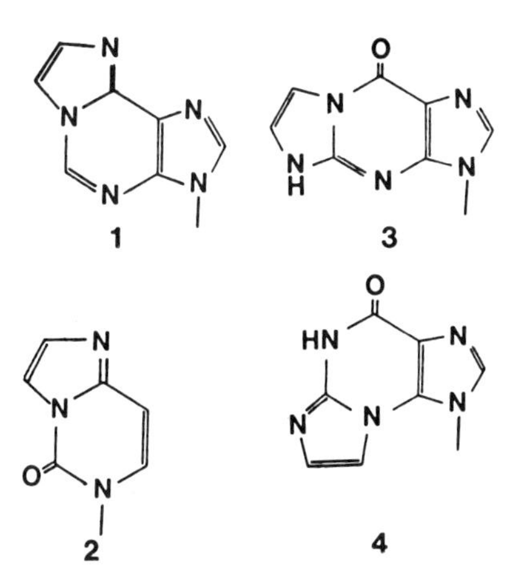

FIGURE 2.14. Haloacetaldehyde reaction products of adenine (1), cytosine (2), and guanine (3 and 4). Guanine products are obtained at a much lower yield than those of adenine and cytosine. Note that the reactive groups in bases are those which normally participate in hydrogen bonding within the double helix; therefore, the depicted products can form only if the bases belong to a single-stranded stretch of nucleic acids. The reaction specificities of haloacetaldehydes remain unaltered under a wide variety of reaction conditions in vitro and inside the cell. Modified products can be detected using a single-strand-specific nuclease S1 (since modified bases can no longer base pair), by chemical cleavage, or by primer extension analysis. From Kohwi-Shigematsu and Kohwi (1992). Reprinted with permission.

be cleaved at modified residues with piperidine treatment alone, and modification sites can then be revealed in a sequence gel. However, the modification signal is often enhanced when DNA is subjected to the Maxam–Gilbert sequencing reaction and modification sites are deduced from additional, modification-dependent cleavages found in the sequence ladder. The reaction specificities of haloacetaldehydes remain unaltered under a wide range of reaction conditions (i.e., pH 5–8, high salt concentrations, temperatures up to 50°C, and the presence of various metal ions). Haloacetaldehydes have been used to recognize intramolecular triplexes in a number of plasmid DNA molecules in vitro (Hanvey et al., 1988b; Kohwi and Kohwi-Shigematsu, 1988, 1991, 1993; Kohwi, 1989a; Kohwi-Shigematsu and Kohwi, 1991; Pestov et al., 1991; Dayn et al., 1992). Since haloacetaldehydes relatively easily penetrate the cellular membrane, incubation of cells in media containing, for example, CAA allows one to probe the presence of unpaired bases in DNA inside the cells. Thus, the presence of H (H*) DNA in plasmid was tested in *E. coli* cells (Kohwi et al., 1992; Ussery and Sinden, 1993). Glycidaldehyde has a somewhat different specificity in that it recognizes unpaired guanines and cytosines (Kohwi, 1989b).

Osmium Tetroxide (OsO_4)

OsO_4 is substantially more reactive to single-stranded DNA than to double-stranded DNA (Palecek, 1992a). It can be used as a site-specific probe for single-stranded pyrimidines, where it adds to the C5=C6 double bond in the presence of pyridine to form osmate esters (Figure 2.15). In the presence of suitable tertiary amines, such as pyridine and 2,2′-bipyridine, osmium tetroxide forms stable complexes with pyrimidine bases, with significantly increased rate of formation of esters. OsO_4 may be used over a wide range of conditions with respect to concentration, pH, temperature, and buffer composition. The site of DNA modification by OsO_4 can be determined by several methods (Palecek, 1991), among which cleavage by hot piperidine and sequencing according to Maxam and Gilbert are the most popular for probing triplexes in vitro (Vojtiskova and Palecek, 1987; Hanvey et al., 1988b; Vojtiskova et al., 1988; Collier and Wells, 1990; Pestov et al., 1991; Klysik, 1992; Panyutin and Wells, 1992). The ability of OsO_4 and pyridine to penetrate into the cells made it possible to use them in probing the triplex structure in *E. coli* cells (Karlovsky et al., 1990; Palecek, 1992b; Ussery and Sinden, 1993).

Potassium Permanganate ($KMnO_4$)

Potassium permanganate in aqueous solutions oxidizes pyrimidine bases and, to some extent, guanine. The reactivity of cytosine is an order of magnitude lower than that of thymine (Kochetkov and Budovskii, 1972); therefore, diluted permanganate solutions are used in specific modification

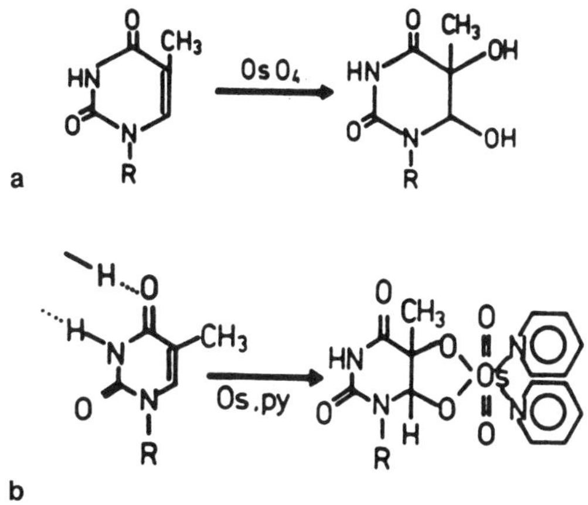

a

b

FIGURE 2.15. Formation of the adducts between nucleic acid bases and osmium tetroxide alone (a) and osmium tetroxide, pyridine (b), which is significantly faster. The reactivity of thymine is increased 100-fold in the presence of pyridine, and 10,000-fold in the presence of 2,2′-bipyridine. Since in the B DNA double-helix the target C5 = C6 double bond of thymine located in the major groove is not accessible to the bulky electrophilic osmium probe, the latter reacts preferentially with single-stranded DNA. Reprinted with permission from Palecek (1991). © CRC Press, Boca Raton, Florida.

of thymines incorporated in DNA. Under certain conditions, the primary reaction product is the piperidine-labile 5-hydroxy-6-keto derivative (Rubin and Schmid, 1980). This reaction is sterically inhibited by stacking interactions in B DNA (Glover et al., 1988, 1990). In the looped portions of an intramolecular triplex, thymines are accessible to $KMnO_4$. This reaction is used to map single-stranded regions of triplexes in a sequencing gel after hot piperidine treatment (Glover et al., 1990; Belotserkovskii et al., 1992; Jayasena and Johnston, 1992a; Panyutin and Wells, 1992; Malkov et al., 1993a).

Diethylpyrocarbonate (DEPC)

DEPC carbetoxylates the N7 position of purines (predominantly adenines), resulting in the opening of the imidazole ring (Figure 2.16) and phosphate backbone scission after piperidine treatment (Herr, 1985; Johnston and Rich, 1985; Kohwi-Shigematsu and Kohwi, 1992). DEPC has relatively low reactivity toward B DNA and enhanced reactivity to purines in single-stranded DNA. Because of this reaction specificity, DEPC is widely used in structural studies of triplexes (Hanvey et al., 1988a; Johnston, 1988; Voloshin et al., 1988; Htun and Dahlberg, 1989; and numerous later works). DEPC was also used to probe for possible triplex

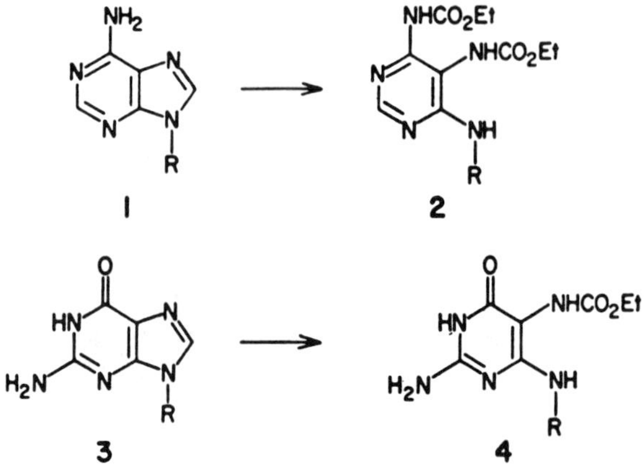

FIGURE 2.16. Diethyl pyrocarbonate modification of adenine and guanine. Unpaired adenine (1) reacts with DEPC to give the ring-opened dicarbetoxylated derivative (2) as the major product. Guanine (3) reacts with DEPC to give the ring-opened product (4) with a severalfold lower yield than adenine. From Kohwi-Shigematsu and Kohwi (1992). Reprinted with permission.

structures in the *Drosophila melanogaster hsp26* gene in vivo (Glaser et al., 1990).

Hydroxylamine and Methoxylamine

The much higher reactivity of hydroxylamine to single-stranded, as compared to double-stranded, DNA and its recognition of structural discontinuities in double-stranded DNA make it a sensitive detector of structures different from the regular B form (e.g., single strands in H DNA) (Johnston, 1992). Reactions of hydroxylamine at low (0.1 *M*) and high (>1 *M*) concentrations result in different products (reviewed by Soyfer, 1969, 1975; Johnston, 1992b). At high concentrations, a reaction of hydroxylamine with cytosine results in a mainly backbone-labilizing derivative of the latter (Figure 2.17). Although the optimum hydroxylamine reaction rate is at pH 6, this reaction can be successfully used at lower pH values where the extrusion of the H form actually takes place.

Methoxylamine has a reaction specificity similar to that of hydroxylamine; however, it seems less sensitive to slight fluctuations of the DNA structure (Johnston, 1988). The major drawbacks of hydroxylamine and methoxylamine are their requisite high concentrations (4 *M*), so the possibility of changes in the DNA structure under study due to the relatively high ionic strength should be checked, as well as the need to titrate amines to achieve the necessary pH. Some additional work to refine the optimal use of these reagents is required (Johnston, 1988).

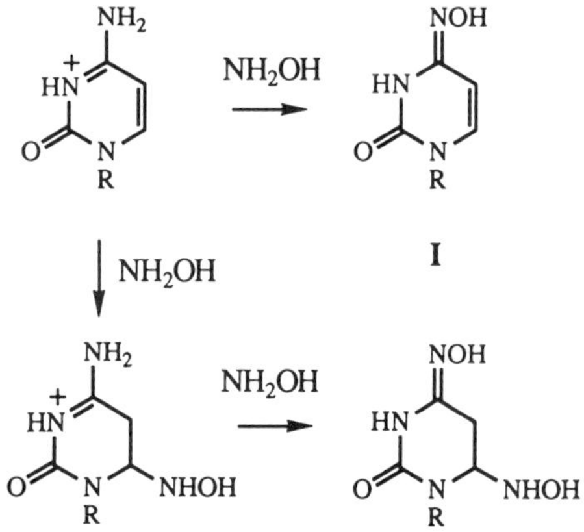

FIGURE 2.17. Reactions of hydroxylamine with cytosine residues. In the presence of oxygen, low concentrations of hydroxylamine result in unspecific reactions with all four bases and extensive cleavage of the polynucleotide chain. At 1 *M* concentration, hydroxylamine mainly yields substance I; but at 7 *M* concentration, the major reaction product is substance III, which is susceptible to ring opening and strand scission with hot piperidine. From Johnston (1992b). Reprinted with permission.

Dimethyl Sulfate (DMS) Footprinting

DMS has been widely used as a Maxam–Gilbert sequencing reagent. In B DNA, it methylates predominantly the N7 of guanine and, to a lesser extent, the N3 of adenine, which are not involved in Watson–Crick hydrogen bonding (Figure 2.18). In the triplex, the N7 of duplex guanines is involved in Hoogsteen hydrogen bonding and thus does not react with DMS. DMS footprinting determines specific guanine residues inaccessible to DMS and localizes triplex structures at the sequence level (Cantor and Efstratiadis, 1984; Pulleyblank et al., 1985; Hanvey et al., 1988a; Voloshin et al., 1988, 1992; Young et al., 1991; Hartman et al., 1992; Jayasena and Johnston, 1992a, 1993; Klysik, 1992; Soyfer et al., 1992; Beltran et al., 1993; Malkov et al., 1993a,b; Milligan et al., 1993).

4,5′,8-Trimethylpsoralen (Me_3psoralen)

Me_3psoralen preferentially binds to 5′-TA residues by intercalation when they are involved in duplex B DNA and does not appreciably bind to single-stranded DNA (Sinden and Ussery, 1992; Ussery et al., 1992). In general, the levels of dark binding (in the absence of irradiation) are

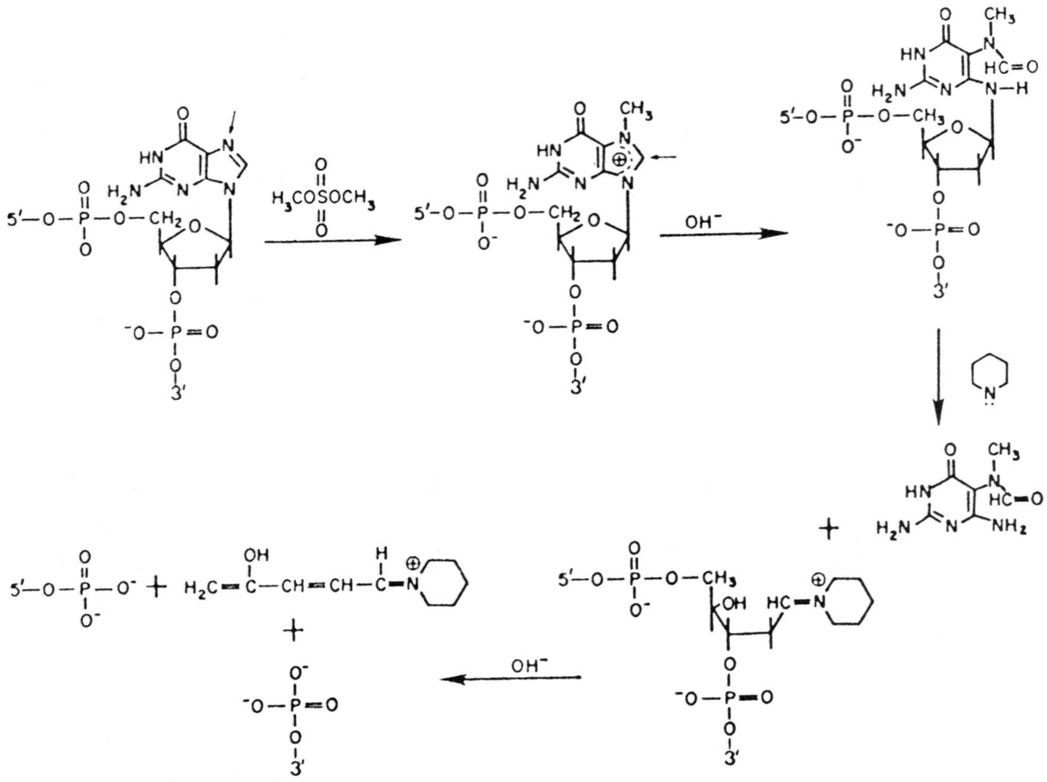

FIGURE 2.18. Reactions of guanine with dimethyl sulfate (DMS) and piperidine, which break DNA at accessible guanines. DMS methylates the guanine N7 when it is not involved in hydrogen bonding with the third strand (or other partner). Piperidine then displaces the ring-opended 7-methylguanine and catalyzes β-elimination of both phosphates from the sugar, which results in a strand break. Some adenines may also be DMS-modified at the N3 position, but piperidine treatment does not lead to strand scission at these adenines. From Maxam and Gilbert (1980). Reprinted with permission.

sufficiently low (15 Me_3psoralen molecules per 1,000 base pairs at saturation), so the topology of DNA is not perturbed. Under UV irradiation (360 nm), Me_3psoralen forms monoadducts and cross-links (Cimino et al., 1985; Ussery et al., 1992). Binding sites are effectively converted into strand breaks by hot piperidine and can be analyzed in the sequencing gel. Alternatively, a primer extension assay can be used, in which case one can determine the monoadduct formation sites which are the polymerase termination sites. When a specifically designed intramolecular triplex is formed in a plasmid, Me_3psoralen does not react with the 5′-TA dinucleotide in the region of central H form bend and TA dinucleotide in triplex–duplex junctions. This inhibition of binding was used to detect the formation of the PyPuPy triplex in vitro and in *E. coli* cells (Ussery and Sinden, 1993).

Copper–Phenanthroline Footprinting

The nucleolytic activity of 1,10-phenanthroline–copper (as well as EDTA–Fe and metalloporphyrins) is activated by reducing agents (e.g., thiol or ascorbic acid) in the presence of molecular oxygen or hydrogen peroxide (reviewed by Sigman, 1990). Cleavage by a phenanthroline–copper chelate has been used to obtain a sequence-level footprint of the third strand on a DNA double helix (François et al., 1988b; Giovannangéli et al., 1992a). In a triplex, the oligonucleotide-covered region was protected, whereas a strong enhancement of cleavage was observed on the purine-rich strand at the triplex–duplex junction on the 3′ side of the homopurine oligonucleotide (François et al., 1988b). Asymmetric DNA cleavage near the triplex is presumably due to intercalation of phenanthroline at the triplex–duplex junction (François et al., 1989a).

Uranyl Ion Footprinting

The accessibility of phosphates in DNA can be determined by uranyl ion-mediated DNA photocleavage analyzed in a sequencing gel (Nielsen et al., 1988). Using this technique, it has been shown that the phosphates of the pyrimidine strand of the duplex are at least as accessible as in double-stranded DNA, whereas the phosphates of the purine strand are partially shielded from interaction with uranyl ions (Nielsen, 1992). It was hypothesized that, rather than steric hindrance, screening of the phosphate group by polycations (such as spermine) or monovalent cations is the reason for the decrease in the accessibility of the purine strand phosphates to uranyl ions.

Inhibition of Abasic Site Formation

Base protonation in acidic media results in labilization of the glycosidic bond of DNA residues and concomitant creation of abasic sites. For example, when double-stranded DNA is exposed to a pH 4.4 buffer for about 100 hours at room temperature, apurination at a rate of approximately 1 in 200 base pairs occurs (Glover et al., 1990). When a protonated triplex is incubated under the same conditions, reactions resulting in abasic sites are significantly inhibited (Duker, 1986; Duker et al., 1986; Glover et al., 1990). This is probably due to the residue rigidity of the triplex, which may inhibit adoption of a planar transition state at the C1′ atoms of nucleotides (Glover et al., 1990). The various apurination reactivities of the duplex and the triplex were used to reveal at the sequence level the predominant H form isomer at low and high superhelical density (Glover et al., 1990).

Chemical Probing of H DNA in the Cells

Some of the chemicals listed in Table 2.3 have been used to probe triple-stranded structures inside cells. These chemicals include diethylpyrocarbonate (Glaser et al., 1990), osmium tetroxide-2,2′-bipyridine (Karlovsky et al., 1990; Ussery and Sinden, 1993), chloroacetaldehyde (Kohwi et al., 1992; Ussery and Sinden, 1993), and 4,5′,8-trimethylpsoralen (Ussery and Sinden, 1993). It is thought that the penetration of chemicals into the cells during incubation in an appropriate solution does not disturb cell integrity (Palecek, 1991). Except for one case, when the H DNA was not found (Glaser et al., 1990), the existence of an intracellular triplex was shown when the pH was not strictly controlled (Ussery and Sinden, 1993), or under conditions supplemented by a missing triplex-stabilizing factor. For example, to detect the PyPuPy triplex, *E. coli* cells were incubated with added Mg^{2+} (Kohwi et al., 1992). Karlovsky et al. (1990) used osmium tetroxide-2,2′-bipyridine to probe unpaired regions in plasmid DNA in vitro and in *E. coli* cells incubated in a low pH medium. Thus, the intracellular medium as a multicomponent mixture of various molecules does not significantly interfere with reactions of the chemicals with their targets.

Photofootprinting

The relatively new triplex photofootprinting method (Lyamichev et al., 1990) is of especial note, because the implementation of this procedure involves a minimum influence on the system studied. This alone makes it preferable to other probing methods for in vivo studies (see Becker and Grossman, 1992 for review). This method is based on the inhibition of formation of dipyrimidine photoproducts (Figure 2.19) in the duplex portion of the triplex under UV irradiation (Lyamichev et al., 1990, 1991; Tang et al., 1991; Malkov et al., 1992; Soyfer et al., 1992). Occupation of the major groove of DNA by the third strand and formation of Hoogsteen hydrogen bonds significantly hinder internal motions in DNA, which

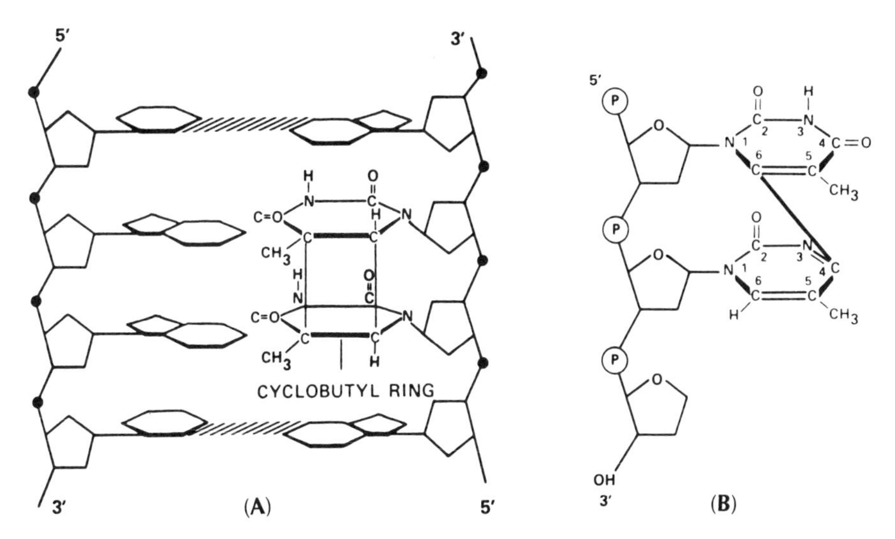

FIGURE 2.19. (A) The cyclobutyl pyrimidine dimer is formed in DNA by covalent interaction of two adjacent pyrimidines in the same polynucleotide chain. Saturation of their 5,6 double bonds results in the formation of a four-membered cyclobutyl ring linking two pyrimidines. To form a dimer, adjacent pyrimidines should adopt a parallel disposition instead of their equilibrium orientation of 34.3° to each other. (B) [6-4] dipyrimidine photoproduct produced by linkage between the C6 position of one thymine and the C4 position of adjacent thymine. In the [6-4] dipyrimidine photoproducts formed in DNA, the 3′ pyrimidine is typically cytosine rather than thymine. From: DNA REPAIR by Friedberg. Copyright © 1984 by W.H. Freeman and Company. Used with permission.

prevents adjacent pyrimidines from adopting positions favorable for UV-mediated dimerization (Lyamichev et al., 1990). Since dimers form primarily in adjacent pyrimidines, the homopurine–homopyrimidine triplex forming regions are particularly sensitive to UV light. To determine the sites of [6–4] dipyrimidine photoadducts, the labeled DNA fragment is treated with hot piperidine to break the labile glycosidic bond of the 3′-Py photoproduct, which subsequently leads to a strand break. In order to map the dipyrimidine cyclobutane dimers, the labeled DNA fragment is subjected to enzymatic treatment with endonuclease V from T4 phage (Tang et al., 1991) or endonuclease activity from *Micrococcus luteus* (Lyamichev, 1991; Malkov et al., 1992), which also results in a single-strand break. The distribution of different photoproducts is revealed separately in a sequencing gel. Figure 2.20 demonstrates the stabilization of the intermolecular PyPuPy triplex by Mg^{2+} and spermidine, according to the photofootprinting data (Lyamichev et al., 1991; Soyfer et al., 1992). The photofootprinting method has been applied to the study of triplex formation under a variety of conditions (Lyamichev, et al., 1990,

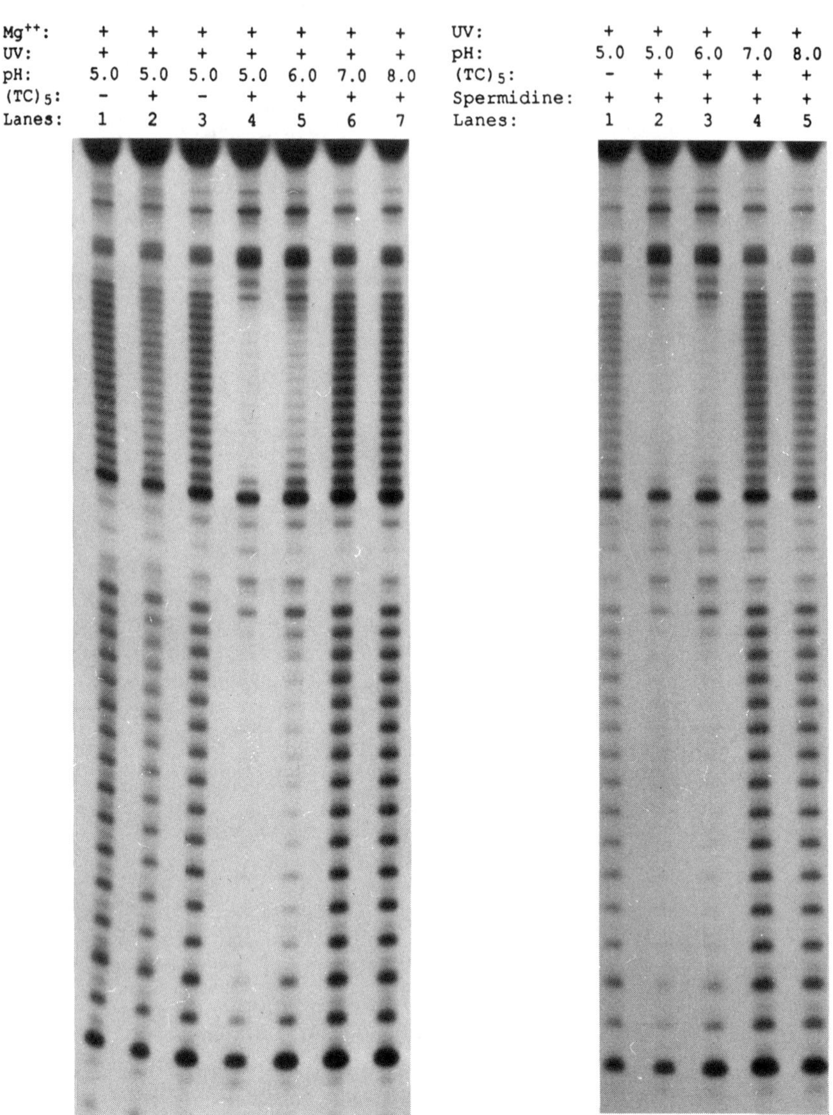

FIGURE 2.20. The photofootprinting of intermolecular PyPuPy triplexes. Hot piperidine treatment results in strand breaks at the points where [6-4] thymine-cytosine photoproducts have been formed and has no effect if DNA has not been modified. Lanes 1 and 3 in panel a and lane 1 in panel b show the pattern of photomodification when duplex DNA containing two $d(TC)_{16} \cdot d(AG)_{16}$ inserts is UV-irradiated. When the $d(TC)_5$ oligonucleotide binds as a third strand to its duplex target, photoproduct formation in the PyPu tract is inhibited, resulting in the disappearance of the bands in lanes 4,5 (panel a) and lanes 2,3 (panel b). The photofootprinting data show that Mg^{2+} and spermidine stabilize the triplex at higher pH (compare lanes 2 and 4,5 in panel a, and lanes 1 and 2,3 in panel b). From Soyfer et al. (1992). Reprinted with permission.

1991; Tang et al., 1991; Klysik, 1992; Malkov et al., 1992; Soyfer et al., 1992; Kovalsky et al., 1993), but, as of this writing, not in vivo.

A Short Overview of Chemical and Enzymatic Probes

A considerable range of chemical probes is now available for the study of different DNA structures, including triplexes. They are potentially applicable to detecting unusual DNA structures in vivo. The most important advantage of their use is the possibility of locating triplex structures at the sequence level. Chemical agents are small and can diffuse to various parts of the DNA structure. Chemical modifications do not result in immediate strand breaks, making possible simultaneous reactions at multiple sites of a supercoiled DNA molecule. The submicrogram amounts of DNA required for these experiments are significantly lower than those necessary for physical studies. With a successful combination of chemical probes, the entire spatial structure of local sites in DNA molecules with unusual conformations, including triple-stranded regions, can be determined. Although enzymes may not be ideal DNA probes, for many reasons, they give information in addition to that from chemical modification, and their use is desirable to avoid confusion which can arise on the basis of chemical probing alone (see below).

A noteworthy problem in chemical and enzymatic probing is that the probing agent reacts with the DNA structure and may therefore disturb the object of study. This may lead to a discovery of a structure that does not exist in the absence of the probe. For instance, single-strand-specific nucleases bind to their targets and could potentially stabilize conformations that contain single-stranded regions. Intercalation of chemical probes could also change the winding state of DNA and shift the conformational distribution. In cases where there is an equilibrium between distinct DNA structures, the reaction of one conformation with a probe efficiently eliminates it from an equilibrium distribution, resulting in the shift of equilibrium to that conformation ("driven equilibrium"). This will result in an overestimation of the proportion of one conformer (Lilley, 1992).

The greatest care should be exercised in designing probing experiments and in interpreting their results. Significant structural perturbations can be avoided using short incubation periods and low reagent concentrations in order to minimize the number of modified sites. Because it is unsafe to draw a conclusion from the data on a single probe, chemical probing experiments with additive structural information should be used. For example, if OsO_4 or $KMnO_4$ is used to detect unpaired thymines, then DEPC should be used to probe the conformational state of adenines. It is desirable to use some unrelated techniques described under Physical Methods for Triplex Study to confirm the formation of the triplex structure.

TABLE 2.4. Reactions using third-strand oligonucleotide-bound agents.

Agent	Information
Psoralen	UV cross-linking at target site, strand break with piperidine
Proflavin derivatives	UV cross-linking at target site
Azidophenacyl	UV cross-linking at target site
Ellipticin	Photoinduced cleavage of two strands
Fe–EDTA	Deoxyribose, duplex scission near target site
1,10-Phenantroline–Cu	Deoxyribose, duplex scission near target site
Staphylococcal nuclease	Duplex scisson near target site
N-Bromoacetyl	G alkylation, strand break with piperidine
N4,N4-ethano-5-methyldeoxycytidine	Cross-linking at target site
5′-p-[N-(2-chloroethyl)-N-(methylamino)]benzamide	G alkylation, strand break with piperidine
Halogenated base analogs	Cross-linking and strand breaks target site

Another significant drawback of using chemical probes is their toxicity. Haloacetaldehydes and glycidaldehyde are carcinogenic; DEPC produces carcinogenic urethane in a reaction with ammonia, and OsO_4 vapors have an irritating effect on eyes, mucous membranes, etc. (Palecek, 1991).

Site-Directed Agents

Similar to the chemical and enzymatic footprint reactions described above, in which triplex formation inhibits the cleavage of the duplex part of DNA, some other agents could potentially be used for triplex probing. Actually, many chemical agents have been applied in site-specific cleavage of duplex DNA through coupling the reactive moiety to the triplex-forming oligonucleotides (Table 2.4). Their action thereby acquires considerable sequence specificity.

FIGURE 2.21. Scission of duplex DNA with an oligonucleotide–EDTA–Fe conjugate. The ribbon model and the sequence of the local triplex between oligonucleotide–EDTA–Fe and its target sequence are shown on the right. The EDTA–Fe moiety is coupled to the thymine residue (*T) via a flexible linker; therefore the cleavage site spans several internucleotide linkages (shown on the backbone by circles whose size is proportional to the extent of cleavage; see also the primary data on a sequencing gel on the left). Substitution of some of the bases (X) in the third strand results in complexes of different stability and different strand cleavage efficiency; compare lanes for X = T, X = A, and X = C, where three different oligonucleotide–EDTA–Fe concentrations, 0.1 (lanes 3,6,9), 0.5 (lanes 4,7,10), and 1.0 μM (lanes 5,8,11), were used. Lane 1: products of an A-specific cleavage reaction. Lane 2 (control): intact 3′ end-labeled restriction fragment after treatment according to the cleavage reactions in the absence of oligonucleotide–EDTA–Fe. Reprinted with permission from Beal and Dervan (1991) *Science* 251, 1360–1363. © 1991 AAAS.

Nuclease-Like Oligonucleotides

Some of the DNA-cleaving agents attached to oligonucleotides utilize an oxidative degradation of deoxyribose (Fe–EDTA and 1,10-phenanthroline–Cu) with concomitant strand breaks. The yields of double-strand cleavage do not exceed 25% for the Fe–EDTA reaction (Moser and Dervan, 1987; Strobel et al., 1988). In spite of relatively low yields, the reaction of duplex DNA with Fe–EDTA attached to the third-strand oligonucleotide (Figures 2.21 and 2.22) is adequate for addressing

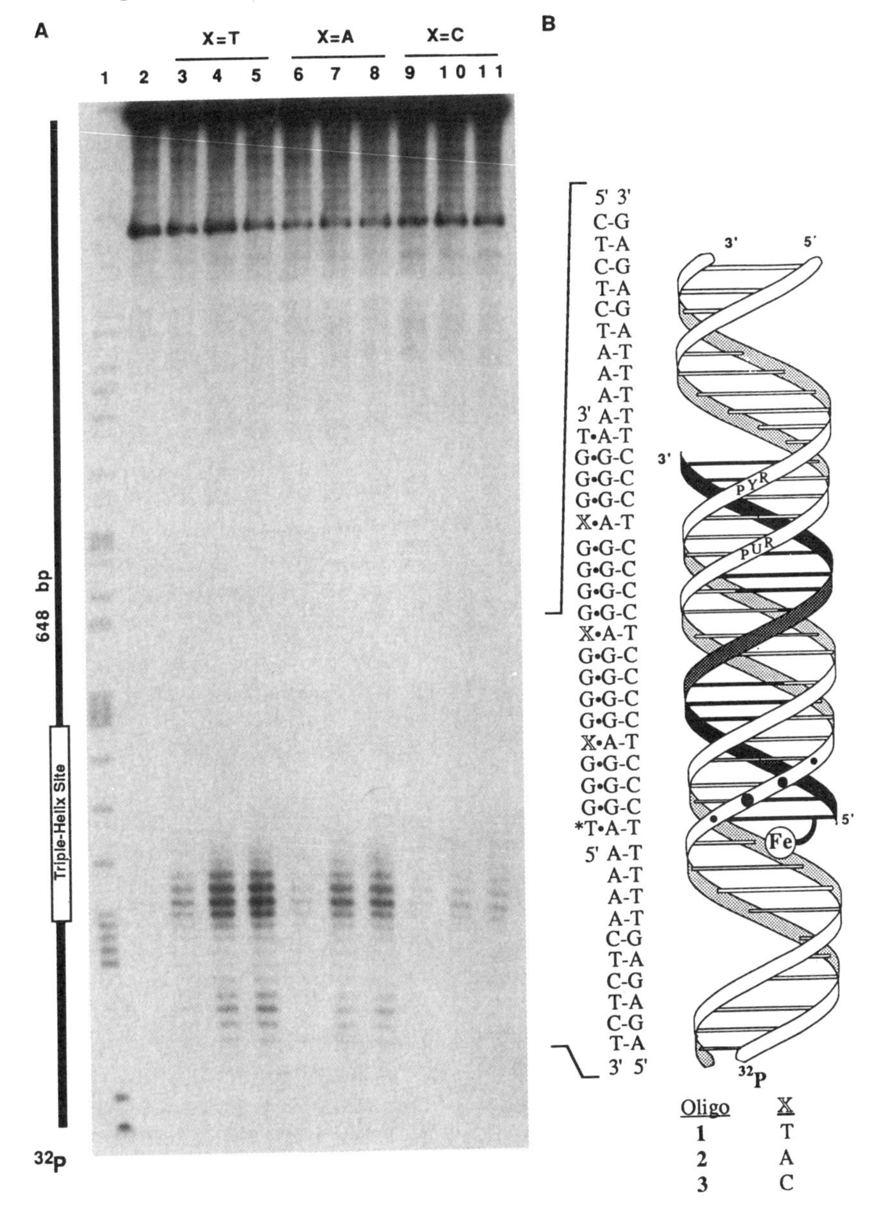

questions related to molecular recognition, including binding specificities, site size, and orientation (Moser and Dervan, 1987; Strobel et al., 1988; Griffin and Dervan, 1989; Povsic and Dervan, 1990; Beal and Dervan, 1991, 1992a,b; Horne and Dervan, 1991; Griffin et al., 1992; Kiessling et al., 1992; Stilz and Dervan, 1993; Shimizu et al., 1994), as well as kinetic and thermodynamic characteristics (Singleton and Dervan, 1992a,b; Colocci et al., 1993; Distefano and Dervan, 1993; Han and Dervan, 1993).

In the case of 1,10-phenanthroline–Cu, reaction yields of up to 70% have been reported (François et al., 1989a). The 1,10-phenanthroline–Cu complex (Figure 2.22) attached to the third-strand oligonucleotide was used to study the specificity of triplex recognition (Jayasena and Johnston, 1992b, 1993) and as a prototype of an artificial nuclease (François et al., 1988a, 1989a,b).

Photoactive Groups Attached to Oligonucleotides

The idea of attaching some functional moiety to an oligonucleotide capable of binding as a third strand was also used in the design of photochemical modification of a duplex target site (Claude Hélène's idea of attachment of psoralen to oligonucleotide, as well as V. Demidov's and V. Soyfer's suggestion on the usage of photoactive nucleotides in the content of oligonucleotides, unpublished data; see also Soyfer et al., 1994). A number of photoactivatable groups covalently linked to third-strand oligonucleotides can cross-link to the duplex DNA or produce photoinduced strand cleavage. Among them are psoralen, azidophenacyl, porphyrins, and proflavin derivatives (Figure 2.22), all of which produce cross-links at the target site (Praseuth et al., 1988a,b; Le Doan et al., 1990; Takasugi et al., 1991). These substances intercalate at the triplex–duplex junction and, under UV irradiation, can cross-link both duplex strands (Thuong and Hélène, 1993). Under piperidine treatment, the cross-links are converted into strand breaks, which can be analyzed at the sequence level to give information about triplex specificity, stability, etc. However, the photoactivatable groups have been largely employed in other applications. In particular, psoralen cross-linking at target sites has been used to prolong the gene repressive effect of the triplex (Duval-Valentin et al., 1992; Giovannangéli et al., 1992b; Grigoriev et al., 1993), and for site-specific mutagenesis (Havre et al., 1993) (see also Chapter 6).

Recognition of DNA by oligonucleotides covalently linked to ellipticine and concomitant photoinduced cleavage and cross-linking has been tested (Perrouault et al., 1990; Le Doan et al., 1991). Cross-linking of ellipticine with DNA was more efficient when DNA was single-stranded; by contrast, when the target was duplex DNA, the reaction resulted largely in strand cleavage. This allowed the suggestion that ellipticine–oligonucleotides act as artificial endonucleases (Perrouault et al., 1990) (see Chapter 6).

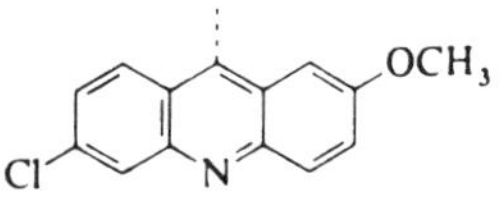

acridine (Acr)

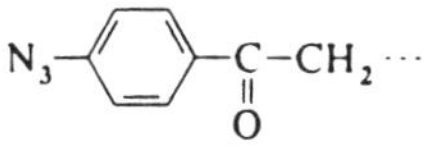

azidophenacyl (N_3-Phe)

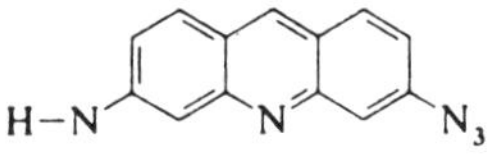

azidoproflavine (N_3-Pf)

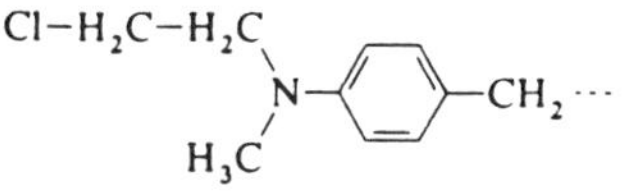

chloroethylamine derivative (ClEtN)

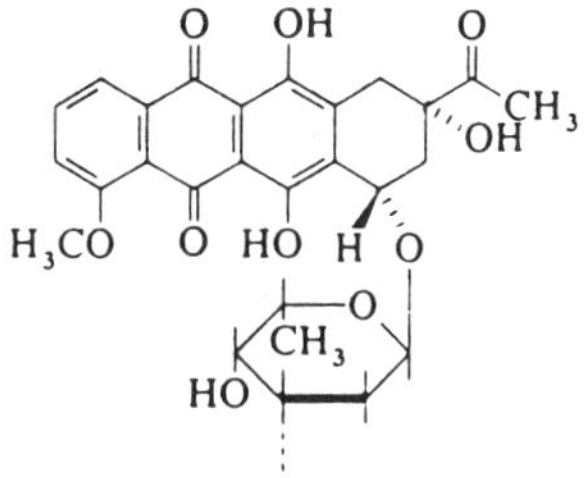

daunorubicine (Dr)

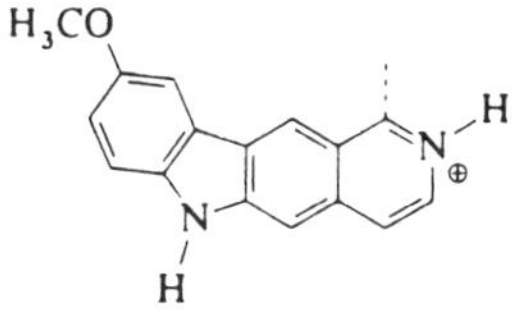

ellipticine (Elli)

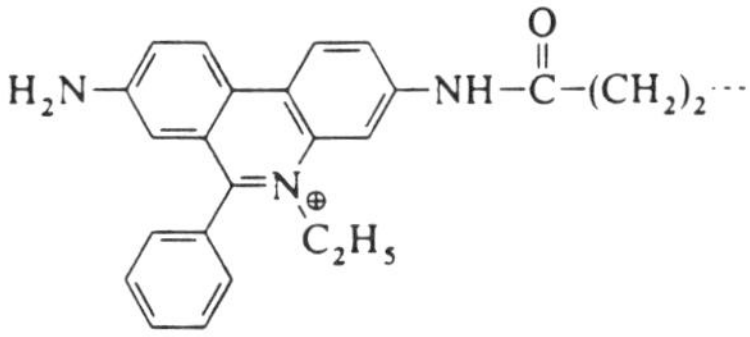

ethidium (Etd)

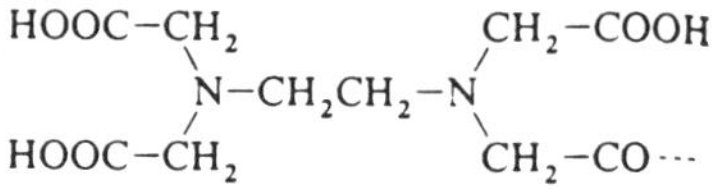

ethylenediaminetetraacetic (EDTA) group

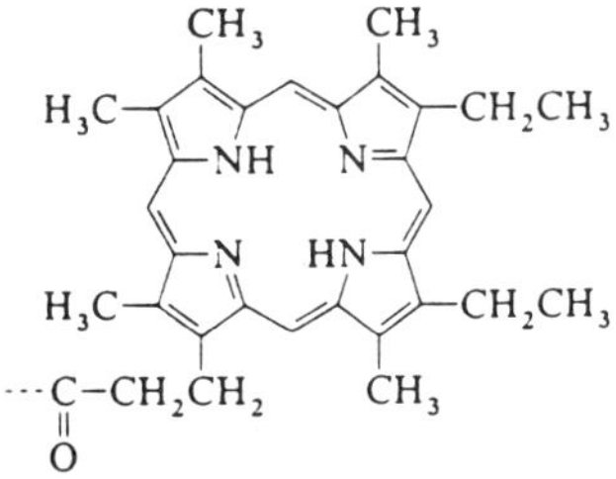

methylpyrroporphyrin XXI (MPPo)

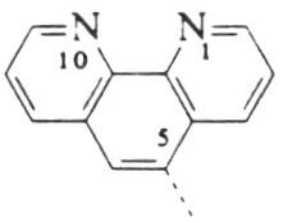

orthophenanthroline (OP)

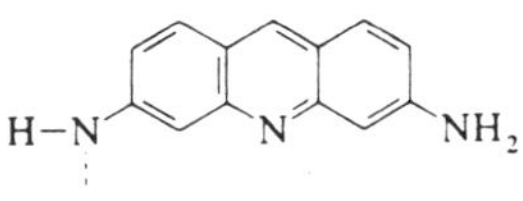

proflavine (Pf)

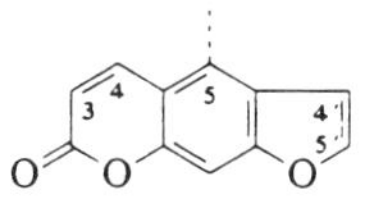

psoralen (Pso)

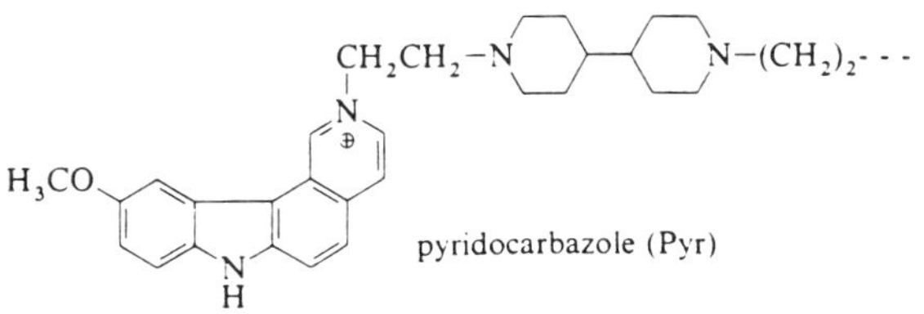

pyridocarbazole (Pyr)

FIGURE 2.22. Structures of different ligands tethered to triplex-forming oligonucleotides. The dashed lines indicate the ligand–oligonucleotide linker. From Thuong and Hélène (1993). Reprinted with permission.

Porphyrins linked to oligonucleotides produce various types of photodamage to target DNA (Le Doan et al., 1990). The observed reactions were cross-linking of oligonucleotides to their target sequence and oxidation of guanine bases dependent on the distance between guanine and porphyrin macrocycle. These reactions accounted for 60% of total photodamage of target site in the complex.

Photoactive Oligonucleotides

Third-strand oligonucleotides can also contain some base analogs (thio- or halogenated) that can form normal Hoogsteen hydrogen bonds with purines in duplexes (Figure 2.23) and serve as photosensitizers, or mediators, that absorb long-wavelength UV or visible light and stimulate photoreactions in ordinary bases. This stimulation may be due to either

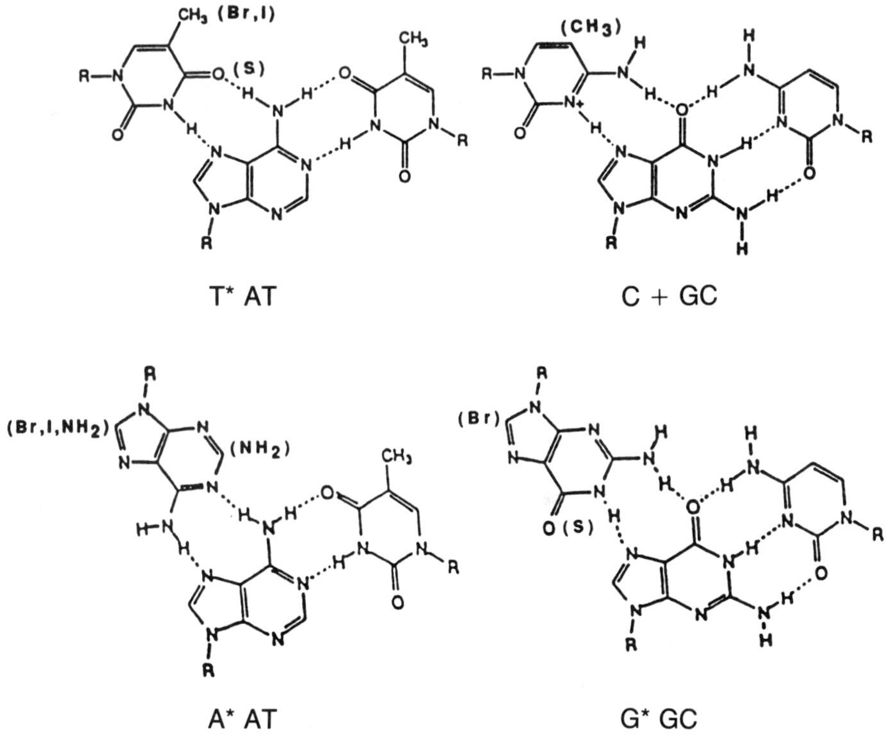

T* AT

C + GC

A* AT

G* GC

FIGURE 2.23. Several base substitutions are shown in the canonical base triads for the PyPuPy and PyPuPu triplexes. These substitutions render modified bases which absorb beyond usual DNA absorbance spectra at longer wavelengths. Thus, at $\lambda > 300\,\text{nm}$ only modified bases in the third strand may absorb energy and transfer it to the nearby bases in duplex DNA, producing a number of photomodifications (strand breaks, cross-links, dimers, etc.).

the direct transfer of excitation energy absorbed by bases or the indirect photodynamic action of reactive particles that form in the solution (see Kochevar and Dunn, 1990, for a review). This stimulation results in cross-linking and strand breaks at the target site (Demidov et al., 1992). Such completely oligonucleotide sensitizers have several advantages over oligomers containing photomodifying groups described in the previous paragraph. It is possibile to use corresponding derivatives of various base analogs in automatic oligonucleotide synthesis. Multiple photomodifying moieties can be inserted into various positions of the synthesized chain. Due to the formation of hydrogen bonds with the purine bases, such base analogs contribute to triplex stability. The hydrogen-bonded state of base analogs makes them undiffusible; therefore, their photomodifying effect should be highly localized.

Base-Specific Oligonucleotide-Bound Chemicals

Several other chemical reactions can be used for structural triplex studies. Originally, they were used in site-directed genome modification and rare chromosome DNA cutting (see Chapter 7). Highly efficient cross-linking (with a yield greater than 95%) of triplex-forming oligonucleotides containing the modified nucleoside N4,N4-ethano-5-methyldeoxycytidine to the N7 of a specific guanine in a double-stranded DNA target occurs in 16-h reactions (Shaw et al., 1991). Oligonucleotide derivatives containing 5′-p-(N-2-chloroehyl-N-methylamino)benzylamide, which alkylate the N7 of guanines, were shown to modify the double-stranded $d(G)_{18} \cdot d(C)_{18}$ target in linear DNA (Fedorova et al., 1988) and, due to Watson–Crick pairing, the looped part of the intramolecular triplex in supercoiled DNA (Vlassov et al., 1988). The most efficient chemical reagent, N-bromoacetyl (Figure 2.24), is capable of yielding a highly localized cross-link with a reaction yield up to 96% (Povsic and Dervan, 1990; Povsic et al., 1992). Reaction of the electrophilic carbon of N-bromoacetyl with the N7 of guanine adjacent to the local triple helix results in covalent attachment of the oligonucleotide to the target sequence. Under hot piperidine treatment, a single-strand break occurs. Two oligonucleotides bound to adjacent inverted purine tracts on double-stranded DNA by triple-helix formation alkylated single guanine positions on opposite strands. In such a way, a yeast chromosome, 340,000 base pairs in size, was cleaved at a single site with an 85% to 90% yield (Povsic et al., 1992).

Oligonucleotide–Enzyme Conjugates

Third-strand binding and its relative affinity can be determined using a conjugate of a third-strand oligonucleotide with a DNA cleaving enzyme. Pei et al. (1990) designed such a hybrid of an oligonucleotide with staphylococcal nuclease, which produces double-strand scissions near the target site. A Lys-84 → Cys mutant staphylococcal nuclease

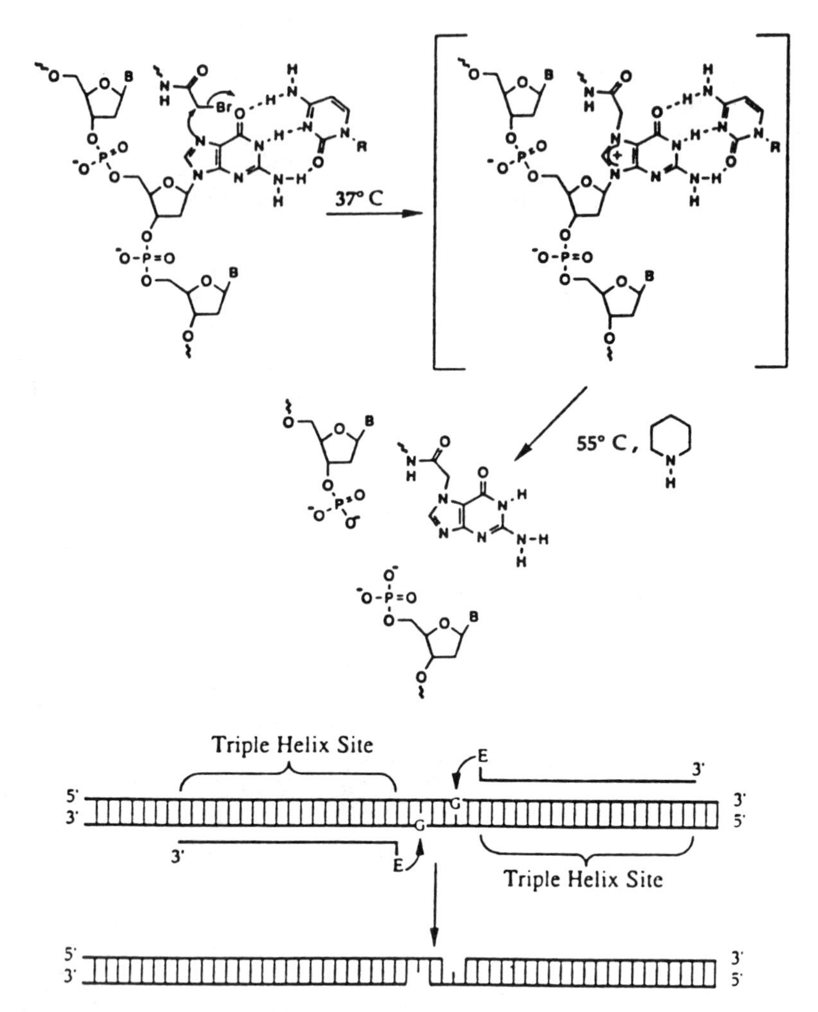

FIGURE 2.24. Highly efficient base-specific duplex DNA cleavage. Upper panel: Triplex formation brings the N-bromoacetyl electrophile in the major groove proximal to a GC base pair in the duplex target site. Alkylation at guanine N7 followed by depurination leads to a backbone cleavage with 5′ and 3′ free phosphate termini. Lower panel: Oligonucleotides carrying an electrophile (E) at the 5′ end bind to adjacent inverted binding sites on duplex DNA via triplex formation and alkylate at single guanine (G) positions on opposite strands. Upon depurination, double-strand cleavage results in single-strand overhangs with sequences suitable for ligation. Reprinted with permission from Povsic et al. (1992). © 1992 American Chemical Society.

was selectively linked to the 5′ and/or 3′ terminus of a thiol-containing polypyrimidine oligonucleotide via a disulfide bond. This semisynthetic nuclease, hybridized to the PuPy tract of plasmid DNA, produced double-strand cleavage at a length of several base pairs with an efficiency greater than 75%.

A Short Overview of Site-Directed Agents

Several kinds of agents have been coupled to triplex-forming oligonucleotides in order to produce localized modification of the duplex in or near the triplex region. Corresponding reactions have been used to probe the sequence specificity and stability of intermolecular triplexes. It should be kept in mind that some of these agents could contribute to oligonucleotide binding at target sites; therefore, some corrections should be allowed for in determining specificity and stability of triplexes. For example, 1,10-phenanthroline intercalates the duplex–triplex junction and makes the triplex more stable. This property is very useful when a site-directed modification of DNA is desirable. Enhancement of triplex stability by ligand intercalation, and even oligonucleotide cross-linking with the duplex, prolong the inhibition of undesirable gene expression.

Site-directed reactions with duplex DNA suggest other kinds of genetic manipulations. Triplex-forming oligonucleotides carrying some active groups are potentially useful for site-directed mutagenesis and point DNA cutting. These new approaches are relatively poorly studied, and a great deal of effort should be devoted to determining the product specificity of modifying agents and the size of affected regions of duplex DNA, and to increasing the efficiency of corresponding reactions.

Conclusion

There are a large number of methods available to study unusual structures of nucleic acids, including triplexes. In the study of oligomeric and polymeric triple-stranded models, physical techniques provide thermodynamic parameters and detailed structural information. Chemical probing and gel-electrophoretic analysis of the distributions of modification along the sequence allow the reconstruction of the spatial disposition of the elements in the structure under investigation. Finally, chemical and photochemical reactions of active groups attached to oligonucleotides form the basis for various genetic and biotechnological applications of triplex methodology.

3
General Features of Triplex Structures

Basic Types of Triplexes

Nucleotide Sequence Requirements

Generally, a triplex consists of a duplex, where the base pairs are formed via Watson–Crick hydrogen bonds, and a third strand, whose bases form hydrogen bonds with one base of each base pair of the duplex. In the earlier studies of polymer or oligomer triplexes, at least one of the strands was homopurine, indicating an important role of hydrogen-bonding capabilities of purine bases in the formation of the structure (Felsenfeld et al., 1957; Rich, 1958a,b; Lipsett, 1963, 1964; Miles, 1964; Sasisekharan and Sigler, 1965; Riley et al., 1966; Morgan and Wells, 1968; Thiele and Guschlbauer, 1968, 1971; Marck and Thiele, 1978; Lee et al., 1979). The underlying basis for triplex formation is that purines have potential hydrogen-bonding donors and acceptors that can form two hydrogen bonds with incoming third bases. By contrast, pyrimidine bases already involved in the duplex can form only one additional hydrogen bond with incoming third bases. The hydrogen bonds of such a type are traditionally called Hoogsteen bonds after their discovery in adenine–thymine cocrystals by Hoogsteen in 1959. To form a more stable structure, the third strand bases bind to the purine bases of the duplex. If purine bases are randomly distributed between two strands, consecutive third-strand bases should switch from one strand of the duplex to the other, resulting in a structural distortion of the sugar-phosphate backbone and lack of stacking interactions. This is energetically unfavorable, therefore, the duplex appropriate for triplex formation contains purine bases only in one strand. Thus, the precondition of triplex formation is the presence of a homopyrimidine (Py) sequence in one strand of the duplex and a complementary homopurine (Pu) sequence in the opposite strand (PyPu tract). Attempts to form triple-stranded helices from alternating purine–pyrimidine strands were unsuccessful (Morgan and Wells, 1968). Two hydrogen bonds with the duplex purine strand can be formed by both pyrimidine and purine bases of the third strand (Felsenfeld et al., 1957;

Rich, 1958a,b; Lipsett, 1963, 1964; Miles, 1964; Morgan and Wells, 1968; Thiele and Guschlbauer, 1968, 1971; Marck and Thiele, 1978; Broitman et al., 1987). Therefore, the third strand in a triplex may be either Py or Pu. Based on this sequence requirement, triplexes can be divided into two types: pyrimidine–purine–pyrimidine (PyPuPy) type and pyrimidine–purine–purine (PyPuPu) type (Table 3.1). As was mentioned earlier in describing the triplex notation, we always place the pyrimidine strand of the duplex at the first position of the triplex, the purine strand of the duplex at the second position, and the pyrimidine or purine third strand at the third position. Similarly, in the base triads, the first letter corresponds to the base in the pyrimidine (or pyrimidine-rich) strand of the duplex, the second letter corresponds to the base in the purine (or purine-rich) strand of the duplex, and the third letter (often after the * sign) corresponds to the base in the third strand.

Intermolecular and Intramolecular Triplexes

Triplexes can be also divided on another basis into intramolecular and intermolecular types (Table 3.1).

Typical intermolecular triplexes can be formed when the polymeric or oligomeric third strand of appropriate sequence binds (through Hoogsteen bonding) to the double-stranded PyPu tract with DNA of RNA geometry. Two types of intermolecular triplexes can be formed, PyPuPy and PyPuPu, depending on the pyrimidine or purine content of the forthcoming oligonucleotide. Some special types of intermolecular triplexes will be described later in this chapter.

Common intramolecular triplexes are formed in DNA sequences containing a PyPu tract with mirror repeat symmetry (Lyamichev et al., 1985,

TABLE 3.1. Basic types of DNA triplexes.

Pyrimidine–purine–pyrimidine (PyPuPy) triplexes	
Generally are pH-dependent	
Intermolecular	Intramolecular (H form)
Form with oligonucleotides containing C or T or their combination, i.e., based on CG*C and/or TA*T triads	Form in DNA with mirror PyPu repeats; formation is facilitated by superhelical stress; CG*C and TA*T triads
Pyrimidine–purine–purine (PyPuPu) triplexes	
Generally are pH-independent	
Require multivalent cations	
Intermolecular	Intramolecular (H* form)
Form with purine-rich oligonucleotides, based on formation of CG*G and /or TA*A triads and a small proportion of TA*T triads	Form in DNA with mirror PyPu repeats under superhelical stress; CG*G and TA*A triads

From Soyfer et al., 1992.

1986; Mirkin et al., 1987). Data from 2-D electrophoresis and S1 nuclease susceptibility showed the formation of unusual structures in supercoiled plasmid containing such tracts (Lyamichev et al., 1985, 1986). Further analysis showed that any deviation from a mirror repeat in the PyPu tract makes formation of an unusual structure extremely unfavorable (Mirkin et al., 1987). Based on these data, Maxim Frank-Kamenetskii and his coworkers proposed an explicit model for the intramolecular triplex in the PyPu tract (Lyamichev et al., 1986). When one half of this PyPu tract is being unwound, one of the unpaired strands bends around the point of symmetry and interacts with the other half of the region, forming the Hoogsteen base pairs (Lyamichev et al., 1986). The other strand of the unwound region is unpaired. This model provided the first reasonable explanation for the susceptibility of the PyPu tracts to nuclease S1 digestion when DNA is exposed to acidic pH. Because both purine and pyrimidine strands may be involved in the formation of triple-stranded regions, PyPuPy and PyPuPu intramolecular triplexes may exist (Figure 3.1). In the first case, the Py strand of an unwound region folds back on the duplex, forming Hoogsteen hydrogen bonds and leaving part of the Pu strand unpaired. In the second case, the Pu strand forms Hoogsteen-like hydrogen bonds with the duplex, leaving part of the Py strand unpaired. Generally, the pyrimidine third strand contains both thymines and cytosines. In the case of cytosine, stable base triads with two Hoogsteen bonds can be formed if the N3 atoms of cytosines are protonated,

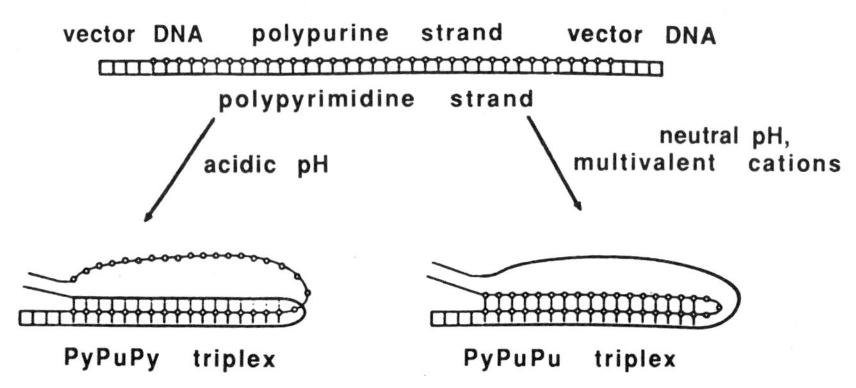

FIGURE 3.1. Two types of intramolecular triplexes formed by the PyPu tracts depending on pH and multivalent cations. Open circles represent the purine strand of the tract, whereas thick lines stand for the pyrimidine strand. In the PyPuPy triplex, half of a pyrimidine strand folds back on the PyPu duplex. Since the cytosines in the third strand must be protonated, this structure requires acidic pH. In the PyPuPu triplex, half of a purine strand is involved in Hoogsteen hydrogen bonding with the PyPu duplex, the process being promoted by multivalent cations (e.g., divalent metal cations). From Panyutin and Wells (1992). Reprinted with permission.

which is possible in mildly acidic conditions (pH ~ 4–5) (Lipsett, 1963, 1964; Howard et al., 1964). The intramolecular PyPuPy triplex was discovered first and was termed the "H form" to indicate the role of an elevated concentration of H^+ ions in the media in which this triplex can exist (Lyamichev et al., 1985, 1986). The term "H form" has been adopted in the scientific literature.

The formation of the similar intramolecular PyPuPu triplex in supercoiled DNA does not require H^+ ions, but it does require the presence of multivalent cations. It was termed the "H* form" to distinguish it from the H form (Bernues et al., 1989).

The original hypothesis about the structure of the intramolecular triplex in supercoiled DNA was based on stereochemical considerations. It was later confirmed by numerous chemical probing experiments that showed the structural states of all individual bases (Vojtiskova and Palecek, 1987; Hanvey et al., 1988b; Htun and Dahlberg, 1988; Johnston, 1988; Kohwi and Kohwi-Shigematsu, 1988; Vojtiskova et al., 1988; Voloshin et al., 1988).

An intramolecular triplex can be formed in an oligonucleotide that folds back twice. This structure will be described at the end of this chapter.

Molecular Details of Triplex Structure

Some general factors were found to determine overall structures of both PyPuPy and PyPuPu triplexes (Table 3.2). For a comprehensive description of the geometric parameters of conceivable molecular models of triplexes, the reader is referred to recent publications (Cheng and Pettitt, 1992a,b; Piriou et al., 1994).

Major-Groove Location of the Third Strand

It has long been recognized that nucleic acid base pairs of double-helical molecules retain hydrogen-bonding capabilities in both the minor and the major grooves of the double helix (see Hélène and Lancelot, 1982, for review). Figure 3.2 shows potential hydrogen bond donor (D) and acceptor (A) sites within both the major and the minor grooves after the Watson–Crick base pairs have formed. The minor groove in the double helix is not wide enough to accommodate the third strand in a manner enabling it to form stable hydrogen bonds with bases incorporated in the duplex. By contrast, the third strand can be placed in the major groove of the DNA where purine bases can form normal Hoogsteen hydrogen bonds (see Cheng and Pettitt, 1992a, for review). The DNA major groove is deep enough, and upon the binding of the third strand, the

TABLE 3.2. Structural characteristics of triple-helical structures.

Feature	Methods	Description
Third-strand base composition	Various	Homopyrimidine (PyPuPy triplex); purine-rich (PyPuPu triplex)
Third-strand orientation	Various	Antiparallel to the like in the duplex
Glycoside bond conformation		PyPuPy and PyPuPu: all *anti*
Intramolecular triplex isomers	Gel electrophoresis	H-y3 and H-r3 predominate in supercoiled DNA, H-y5 and H-r5 also possible at low superhelicity
Stabilization (PyPuPu triplex)	Various	Divalent metal cations; polyamines; supercoiling (intramolecular triplex)
Strand conformation		
Poly(dT)poly(dA)poly(dT)	X-ray	12-fold helix, N-type sugar, A form
poly(dT)poly(dA)poly(dT)	IR	S-type sugar, B form
Various homopolymer comlexes	IR, Raman IR, Raman	N- and S-type sugars depending on exact combination of ribo and deoxyribo strands
PyPuPy triplex of mixed sequence	NMR	Predominantly S sugar
PyPuPu triplex, mixed sequence	NMR	Predominantly S sugar
Minimum length		5 triads (intermolecular) 15-bp length of PuPy tract (H form)
Maximum length		No limitations for intermolecular triplex except for 150 bp for TA*A triplex Mutiple conformers formed in long tracts for (H and *H forms)

resulting triplex has a diameter only a few angstroms larger than the 20 Å diameter of the starting duplex (Laughton and Neidle, 1992a).

Base Triads in Nucleic Acids

Figure 3.3 shows the base triads which can be formed with natural bases. Base triads involving thymine, cytosine and guanine in the third strand may exist in two isomeric configurations (Lipsett, 1964; Miles, 1964; Davis, 1967; Morgan and Wells, 1968; Beal and Dervan, 1991; Sun et al., 1991a; Cheng and Pettitt, 1992b; Giovannangéli et al., 1992a; reviewed by Cheng and Pettitt, 1992a; Thuong and Hélène, 1993). In one such configuration (Hoogsteen base pairing), the third bases are oriented according to the hydrogen bonding scheme discovered by Hoogsteen (Hoogsteen, 1959). In the other configuration (reverse-Hoogsteen base pairing), third bases are rotated 180° relative to their position in the previous scheme.

Common PyPuPy triplexes include CG*C$^+$ and TA*T (UA*U) triads. These triads were suggested in the early investigations of polymer and oligomer triplexes based on stereochemical considerations and data from

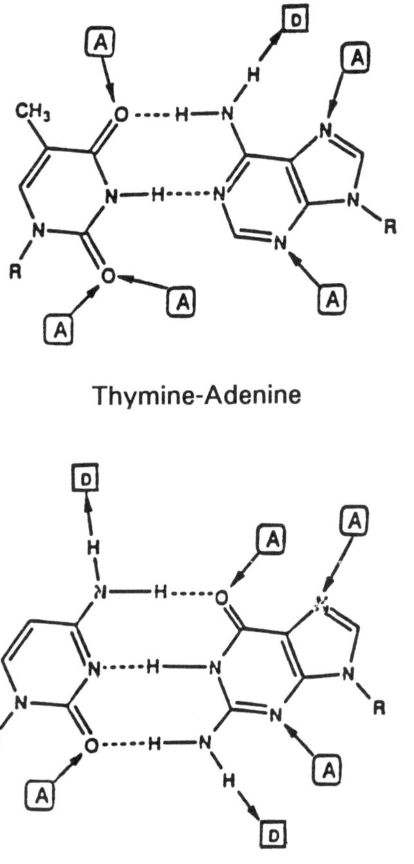

Thymine-Adenine

Cytosine-Guanine

FIGURE 3.2. Hydrogen bond donors and acceptors in the Watson–Crick base pairs. Part of the hydrogen bond-forming capability has been used to make two bonds in the TA pair and three bonds in the CG pair. The rest of the donors (D) and acceptors (A) are exposed in the major and minor grooves (oriented top and bottom, respectively). At the major groove side, both adenine and guanine have two unused hydrogen-bonding capabilities appropriate for Hoogsteen hydrogen-bonding. Reprinted with modifications from *Prog. Biophys. Mol. Biol.* 39, Hélène, C. and G. Lancelot. Interactions between functional groups in protein-nucleic acid associations, pp. 1–68. © 1982, with permission from Elsevier Science Ltd, The Boulevard, Langford Lane, Kidlington OX5 1GB, UK.

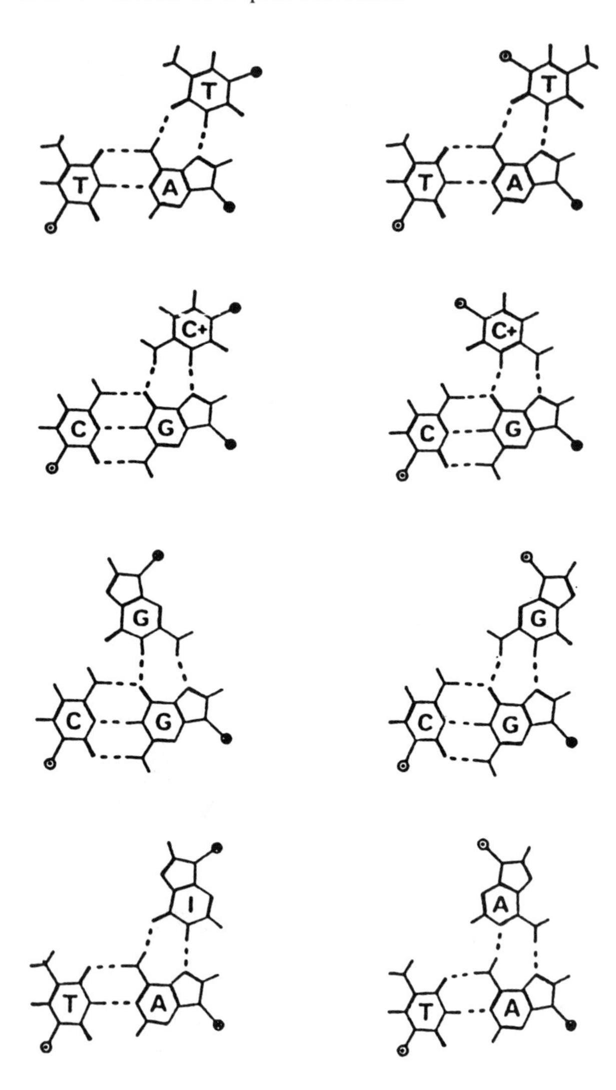

FIGURE 3.3. Base triads formed between the Watson–Crick TA and CG base pairs and T, protonated C (C^+), G, A, and I. The Watson–Crick pairs are oriented horizontally, and the incoming third bases make Hoogsteen bonds from the top. The third-strand bases are in the *anti* conformation with respect to the sugars; therefore, the left column triads correspond to the Hoogsteen hydrogen-bonding pattern and a parallel orientation of the third strand and the purine strand of the target duplex. Similarly, in the right column the triads have a reverse-Hoogsteen hydrogen-bonding pattern, and the third strand and the purine strand of the duplex are in an antiparallel orientation. From Thuong and Hélène (1993). Reprinted with permission.

infrared studies (Felsenfeld et al., 1957; Lipsett, 1964; Miles, 1964; Morgan and Wells, 1968). They are isomorphous, that is, they can be superimposed one over another so that the positions and orientations of corresponding glycosidic bonds, as well as the positions of the C1′ atoms, practically coincide. The PyPuPu triplexes include CG*G and TA*A triads (Lipsett, 1964; Marck and Thiele, 1978; Broitman et al., 1987) which are not truly isomorphous, so the conformation of the sugar–phosphate backbone is not perfectly regular along the third strand. Figure 3.4 shows the arrangements of bases and their glycoside bonds derived from molecular modeling of $(dT)_{10} \cdot (dA)_{10} \cdot (dT)_{10}$ and $(dC)_{10} \cdot (dG)_{10} \cdot (dG)_{10}$ (Sun et al., 1991a). Using DNase I footprinting, affinity cleavage, NMR spectroscopy, and molecular modeling, it was also shown that the PyPuPu triplex can accommodate a number of TA*T triads at the expense of some distortion in optimum base pairing and/or backbone conformation (Cooney et al., 1988; van Vlijmen et al., 1990; Beal and Dervan, 1991; Radhakrishnan et al., 1991a).

In addition to the "canonical" triads described above, one can test some other base combinations. All 32 possible base combinations with their geometric parameters have been collected in a database (Piriou et al., 1994). A number of modified bases can participate in triads, as was discussed in the historical background to polymer triplexes (Chapter 1). Less common triads formed from natural bases and triads, including synthetic base analogs, will be covered in section "Extension of triplex recognition schemes" in Chapter 4.

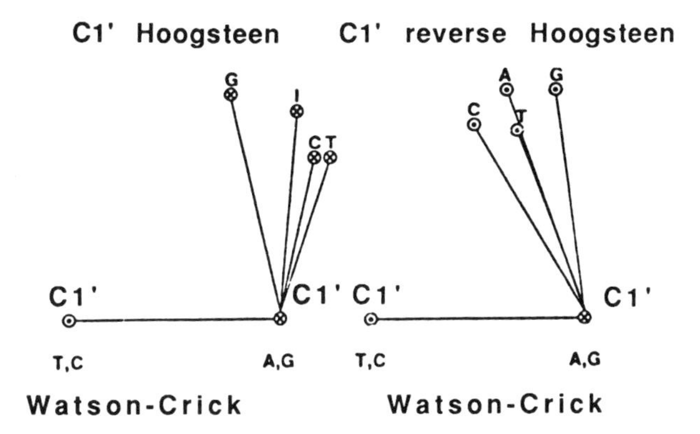

FIGURE 3.4. Isomorphism of base triads. The positions of the C1′ atoms of the three nucleotides which belong to each triad show that only the Hoogsteen-type triads TA*T and $CG*C^+$ are isomorphous. The alternation of all other possible triads within Hoogsteen or reverse-Hoogsteeen triplexes results in some irregularities in the geometry of the sugar–phosphate backbone. Reproduced from Sun et al. (1991a). Reprinted with permission.

Orientation of the Third Strand

The bases of the third strand in the left part of Figure 3.3 form Hoogsteen hydrogen bonds and have *anti* conformation relative to the sugar ring (see Figure 1.5, Chapter 1), which corresponds to a parallel orientation of the third strand and the purine strand of the duplex. For the PyPuPy triplex, this means an antiparallel orientation of pyrimidine strands. Such a strand orientation in the poly(dT) · poly(dA) · poly(dT) triplex was suggested by Arnott et al. (1974) on the basis of X-ray data. This strand orientation also followed from the H DNA model (Lyamichev et al., 1986); for intermolecular triplexes, from the optical melting curves for complexes with parallel and antiparallel oriented third-strand oligonucleotides (Manzini et al., 1990); and from experiments in which orientation was directly determined from an analysis of damaged sites produced by cleaving or photocross-linking reagents attached at the end of oligonucleotides (Le Doan et al., 1987; Moser and Dervan, 1987).

The formation of reverse-Hoogsteen hydrogen bonds by the third bases and their *anti* conformation relative to the sugar ring (and, therefore, the antiparallel orientation of the third purine strand and the purine strand of the duplex) is consistent with the H* DNA model (Kohwi and Kohwi-Shigematsu, 1988). The *anti* orientation of the third-strand purine bases relative to the sugar–phosphate backbone was shown experimentally using the NMR technique (Radhakrishnan et al., 1991a, 1993). The thymine base in the purine-rich third strand also has an *anti* conformation relative to the glycosidic bond. A controversy about the strand orientation for this type of triplex arose when Cooney et al. (1988), based on the data of DNase footprinting, proposed a parallel orientation of the homopurine strand of the duplex and the purine-rich (G- and T-containing) third strand. Later the same researchers showed an antiparallel orientation of the third strand using band-shift assay and target DNA cleavage by Fe–diethylenetriaminepentaacetic acid attached at either one or another ends of triplex-forming oligonucleotides (Durland et al., 1991). The same conclusion was drawn by other researchers who mapped triplex-forming oligonucleotides on DNA by Fe–EDTA affinity cleavage (Beal and Dervan, 1991) and photofootprinting (Frank-Kamenetskii et al., 1991; Soyfer et al., 1992).

Experimental data on third-strand orientations are consistent with theoretical predictions. Molecular dynamics simulations, calculations of internal potential energy, and solvation energy for the PyPuPy triplex composed of TA*T triads showed a preference for Hoogsteen hydrogen bonds and an *anti* conformation of its constituent thymines (Cheng and Pettitt, 1992b). This means an antiparallel orientation of the third Py strand and the Py strand in the duplex. Although an *anti* conformation of purines in polynucleotides is preferable to their *syn* conformation, the rotational barrier is relatively low (Saenger, 1984). Molecular dynamics simulations show that, for the PyPuPu triplex composed of CG*G triads, both Hoogsteen hydrogen bonds (*syn* conformation of third-strand guanines) and reverse-Hoogsteen hydrogen bonds (*anti* conformation of third-strand guanines) are possible if the solvent effects are not taken into account (Cheng and Pettitt, 1992b; Laughton and Neidle, 1992b). However, solvation of the bases makes the *anti* conformation preferable. NMR data on the PyPuPu triplexes containing CG*G, TA*A, and TA*T triads show an antiparallel orientation of the duplex purine strand and the purine-rich third strand and *anti* conformations of the glycosidic bonds (Radhakrishnan et al., 1991a, 1993; Radhakrishnan and Patel, 1994a,b). Thus, an antiparallel orientation of similar strands and *anti* glycosidic bonds was observed in both the PyPuPy and the PyPuPu triplexes.

Overall Conformations of Triplexes

For more than a decade, the molecular details of triple-stranded complexes were considered similar to those in the model of Arnott et al. (Arnott and Bond, 1973; Arnott and Selsing, 1974; Arnott et al., 1976). From the X-ray fiber analysis of poly(U) · poly(A) · poly(U), poly(I) · poly(A) · poly(I) and poly(dT) · poly(dA) · poly(dT), the strands involved in the triplex were believed to have similar A form structures with 12 triads per turn of helix and C3′-endo furanose rings. Studies over the last several years have shown that many triple-stranded structures (Table 3.2) differ from this canonical A form. Overall strand conformations depend on the strand sequence and constituent backbone types. The available data show that strands in triplexes can have characteristics typical for either A or B conformations or a mix of characteristics for both A and B conformations.

First, using infrared spectroscopy, it was shown that the polynucleotide helix poly(dT) · poly(dA) · poly(dT) has the C2′-endo sugar pucker characteristic of the B form structure (Liquier et al., 1991; Howard et al., 1992), instead of C3′-endo sugar pucker in the model by Arnott et al. (Arnott and Selsing, 1974; Arnott et al., 1976). In this particular case, the B form geometry could be explained by the propensity of poly(dA) · poly(dT) for the B′ form over the A form (Alexeev et al., 1987). However, FTIR spectra show that other PyPuPy triplexes also contain features of the B form structure. For example, completely deoxyribose triple helices (where the backbone can be adjusted to A and B forms) composed of C and G contain both C3′-endo and C2′-endo type sugars. Presumably, the purine strand is in the B form conformation (Akhebat et al., 1992). Homopolymer PyPuPy triplexes that contain C and G bases, either prepared starting with a double helix containing only riboses (restricted to the A form geometry) or containing a ribose-type third strand, all have

C3′-endo (A form) type geometries of sugars (Akhebat et al., 1992). A complex sugar pucker pattern was found for the A-, T-, and U-containing PyPuPy triplexes (Liquier et al., 1991). Only the C3′-endo type of the sugar pucker is detected in poly(rU) · poly(rA) · poly(rU) and poly(dT) · poly(rA) · poly(rU). Both the C3′-endo and the C2′-endo type sugar puckers were detected in poly(rU) · poly(rA) · poly(dT), poly(rU) · poly(dA) · poly(dT), and poly(dT) · poly(rA) · poly(dT) triplexes. Similar structural studies of the homopolymer PyPuPu triplexes show that the poly(dC) strand has C2′-endo type sugars, the poly(dG) strand has C3′-endo sugars in the duplex, and the third strand has sugar conformations of either C3′-endo in poly(rG) or C2′-endo in poly(dG) (Ouali et al., 1993).

The X-ray data from single crystals of oligonucleotide triplexes show that the diffraction patterns correspond to the B form helices. In addition, the sugar puckers are in the C2′-endo family. Yet, the $(dT)_{12} \cdot (dA)_{12} \cdot (dT)_{12}$ triplex has the nonstandard values of 13 triads per turn and 3.26 Å for a rise per residue. The PyPuPy triplex with a heterogeneous sequence has 12 triads per turn and 3.20 Å pitch (Liu et al., 1994).

Second, binding of distamycin specific for B form DNA indicates that the minor groove in the poly(dT) · poly(dA) · poly(dT) triplex has a structure typical of the B form rather than of the A form (Howard et al., 1992).

Third, the molecular dynamics simulation of the PyPuPy triplex, starting from the A form model of Arnott et al. (1976), shows that the equilibrium structure significantly differs from the A form, especially in sugar conformation and phosphate group orientation for the purine strand (Laughton and Neidle, 1992a). Another difference is that a wider major groove is predicted by the theoretical model (Laughton and Neidle, 1992a). The molecular dynamics simulation of the poly(dT) · poly(dA) · poly(dT) triplex, starting from either the A or the B form, shows that it is relatively difficult to accommodate the third strand in the A form structure and relatively easy in the B form (Sekharudu et al., 1993). In the equilibrium structure, the sugar pucker, the width of the major groove, and the base tilt are similar to those in B DNA, whereas the axial base displacement and helical twist resemble those of A DNA. Molecular modeling, based on molecular mechanics energies (Laughton and Neidle, 1992b), shows that strands in the PyPuPu triplex adopt conformations similar to those in the PyPuPy triplex, namely, the A-like conformation of the purine strand in the duplex and the B-like conformation of other strands. The A-like conformation of the purine duplex strand and conformations intermediate between A-form and B-form for duplex pyrimidine and third purine strands have been predicted in molecular dynamics and molecular mechanics simulations (Cheng and Pettitt, 1995; Weerasinghe et al., 1995). Similar conformational trends for individual strands were hypothesized on the basis of resonance-Raman studies of triplexes (Gfrorer et al., 1993).

Fourth, photofootprinting data (Lyamichev et al., 1990; Soyfer et al., 1992) show that the duplex part of the DNA triplex can have some features related to the A form. Protection from photomodification for pyrimidines in the triplex is similar to the decrease in photomodification yield for A form DNA (Becker and Wang, 1989). In addition, third-strand binding restricts internal base motions in the triplex, disfavoring pyrimidine positions necessary for dimerization. Thus, the triplex also differs from canonical DNA structures in the decreased mobility of its constituent bases (Lyamichev et al., 1990).

Fifth, NMR studies of more complex oligonucleotide triplex-forming sequences showed that the majority of sugars in the PyPuPy triplexes have C2′-endo puckers, and the entire structure is closer to an underwound B DNA than to an A DNA (Radhakrishnan et al., 1991b, 1992a,b; Macaya et al., 1992a,b). Studies of intramolecular oligonucleotide PyPuPu triplexes showed a similar overall structure but also revealed distinct structural differences between the helical rise and axial twist for the GG, GT, and TG steps in the third strand (Radhakrishnan and Patel, 1993a). Thus, there exists a heterogeneity of conformation, not only between the different strands of the same triplex, but also along the same strand.

Sixth, the susceptibility to chemical probing of the boundary between the B form duplex and the oligonucleotide-formed triplex indicates a change in duplex structure at the site of the junction (Collier et al., 1991a; Hartman et al., 1992). Altered duplex conformation does not extend far beyond the triplex structure. The triplex–duplex junction is highly localized, because it does not interfere with protein binding at immediately adjacent sites (Hanvey et al., 1990; Huang et al., 1992). This conclusion is supported by the data of Colocci et al. (1993), who did not find propagation of triplex-modified duplex structures further than one base pair at the adjacent sequence.

Thus, depending on the sequences involved and the polynucleotide backbone type, triple helices may have different geometries, although in the recent review by Frank-Kamenetskii and Mirkin (1995) the authors declared that "the duplex within the triplex adopts a B-like configuration; the helical twist in the triplex, however, is significantly smaller than that for B-DNA" (p. 73).

Stabilization Common to Intramolecular and Intermolecular Triplexes

Neutralization of phosphate groups should reduce the repulsion of negatively charged sugar–phosphate backbones, facilitate the approach of the third strand to the duplex target, and stabilize the triplex structure that has been formed. The triplex-stabilizing effects of several types of cations have been studied.

Metal cations are known to have triplex-stabilizing effects. Monovalent cations at concentrations of approximately $0.1\,M$ are ordinarily used in triplex studies. The stability of PyPuPy triplexes containing AU*U and TA*T triads increases when Na^+ or K^+ concentration increases beyond $0.1\,M$, as shown by calorimetric and UV-melting experiments (Blake et al., 1967; Krakauer and Sturtevant, 1968; Durand et al., 1992a). Divalent cations stabilize this triplex even more effectively. Concentrations of Mg^{2+} or Mn^{2+} of $10\,mM$ are sufficient for the complete formation of the poly(U) · poly(A) · poly(U) triplex (Felsenfeld et al., 1957; Felsenfeld and Rich, 1957). This trend suggests a significant role of the stabilizing mechanism, different from a simple bulk screening of phosphates by positive ions, where increasing the ionic strength should provide a stabilizing effect. Cations bound to negatively charged phosphate groups appear to be responsible for this aspect of the triplex-stabilizing effect. Mg^{2+} or Mn^{2+} have lower ionic radii and higher charges (CRC Handbook of Chemistry and Physics, 1993), and, therefore, can create a higher neutralizing charge density than Na^+ or K^+. Triplexes containing protonated bases (see below) are also stabilized by increasing ionic strength (Shea et al., 1990; Völker and Klump, 1994). However, the relative stabilities of CG*C^+ and TA*T triads differ at low and high salt concentrations. The CG*C^+ triad is more stable than the TA*T triad below $0.4\,M$ Na^+. Above $0.4\,M$ Na^+, the TA*T triad is more stable than the CG*C^+ triad (Völker and Klump, 1994).

The formation of intramolecular and intermolecular PyPuPu triplexes requires the presence of some multivalent cations including polyamines, and di- and trivalent metal cations (Cooney et al., 1988; Kohwi and Kohwi-Shigematsu, 1988; Bernues et al., 1989, 1990; Kohwi, 1989b; Collier and Wells, 1990; Lyamichev et al., 1990, 1991; Frank-Kamenetskii et al., 1991; Panyutin and Wells, 1991; Dayn et al., 1992b; Kang et al., 1992a; Soyfer et al., 1992). These ions have several types of binding sites on polynucleotides, and therefore a complex interplay of stabilizing interactions appears to exist (see Chapter 5 for more details).

In some cases, the formation of triplexes requires specific pH values. This requirement stems from the necessity to protonate the third-strand bases so that they can form two Hoogsteen hydrogen bonds with the duplex purines. For example, under acidic conditions, cytosine and adenine are protonated ($pK_a = 4.3$ and 3.8, respectively). Thus, acquired protons may participate in the hydrogen bonds with duplex guanines, allowing the formation of CG*C^+ (Lipsett, 1963, 1964) and CG*A^+ (Malkov et al., 1993a) triads. Interaction of protonated bases with the negatively charged phosphates of the target duplex is another stabilizing (or triplex-promoting) contribution. An acidic environment promotes the formation of protonated triplexes both between the PyPu tract and an external third strand and within the supercoiled plasmid. The pH dependence of triplex formation is strong (Lyamichev et al., 1985; Soyfer et al., 1992; see Frank-Kamenetskii, 1992, for review), and in the absence of other triplex stabilizers, the upper pH limit for triplex formation is about 5.0 to 5.5. The contribution of the base protonation as a factor providing a second Hoogsteen hydrogen bond may be of key importance during triplex formation and less significant in maintaining a preformed triplex structure if other compensating stabilizing factors are available. For instance, when triplexes have been formed under acidic conditions, they may be quite stable in neutral pH media containing spermine as an external triplex stabilizer (Hampel et al., 1991).

Factors That Destabilize Triplexes

Some sequence peculiarities have triplex-destabilizing effects. They include, for example, the presence of imperfectly matched triads, abasic sites, and extra bases inside the triplex-forming sequence. The consequences of these defects on the triplex structure are discussed in this chapter under the Thermodynamics of Triplexes.

The repulsion of adjacent protonated cytosines can counterbalance the stabilizing stacking interactions of the third-strand bases and result in weak triplex formation (Lee et al., 1984). In the case of intramolecular PyPuPy triplexes, the contributions of the stacking interactions, the effect of the loop (which does not allow easy detachment of the third strand), and the significant enhancement of the stability of CG^*C^+ triads at acidic pH outweigh the unfavorable electrostatic repulsion of protonated cytosines. A triplex can be formed in $(C)_n \cdot (G)_n$ sequences (Lyamichev et al., 1987; Kohwi and Kohwi-Shigematsu, 1988). Moreover, according to the data from chemical probing by osmium tetroxide and UV melting, the stability of triplexes diminishes with reduction in the percentage of GC in the triplex-forming sequences (Hanvey et al., 1989; Völker and Klump, 1994). In the case of intermolecular triplexes, a decrease in the relative stabilities of triplexes, as measured by the affinity cleaving technique, accompanied an increase in the number of adjacent CG^*C^+ triads at neutral pH (Kiessling et al., 1992) and even at acidic pH (Jayasena and Johnston, 1992b). A similar trend was observed for oligonucleotides folded into intramolecular triplexes (Völker and Klump, 1994). At neutral pH, this effect might be caused either by destabilizing interactions between adjacent protonated cytosines that overweigh stabilizing factors, or by incomplete protonation of the adjacent third-strand cytosines, resulting in the reduced stability of the CG*C triad with one hydrogen bond (Kiessling et al., 1992). It is difficult to estimate the relative destabilizing contribution of adjacent cytosines at acidic pH from the affinity cleavage data: the cleavage efficiency of EDTA–Fe used for affinity cleaving decreases sharply below pH 7 (Moser and Dervan, 1987). In the UV-melting experiments, the electrostatic repulsion between adjacent protonated cytosines at acidic pH was less pronounced than at neutral pH (Völker and Klump,

1994). Thus, the triplex–destabilizing effect of contiguous CG*C^{+} triads is more significant at neutral pH and less significant at acidic pH.

The effect of 5′-phosphorylation on the stability of the oligomer PyPuPy DNA triplex has been demonstrated by both gel electrophoresis and UV melting (Yoon et al., 1993). The order of stability is (from greatest to least): no phosphate on all of the strand > phosphate on both pyrimidine strands > phosphate on purine strand > phosphate on all three strands. The differential stability of triple helix species is believed to stem from an increase in rigidity due to steric hindrance from the 5′-phosphate.

Specific Features of Intramolecular Triplexes (H and H Forms)*

Relaxation of some torsional stress in closed circular DNA may result from the formation of non-B form DNA structure (reviewed by Wells et al., 1988; Frank-Kamenetskii, 1990; Palecek, 1991). Therefore, DNA supercoiling can provide the energy to drive formation and stabilization of H DNA (Lyamichev et al., 1985, 1987; Mirkin et al., 1987; Vojtiskova and Palecek, 1987; Hanvyey et al., 1988a,b; Htun and Dahlberg, 1988; Johnston, 1988; Kohwi and Kohwi-Shigematsu, 1988; Bernues et al., 1989, 1990).

Formation of the H form requires a mirror-repeated nucleotide sequence (H palindrome) with an arbitrary, preferably AT-rich, insert between the two halves of the repeat (Mirkin et al., 1987; Shimizu et al., 1993). The denaturation bubble necessary to unwind the double helix locally and initiate strand rearrangement to form H DNA (see below) is more readily produced in the easily melting AT-rich sequences. The types of sequences flanking the PyPu tract are also important, since it has been shown that having more stable duplexes in GC-rich flanks reduces the rate of H DNA formation (Kang et al., 1992b). Any deviation from the mirror-repeated sequence will result in a significant increase of superhelical tension necessary to extrude the PyPuPy-type triplex (Mirkin et al., 1987; Belotserkovskii et al., 1990). The mirror repeats are characterized by a pseudosymmetry rather than a true symmetry. This results in different probabilities for the occurrence of the two isomers (see Figure 1.29) (Mirkin et al., 1987; Htun and Dahlberg, 1989). The conformer in which the 3′ half of the Py repeat is used as the third strand was denoted H-y3, and the other isomer, in which the 5′ half of the Py repeat is used as the third strand, was denoted H-y5 (Htun and Dahlberg, 1989). Similar conformers (H-r3 and H-r5) can be formed when the purine strand is folded back, and donated as the third strand.

The mirror repeat requirement is less strict for the H* form than for the H form. Dayn et al. (1992b) were successful in designing H* forms consisting of CG*G and TA*T triads. Their sequences were not mirror repeated, because adenine and thymine bases were positioned as inverted repeats. Neither were these the pure PyPu tracts, because thymine residues were interspersed in otherwise purine-rich strands (Figure 3.5). Malkov et al. (1993a) described another variant of the H* form without mirror symmetry, where the bases of a completely purine third strand took part in the CG*G and CG*A$^+$ triads. In the most recent example of an unusual H* form, the structure consists of an equal number of alternating CG*A$^+$ and TA*T triads (Klysik, 1995). In this case, the sequence also does not have mirror symmetry, but a more striking feature is that

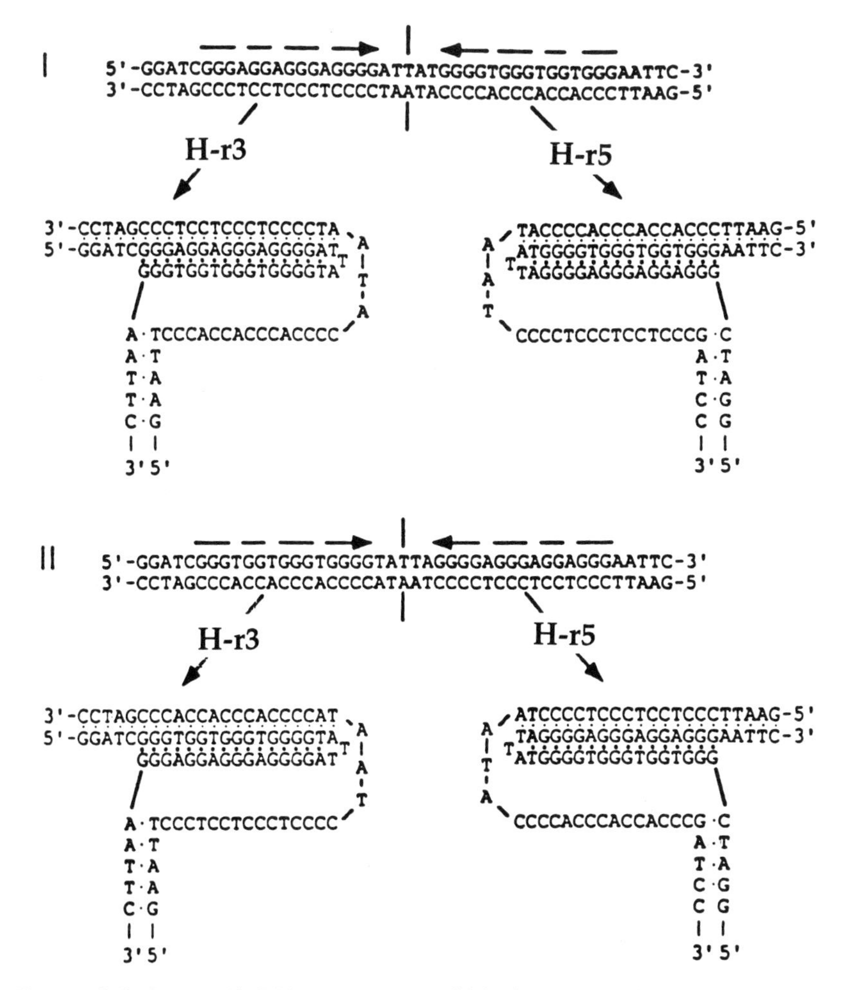

FIGURE 3.5. Some H* DNA structures which do not require an exact mirror symmetry because wherever possible TA*A triads were substituted by TA*T triads. In both sequences I and II, the GC pairs are arranged according to a mirror symmetry (a vertical line shows the pseudosymmetry axis), and the AT pairs are arranged as inverted repeats. Reproduced from Dayn et al. (1992b).

the third strand is neither purine- nor pyrimidine-rich. However, an antiparallel orientation of the duplex purine strand and the third strand allows the classification of this triplex as the PyPuPu type. Some researchers use the terms "parallel and antiparallel triplexes." The "parallel triplex" implies a parallel orientation of the duplex purine strand and the third strand. This is characteristic of the PyPuPy triplex, where two strands (pyrimidine and purine) in the content of the duplex are oriented in opposite directions and a third (pyrimidine) strand is parallel to the duplex purine strand. The "antiparallel triplex" implies an antiparallel orientation of the duplex purine strand and third strand, that is characteristic of the PyPuPu triplex. Klysik (1995) designed a triplex with an equal number of purine and pyrimidine bases in the third strand, and called it an antiparallel triplex.

The nucleotide composition of the PyPu tract may be an important factor. Studies of H DNA formation as a function of G+C content of the PyPu tract (Hanvey et al., 1988a) showed that tracts with 25% to 66% G+C form similar H DNA structures. An $A_{20} \cdot T_{20}$ tract remained in a B-like duplex conformation irrespective of variations in pH and DNA supercoiling. It was shown later that the $A_n \cdot T_n$ tracts are able to form an H DNA structure only at significantly high n [$n = 69$ in the study of Fox (1990)]. The $G_n \cdot C_n$ tracts may adopt H or H* form structures dependent on pH and the presence of divalent metal cations (Hanvey et al., 1988a; Kohwi and Kohwi-Shigematsu, 1988). The H* DNA structure forms in the G+C-rich PyPu tracts in the presence of multivalent cations (see below). At 50% G+C content, the details of locally unwound structure are somewhat uncertain: at neutral pH a whole pyrimidine strand is susceptible to single-strand-specific agents (Panyutin and Wells, 1992; Beltran et al., 1993). Whether a structure corresponds to a mixture of different H* isomers, or to an H* hairpin with Hoogsteen-bonded purine strand and unpaired pyrimidine strand (Beltran et al., 1993), remains to be elucidated. At low G+C content of the PyPu tract (33%), H* DNA structure did not form (Panyutin and Wells, 1992).

Both triple-stranded and single-stranded regions are present in the structure of H DNA. Due to the foldback structure of H DNA, a flexible kink is introduced into DNA which serves as a hinge between flanking duplexes (Htun and Dahlberg, 1988, 1989). The presence of such a hinged structure was experimentally confirmed by Stokrova et al. (1989), who used electron microscopy to observe thick "stems" at the sites where curvature of supercoiled DNA molecules containing the PyPu tracts occurred. Observation of these "stems" correlated with a lowering of the pH of the buffer, which should result in the stimulation of H DNA formation.

No chemical reactivity at the base of the acceptor duplex in H DNA was found (Pulleyblank et al., 1985; Htun and Dahlberg, 1988). This was interpreted as an absence of any discontinuity between the triplex and the adjacent duplex (Htun and Dahlberg, 1988, 1989). It can be reasoned, however, that it is difficult to detect a highly localized helix distortion at the site of the triplex–duplex junction because of the steric hindrance of the third strand leaving the duplex. Experiments on intermolecular triplexes have shown that some duplex irregularity at the triplex–duplex junction really exists (Chomilier et al., 1992; Hartman et al., 1992; Huang et al., 1992; see section "Overall Conformation of Triplexes" in this chapter).

A detailed model of intramolecular triplex formation (Figure 3.6) was elaborated by Htun and Dahlberg (1989). In this model, a six-base-pair denaturation bubble in the PyPu tract allows the duplexes on either side to rotate slightly and to fold back, providing the formation of the first triad. This nucleation event is critical in establishing which of several nonequivalent H DNA isomers is to be formed by any DNA molecule. Subsequently, one half of the PyPu tract accepts a single strand as it is released by the progressive denaturation of the other half of the PyPu tract. Either the 3′ or the 5′ half of the Py strand could function as the third strand in H DNA (H-y3 and H-y5 isomers), with its counterpart strand being unstructured. Since the initial denaturation bubble unwinds the duplex by about one half-turn, the resulting two duplexes should rotate counterclockwise or clockwise by about one half-turn in order to position the strand to be folded back opposite to the major groove of the accepting duplex. Thus, two rotations would differ in one full turn of the overall negative supercoil relaxation. Both H-y3 and H-y5 isomers appear equivalent when drawn in two dimensions, yet more negative superturns are relaxed by formation of H-y3 than by H-y5 form DNA. Similar reasoning can be used to explain the predominance of the H-r3 isomer over the H-r5 isomer when the purine strand is folded back. The model predicts that when the same negatively supercoiled DNA is folded into the PyPuPy or PyPuPu triplexes, the molecule carrying the H-r3 isomer will be relaxed by about one-fourth of a superhelical turn less, compared to the DNA with the H-y3 isomer. At low negative supercoiling, H-y5 and H-r5 conformations, which are unfavorable at high supercoiling, can be assumed. More recent data (Kang et al., 1992a; Kang and Wells, 1992) indicate that the isomerization into H-y3 or H-y5 forms may also depend on the loop sequence or specific cations (see below).

The issue of the acceptable size of the spacer that interrupts regular PyPu tract participating in an intramolecular triplex has been the subject of several investigations. The bases of this sequence at the U-turn strand position are unpaired. An acceptable strand folding can be provided by as few as two unpaired bases (Veselkov et al., 1993), but in most of the published experimental studies, it is usually provided by three to five bases (Lyamichev et al., 1986, 1991; Vojtiskova and Palecek, 1987; Hanvey et al., 1988a; Htun and Dahlberg, 1988; Johnston, 1988; Kohwi and Kohwi-Shigematsu, 1988). This corresponds to the favorable loop

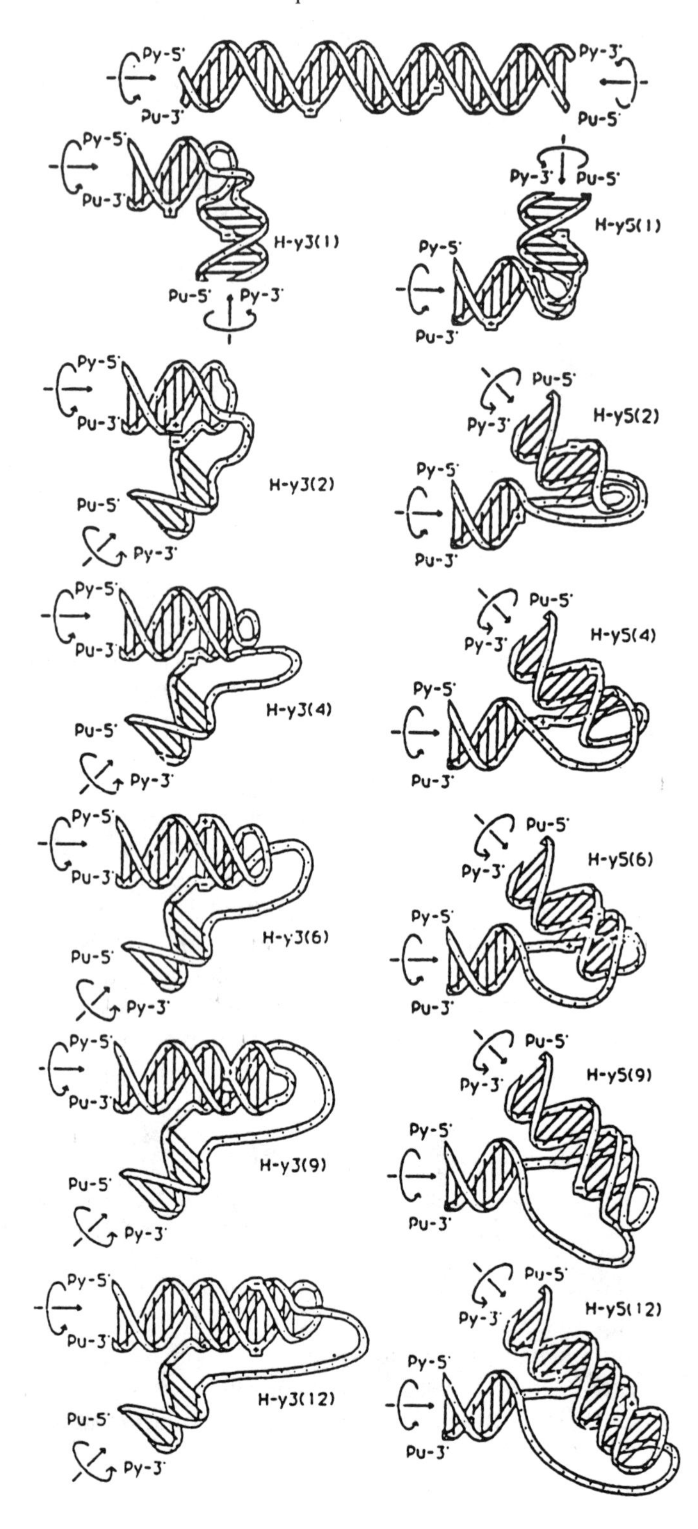
Py-5'
Pu-3'
Py-3'
Pu-5'
H-y3(1)
H-y5(1)
H-y3(2)
H-y5(2)
H-y3(4)
H-y5(4)
H-y3(6)
H-y5(6)
H-y3(9)
H-y5(9)
H-y3(12)
H-y5(12)

lengths of four to five bases determined in molecular mechanics simulations and energy minimization (Harvey et al., 1988). Longer triplex loops increase the energetic cost of triplex formation and, in the case of H DNA, require a higher supercoil energy. For example, elongation of the loop size (four to twelve bases) required increasing the superhelical density for triplex formation (Parniewski et al., 1989; Shimizu et al., 1989). A 46-base spacer with a random sequence inhibited H DNA formation under the same conditions where H DNA with 12 bases in the loop can be formed (Parniewski et al., 1989). If two spatially separated PyPu tracts are brought closer (e.g., under conditions promoting the cruciform in the alternating A-T spacer sequence), an intramolecular triplex structure can be adopted (Klysik, 1992). When two sufficiently long PyPu tracts (24–27 base pairs) are separated by a spacer of increasing length, the triplexes (H and H* forms) can form separately in each of the tracts, provided that the spacer length is of the order of 10 base pairs (Shimizu et al., 1990).

The importance of the spacer sequence was understood long ago. For example, the triplex formation was facilitated when the DNA was designed in such a way, that the AT-rich spacer was inserted between two triplex-forming mirror repeats (Lyamichev et al., 1986; Mirkin et al., 1987). Under such circumstances, the initial unwinding of the central AT-rich region was easier. This agrees with the data from Shimizu et al. (1993), who also showed that as the G+C content of the spacer increases, the triplex requires more superhelical stress to be extruded. The length and base composition of the interrupting sequence influenced the particular triplex isomer (H-r5, H-r3 and H-y3) formed in the tandem $(dG)_{10} \cdot (dC)_{10}$ tract at pH 5 (Kang and Wells, 1992). The loop sequence was also shown to play a crucial role in isomerization of the PyPuPy triplexes in the tandem $(TC)_6CT \cdot (GA)_6AG$ sequence in supercoiled plasmid (Shimizu et al., 1993). Thermal denaturation studies of model triplexes between circular and linear oligonucleotides (see Figure 3.17), which might provide a good approximation of interactions at a tip of H DNA, demonstrate

←

FIGURE 3.6. Model for the formation of H DNA. The rearrangement of the PyPu duplex into the H DNA begins with the initial local melting in the middle of the tract, allowing the duplex to fold back on itself. An unwound pyrimidine strand folds into the major groove, where it forms first Hoogsteen bonds with the purine strand. Depending on which half of the PyPu tract initiates Hoogsteen pairing, structural isomers H-y3 (left) and H-y5 (right) can be formed. Coordinated rotation of the two duplexes allows the winding of the pyrimidine strand as it is released by the denaturation of its parent duplex. Elongation occurs until it is possible to form stable triads and is driven by the energy associated with the negative supercoiling. Reprinted with permission from Htun and Dahlberg (1989) *Science* 243, 1571–1576. © 1989 AAAS.

that interactions between the sequence of the loop in the pyrimidine strand and the sequence of the outgoing purine strand are important for stabilization of the entire structure (Booher et al., 1994; S. Wang et al., 1994).

In addition to superhelical tension, the formation of protonated PyPuPy triplexes is facilitated by mildly acidic conditions (Lyamichev et al., 1985; Mirkin et al., 1987). The PyPuPu triplexes generally require various multivalent cations (Kohwi and Kohwi-Shigematsu, 1988; Bernues et al., 1989, 1990; Panyutin and Wells, 1992). As a result, the formation of H and H* DNA is interdependently governed by low pH, negative supercoiling, bivalent cations, and the increasing length of the PyPu tracts (see, e.g., Lyamichev et al., 1985; Htun and Dahlberg, 1988; Johnston, 1988; Kohwi and Kohwi-Shigematsu, 1988; Bernues et al., 1989; Collier and Wells, 1990; Beltran et al., 1993).

Figure 3.7 shows the interdependence of low pH and superhelical density of plasmid DNA in promoting extrusion of the H form, as derived from the data on 2-D electrophoresis and S1 nuclease digestion (Lyamichev et al., 1985). It is clear that two factors are supplementary: at pH 4, no negative supercoiling was required for H DNA formation, and extreme conditions of negative supercoiling ($-\sigma \sim 0.1$) were required for H DNA formation at neutral pH. These results are consistent with the data on the chemical probing of H DNA (Htun and Dahlberg, 1988; Johnston, 1988).

Figure 3.8 shows the dependence of H* DNA formation on the concentration of Zn^{2+} ions added to the medium, as derived from a quantitative analysis of supercoil relaxation (Beltran et al., 1993). Numerous studies

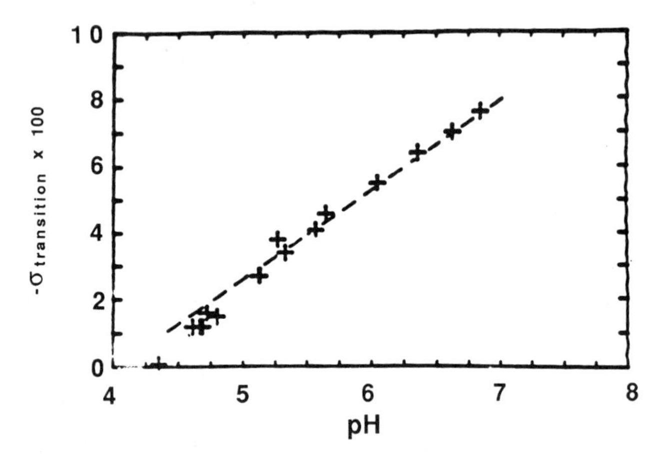

FIGURE 3.7. The pH dependence of negative superhelical density of DNA ($-\sigma_{tr}$) required to extrude a protonated H DNA. The experimental values of $-\sigma_{tr}$ corresponding to the transition midpoint were determined from the 2-D gel electrophoretic data. From Lyamichev et al. (1985). Reprinted with permission.

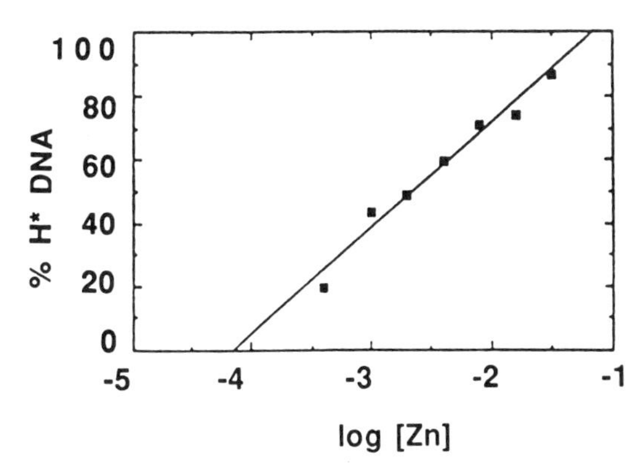

FIGURE 3.8. Yield of H* DNA as a function of Zn^{2+} concentration in the medium. Similar dependencies were also found for Cd^{2+} and Mn^{2+} but with a lower percentage of H* DNA formed. From Beltran et al. (1993). Reprinted with permission.

have shown that H* DNA containing adenine and guanine in the third strand is promoted by the transition metal cations Zn^{2+}, Cd^{2+}, and Mn^{2+}, but not by the alkaline earth cations Ca^{2+} and Mg^{2+} or the other cations Co^{2+}, Ni^{2+}, Cu^{2+}, and Hg^{2+} (Bernues et al., 1989, 1990; Collier and Wells, 1990; Shimizu et al., 1990; Frank-Kamenetskii et al., 1991; Lyamichev et al., 1991; Kang et al., 1992a; Soyfer et al., 1992; Malkov et al., 1993b). It is possible that there are different kinetic barriers for the extrusion of H* DNA in the presence of different cations. For example, preincubation $(GA)_{37} \cdot (CT)_{37}$ containing plasmid at elevated temperature in the presence of Ca^{2+} or Mg^{2+} helps to extrude the triple-stranded structure (Panyutin and Wells, 1992). By contrast, the $(dC)_n \cdot (dG)_n \cdot (dG)_n$ triplex in plasmid DNA is easily promoted by Mg^{2+} and Ca^{2+} ions (Kohwi and Kohwi-Shigematsu, 1988; Lyamichev et al., 1991; Kang and Wells, 1992) and transition metal cations (Kang and Wells, 1992; Panyutin and Wells, 1992).

However, PyPuPu triplexes can be promoted not just by divalent metal cations. According to the data from chemical probing, trivalent cations in lower concentrations than divalent cations (1 m*M* spermidine and Co^{3+} versus 10 m*M* of Mg^{2+}, Ca^{2+}, or Co^{2+}) can be used to stabilize the H* form (Panyutin and Wells, 1992). Yet, when there are several possible conformers, divalent metal cations, spermidine, and Co^{3+} stabilize different structures (Panyutin and Wells, 1992; Soyfer et al., 1992). The interplay of various cations in triplex stabilization will be discussed at more length in Chapter 5.

The interdependence of PyPu tract length, superhelical density, and pH in H DNA formation was studied by 2-D electrophoresis and chemical probing (Htun and Dahlberg, 1989; Lyamichev et al., 1989a,b; Collier and Wells, 1990; Panyutin and Wells, 1992). It was found that one supercoil is relaxed for every 10 or 11 base pairs of the PyPu tract (Htun and Dahlberg, 1989; Collier and Wells, 1990) (Figure 3.9). Therefore, the longer the PyPu tract, the more superhelical relaxation can be obtained by extrusion of the H form. H DNA can be observed in a long $d(GA)_{37} \cdot d(TC)_{37}$ tract at neutral pH and a moderate level of negative supercoiling (Collier and Wells, 1990). Evidently, this is valid for both PyPuPy and PyPuPu triplexes (Lyamichev et al., 1989b; Panyutin and Wells, 1992). A calculation of the free energy of extrusion of the H form, which used experimental data for input parameters, resulted in an estimation of 15 base pairs as the absolute minimum length of the PyPu tract where the H DNA can be formed at any superhelical density (Lyamichiev et al., 1989b).

According to the data from S1 nuclease and chemical probing, in the specific case of long adenine tracts, the production of the H form required a sufficiently long sequence (Fox, 1990). These tracts are known to adopt the B′ conformation, which is different from the standard B form (see Arnott and Selsing, 1974; Alexeev et al., 1987, for a description of the

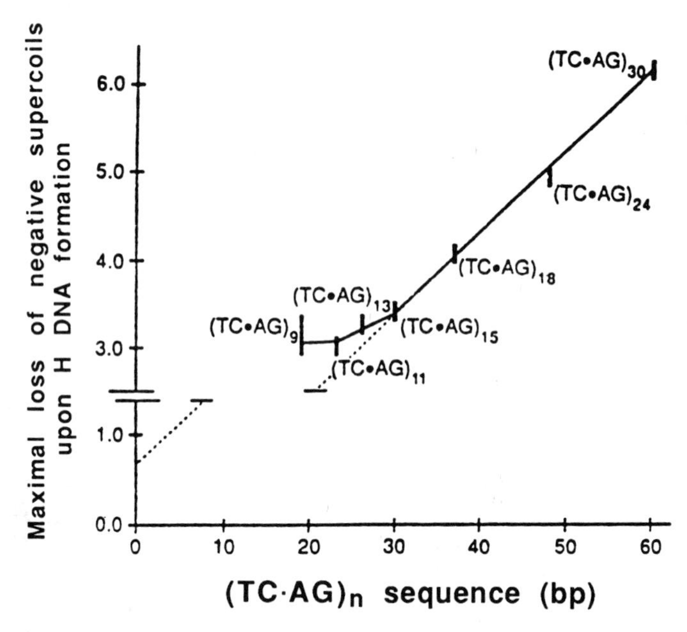

FIGURE 3.9. Dependence of negative supercoil relaxation on the length of the PyPu tract. The loss of negative supercoils was determined for $(GA)_n \cdot (TC)_n$ repeats of various lengths from the 2-D gel electrophoretic data. It follows from the slope of the graph between $(GA)_{15} \cdot (TC)_{15}$ and $(GA)_{30} \cdot (TC)_{30}$ that one supercoil is relaxed for every 11 base pairs of the PyPu tract. Reprinted with permission from Htun and Dahlberg (1989) *Science* 243, 1571–1576. © 1989 AAAS.

structure, and Price and Tullius, 1993, for a later list of references). The H form was detected in $(dA)_{68} \cdot (dT)_{69}$, but not in $(dA)_{23} \cdot (dT)_{23}$ or $(dA)_{33} \cdot (dT)_{33}$ inserts in plasmid DNA. This is probably due to additional interplane hydrogen bonds in the B′ form and a higher rigidity of the B′ form compared to the B form. Consequently, the B′ form is unable to adjust its molecular conformation to that necessary for triplex formation.

In sufficiently long PyPu tracts (>35 base pairs), multiple nucleation events can result in a mixture of several H DNA conformers (Htun and Dahlberg, 1989; Collier and Wells, 1990; Kohwi-Shigematsu and Kohwi, 1991; Panyutin and Wells, 1992; Kohwi and Kohwi-Shigematsu, 1993). For example, chemical probing analysis of intramolecular triplexes with increasing length of triplex-forming sequence and variation of pH, ion concentrations, etc. is complicated by the formation of either H or H* form DNA (Kohwi and Kohwi-Shigematsu, 1988) or by the presence of mixed structures. In some of these structures, either half of the Py or Pu strand can be involved in triplex formation. A bitriplex structure, where the looped strand of H DNA is donated to form the H* DNA with another PyPu tract, is also possible (Kohwi-Shigematsu and Kohwi, 1991; Panyutin and Wells, 1992; Sprous and Harvey, 1992; Kohwi and Kohwi-Shigematsu, 1993). Figure 3.10 shows an example of a bitriplex structure which can be formed in the $d(G)_n \cdot d(C)_n$ tract under conditions appropriate for both H and H* DNA formation (Kohwi-Shigematsu and Kohwi, 1991). An alternative H hairpin structure with a completely unpaired Py strand was suggested on the basis of the chemical modification pattern for the $d(GA)_{22} \cdot d(CT)_{22}$ sequence at neutral pH (Beltran et al., 1993; Martinez-Balbas and Azorin, 1993). However, those results do not exclude the presence of a triplex isomer mixture.

H and H* DNA can accommodate some mismatched triads. There has been no systematic study of the acceptable number and position of such imperfect triads. The studies of the PyPuPy triplexes with point substitutions in the generally mirror-repeated sequence (Mirkin et al., 1987; Voloshin et al., 1988; Belotserkovskii et al., 1990) showed that mismatched triads result in destabilization of the H form, the magnitude of the effect being dependent on the specific type of mismatch. A 42-base-pair PyPu sequence with three consequtive mismatches was shown not to form a large triplex (Hanvey et al., 1989). The data from chemical probing indicate that a mixture of two triple-stranded helices is formed in which the mismatched site is either in the middle of the folded pyrimidine strand or completely out of the triplex zone (Hanvey et al., 1989).

Specific Features of Intermolecular Triplexes

When intramolecular H and H* DNA forms, only three possible strand combinations can exist (triple helix + single strand, double helix, and single strands). When oligo- or polynucleotide chains are interacting, more

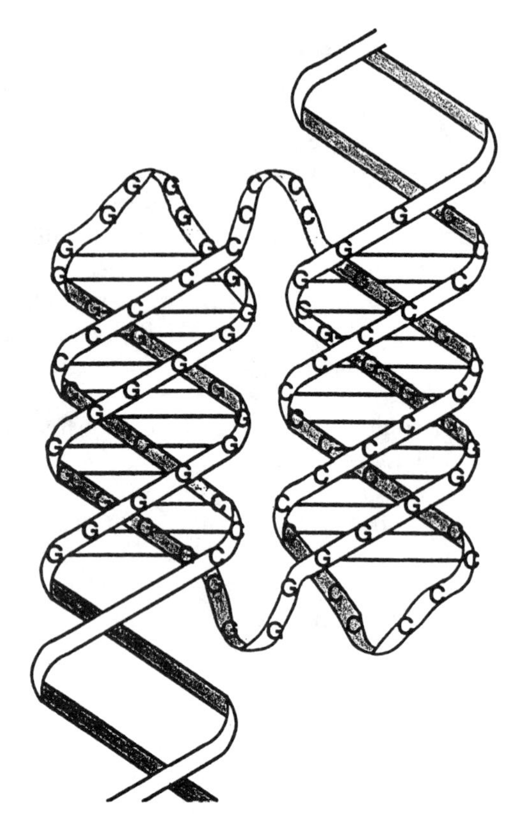

FIGURE 3.10. A model of bitriplex structure formed in long PyPu tracts which is in accordance with the chemical probing data. Reproduced from Kohwi-Shigematsu and Kohwi (1991) *Nucleic Acids Res* by permission of Oxford University Press.

possible combinations may appear. These can exist as single-stranded molecules or can form duplexes, triplexes, or combinations of single-, double-, and triple-stranded molecules. Given sufficient data on melting temperatures and the stoichiometry of helical complexes as a function of ambient conditions, it is possible to construct a phase diagram for polynucleotide double- and triple-stranded configuration.

Such a phase diagram (Figure 3.11) was obtained for the mixture of poly(A) and poly(U) from numerous UV-melting and direct calorimetric measurements (Fresco, 1963; Massoulie et al., 1964a; Stevens and Felsenfeld, 1964; Krakauer and Sturtevant, 1968). Four rearrangement reactions were considered:

$$\text{poly(A)} \cdot \text{poly(U)} = \text{poly(A)} + \text{poly(U)} \quad (1)$$

$$2\text{poly(A)} \cdot 2\text{poly(U)} = \text{poly(U)} \cdot \text{poly(A)} \cdot \text{poly(U)} + \text{poly(A)} \quad (2)$$

$$\text{poly(U)} \cdot \text{poly(A)} \cdot \text{poly(U)} = \text{poly(A)} + 2\text{poly(U)} \tag{3}$$

$$\text{poly(U)} \cdot \text{poly(A)} \cdot \text{poly(U)} = \text{poly(A)} \cdot \text{poly(U)} + \text{poly(U)} \tag{4}$$

Reaction (1) corresponds to a dissociation (melting) of the duplex. Reaction (2) corresponds to a rearrangement of the duplex, resulting in formation of a triplex and a single strand. Reaction (3) describes the melting of the triplex as a whole. Reaction (4) relates to the liberation of the third strand, which leaves the duplex intact.

This diagram was constructed with temperature and Na^+ concentration as variables (Krakauer and Sturtevant, 1968). A four-dimensional diagram would be necessary to account for the varying pH and varying proportions of poly(A) and poly(U); however, the effect of these variables can be described additionally. According to Cantor and Schimmel (1980c), the following explanation of the phase diagram can be presented. The high-temperature region IV corresponds to the existence of single-stranded polymers because of the instability of hydrogen-bonded helical structures. Both double- and triple-stranded helices are less stable at low salt concentrations, and a much higher temperature is needed to dissociate them at

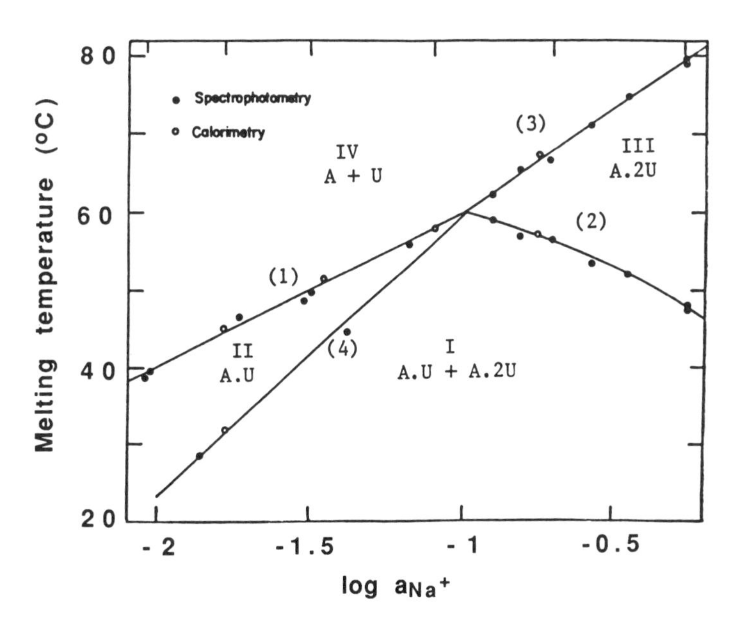

FIGURE 3.11. Phase diagram for the poly(A) + poly(U) system. The dependence of T_m on Na^+ concentration at neutral pH subdivides the plane into four regions where different polymer complexes exist. Transitions between these regions correspond to the reactions: (1) poly(A) · poly(U) = poly(A) + poly(U); (2) 2poly(A) · 2poly(U) = poly(U) · poly(A) · poly(U) + poly(A); (3) poly(U) · poly(A) · poly(U) = poly(A) + 2poly(U); (4) poly(U) · poly(A) · poly(U) = poly(A) · poly(U) + poly(U). Reproduced from Krakauer and Sturtevant (1968).

high salt concentrations. Thermodynamic analysis performed by Cantor and Schimmel (1980b) shows that multistranded nucleic acid structures bind larger amounts of cations per phosphate group than single-stranded structures. This is due to the need to reduce the electrostatic repulsion of strongly charged sugar–phosphate backbones during the formation of double- and triple-stranded complexes. As a result, increasing salt concentration stabilizes the triplex because of the reduction of electrostatic free energy of the complex.

Region II corresponds to the double-stranded helix poly(A) · poly(U). The stoichiometrically excessive single strands poly(A) or poly(U) cannot form stable triplexes and remain unbound in the solution.

Region III corresponds to the stable triplex poly(U) · poly(A) · poly(U). Excessive poly(A) or poly(U) exist as single strands in solution.

Both poly(A) · poly(U) and poly(U) · poly(A) · poly(U) are stable in region I. The mixing curve (Figure 2.1, Chapter 2) has actually been obtained under conditions corresponding to region I, provided that the salt concentration is fixed and the proportion of poly(A) and poly(U) is changed. If the part of poly(U) relative to the sum of poly(U) and poly(A), $X_U < 0.5$, then the mixture contains a double helix poly(A) · poly(U) and a free poly(A). When $0.5 < X_U < 0.67$, the mixture contains poly(A) · poly(U) and poly(U) · poly(A) · poly(U) in such proportion that no single strand is present. When $X_U > 0.67$, the mixture contains poly(U) · poly(A) · poly(U) and single-stranded poly(U). These results are explained by the lower stability of single strands compared with the stability of multistranded complexes (Cantor and Schimmel, 1980b).

The proportion of polynucleotides in the mixture determines possible transitions between regions I and II or III. If ternary complexes exist in the solution, then upon decreasing salt concentrations and transition into region II [reaction (4)], they will melt and result in a double helix and a single strand. On the other hand, if a double helix exists, then upon increasing salt concentration and transition into region III [reaction (2)] it will rearrange into triple helix and single strand.

Reaction (2) is interesting because it shows that at high ionic strength, the existence of a triplex helix and a single strand is more likely than the existence of a double helix. This is explained by the need to screen three interphosphate interactions in the triplex rather than a single one in the duplex. This means that a triple helix and a single strand bind more counterions than two duplexes. As a result, dissociation of the third strand depends on ionic strength more strongly than the usual melting of the double helix.

Figure 3.12 shows the dependence of triple helix melting on increasing Na^+ concentration (Durand et al., 1992a). The ionic strength dependence for the temperature of dissociation of the third strand is steeper compared to the ionic strength dependence for the duplex melting. This results in closer triplex–duplex and duplex–single strand transitions at higher salt

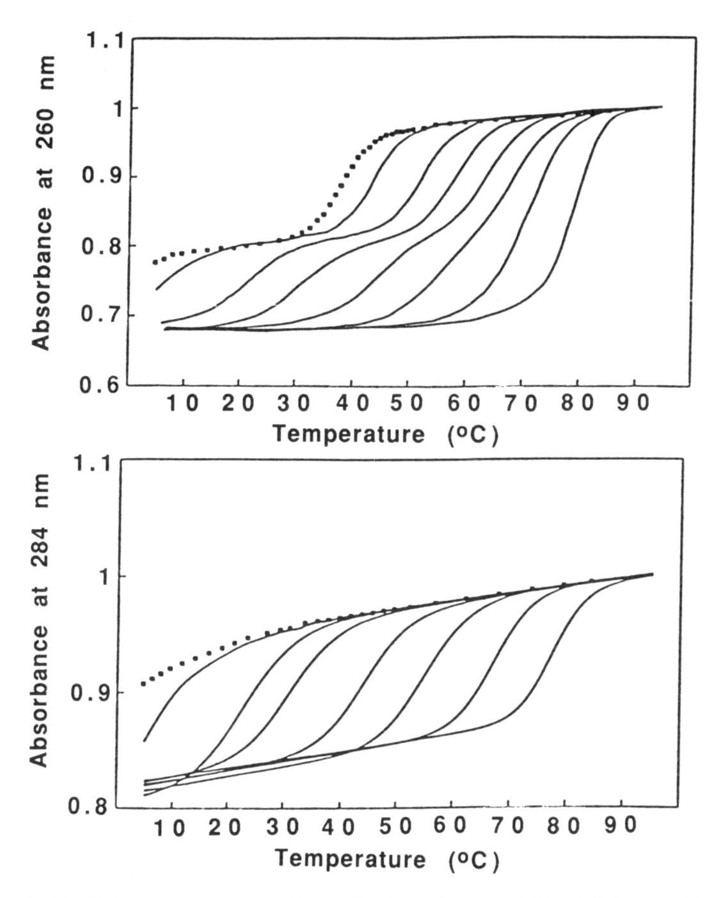

FIGURE 3.12. Salt concentration-dependent melting of the triplex and duplex formed by a model polymer $(dA)_{12}$-x-$(dT)_{12}$-x-$(dT)_{12}$, where x = hexaethylene linker. Shown are the absorbance profiles at 260 nm (where both transitions are seen) and at 284 nm (where only triplex melting contributes to the absorbance change). NaCl concentration increases from left to right through 0, 0.01, 0.05, 0.10, 0.25, 0.50, 1.0 and 2.0 *M*. The difference between the midpoints of the two transitions becomes smaller as NaCl concentration increases, until the complex melts as a whole. Reprinted with permission from Durand et al. (1992a). © 1992 American Chemical Society.

concentrations. For example, at 0.05 *M* Na^+, the third strand dissociates at 21.7°C and the duplex melts at 52.8°C (a difference of 31°C); at 0.5 *M* Na^+, the corresponding values are 56.5° and 70.1°C (a difference of 13.6°C).

The effect of pH is revealed under alkaline conditions, where the uracil bases lose the protons involved in hydrogen bonding with the adenines. Due to that, the process of triplex dissociation proceeds through reaction

(4) and then (1), even at a high salt concentration (Michelson et al., 1967). The critical value of pH, above which such a phenomenon occurs, depends on a salt concentration.

At acid pH, poly(A) forms a helical protonated structure which competes with poly(A) · poly(U) and poly(U) · poly(A) · poly(U). At low salt concentration, between pH 5.4 and 5.0, the dissociation of poly(U) · poly(A) · poly(U) takes place in two steps; at pH below 5.0, the triplex dissociates directly into single strands (Michelson et al., 1967).

According to the studies of polynucleotide–oligonucleotide models, increasing the length of the PyPu tracts makes the triplexes more stable. However, if there is a polymer pyrimidine template, the purine strand may be composed of a number of short covalently unbound blocks, which may be as short as one nucleotide (nucleoside) (Howard et al., 1966b; Huang and Ts'o, 1966; Sarocchi et al., 1970; Hattori et al., 1975). In such dynamic complexes, the polymer pyrimidine strand makes hydrogen bonds with stacked purine bases, forming a specific duplex that does not contain a sugar–phosphate backbone in the purine strand. This duplex presents a target for attachment of another polymer stand. It is clear that such triplexes are relatively unstable and require a large excess of a monomer component. The studies in which the triplexes were formed from polynucleotide and oligonucleotide components of various lengths showed an increase in triplex stability with increasing oligomer length (Lipsett et al., 1960; 1961; Naylor and Gilham, 1966; Cassani and Bollum, 1969; Raae and Kleppe, 1978; Kitagawa and Okuhara, 1987). In accord with this general trend, somewhat different experimental data were reported: (1) a linear dependence of the triplex melting temperature T_m on reciprocal oligonucleotide length $1/n$ (Lipsett et al., 1960), and (2) a linear dependence of $1/T_m$ on $1/n$ (Cassani and Bollum, 1969). The latter dependence is well known for the case of oligonucleotide duplexes (see Cantor and Schimmel, 1980c, for review).

Usually, the length of the third-strand oligonucleotide is not less than 9 or 10 nucleotides (Le Doan et al., 1987; Moser and Dervan, 1987; Lyamichev et al., 1988; Pilch et al., 1990a; Malkov et al., 1993a; Rubin et al., 1993). An insufficient stability of shorter intermolecular triplexes under physiological conditions prompted many researchers to design many oligonucleotide analogs including triad-stabilizing base analogs, nonionic and hydrophobically modified oligomers, oligonucleotides switching from one strand to another at adjacent PyPu tracts, etc. (see Chapter 4, for more detail). In experiments aimed at inhibition of transcription from specific genes (see Chapter 7), purine-rich oligonucleotides 21 to 39 nucleotides in length were used to form PyPuPu-type triplexes (Cooney et al., 1988; Orson et al., 1991; Postel et al., 1991; McShan et al., 1992; Ebbinghaus et al., 1993; Ing et al., 1993; Roy, 1993). In none of these works was the oligonucleotide length optimized. However, it was suggested recently that the best oligonucleotide length for the PyPuPu triplex is about 12 nucleotides: longer oligonucleotides bind slightly weaker, whereas shorter oligonucleotides bind significantly more weakly (Cheng and Van Dyke, 1994). The reason for the lower binding affinity of long oligomers was suggested in the accumulation of structural deformation upon binding of an incompletely fitted duplex and third strand, as was earlier demonstrated for the regular polymer triplexes (Pochon and Michelson, 1965; Broitman et al., 1987).

The influence of a triplex on an underlying double-helical structure can be estimated from interaction of various substances with adjacent double- and triple-stranded regions. This influence is highly localized, because an intermolecular triplex does not interfere with protein binding at the immediately adjacent site (Hanvey et al., 1990; Huang et al., 1992). However, the triplex–duplex junction is susceptible to chemical probing, which indicates some kind of distortion of the regular double helix. The unusual retardation of a double-helical fragment when it binds the third-strand oligonucleotide was interpreted as a result of an induced point bend in DNA (Chomilier et al., 1992). These authors additionally justified their conclusion by the results of molecular modeling, which also predict helix bending.

The formation of intermolecular triplexes does not require the initiating denaturation bubble, as the formation of H (H*) DNA does, and in this respect it should be easier. In reality, it is more difficult, since the third strand can dissociate more easily compared to H (H*) DNA, and the stimulatory effect of supercoiling is absent. Intermolecular PyPuPu triplexes are influenced by various divalent metal cations in a slightly different manner from intramolecular triplexes (see page 121). According to the results of dimethyl sulfate (DMS) footprinting and photofootprinting, a stable complex for $d(C)_n \cdot d(G)_n \cdot d(G)_n$ sequence is formed in the presence of alkali earth (Mg^{2+}, Ca^{2+}) and transition cations (Cd^{2+}, Co^{2+}, Mn^{2+}, Zn^{2+}, and Ni^{2+}), but not Ba^{2+} or Hg^{2+} (Frank-Kamenetskii et al., 1991; Soyfer et al., 1992; Malkov et al., 1993b). By contrast, a stable complex between $d(AG)_5$ and the plasmid carrying two $d(CT)_{16} \cdot d(AG)_{16}$ inserts is formed in the presence of the transition metal ions Cd^{2+}, Co^{2+}, Mn^{2+}, Zn^{2+}, and Ni^{2+}, but not Mg^{2+}, Ca^{2+}, Ba^{2+}, or Hg^{2+} (Figure 3.13). A similar result was obtained for the irregular PyPu sequence and the corresponding purine mirror-repeated oligonucleotide dGAGAAAGGGG. Thus, the presence of one of the transition metal cations (Mn^{2+}, Zn^{2+}, Cd^{2+}, Co^{2+}, and Ni^{2+}) is a universal condition for the formation and/or stabilization of intermolecular PyPuPu triplex. The above data imply that some metal-binding sites in polynucleotides are important for the formation of PyPuPu triplexes. The differential effects of divalent cations on triplex formation can be explained by the unequal enhancement of Hoogsteen hydrogen bonds when cations coordinate the N7 of purines (Potaman and Soyfer, 1994). This hypothesis will be outlined in Chapter 5.

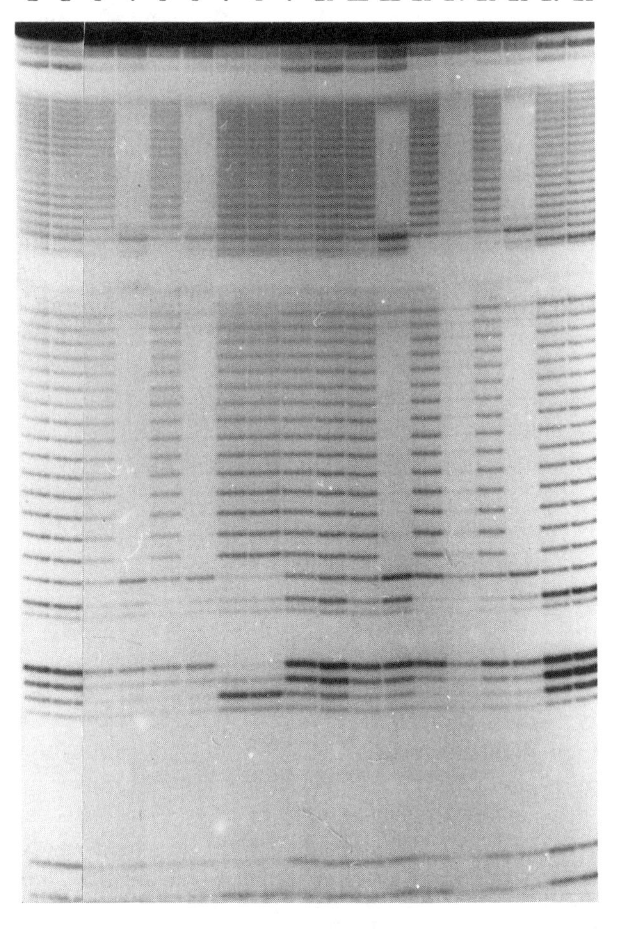

FIGURE 3.13. Effects of divalent metal cations on the formation of the intermolecular PyPuPu triplex between a $d(AG)_5$ oligonucleotide and a fragment of the pTC33.4 plasmid with two $d(CT)_{16} \cdot d(AG)_{16}$ inserts according to dimethyl sulfate footprinting. Triplexes were obtained by incubating the duplex target with the third-strand oligonucleotide in Tris buffer, pH 7.0–7.7, containing either 7 m*M* of the indicated cations (lanes 1–16) or 1 m*M* EDTA (lanes 17 and 18). Reproduced from Malkov et al. (1993b) *Nucleic Acids Res* by permission of Oxford University Press.

However, the PyPuPu triplex is promoted not only by the purine base-coordinating cations. For example, experiments utilizing affinity cleavage probing require the presence of the ethylenediaminetetraacetic acid (EDTA) moiety, which is incompatible with the presence of divalent metal cations (Beal and Dervan, 1991, 1992a,b; Stilz and Dervan, 1993). In this case, 100 μ*M* spermine, which has four positive charges and interacts with phosphates and perhaps the bases, stabilizes the PyPuPu triplex.

In contrast to H DNA, where only natural nucleic acid bases take part in the formation of triads, intermolecular triplexes can utilize a number of base analogs, which can stabilize the whole complex. For example, homopolynucleotides containing 5-halogenouracils form more stable PyPuPy triplexes with poly(A) than homopolynucleotides containing uracil (reviewed by Michelson et al., 1967). Methylation of cytosine at C5 also produces a stabilizing effect on the PyPuPy triplexes (Lee et al., 1984; Povsic and Dervan, 1989; Xodo et al., 1991; Hausheer et al., 1992). Triplexes containing these base-substituted analogs will be considered in more detail in Chapter 4 along with a number of other triplexes containing unusual bases.

DNA–RNA Triplexes

Functional triplexes can be formed from certain combinations of RNA and DNA strands. There are eight possible types of triplexes where each strand is of DNA or RNA type (Han and Dervan, 1993). The variability in these complexes may be in the chemical differences between RNA and DNA (the 2′OH and the thymine methyl groups, respectively) or in the structural differences (the number of bases per helix turn, charge density, width and depth of the major and minor grooves, etc.). As shown for homopolymer or homooligomer PyPuPy triplexes, UA*U triplexes are more stable (having a higher melting temperature under comparable ionic strength) than TA*T triplexes (Massoulie et al., 1964a,b, 1966; Stevens and Felsenfeld, 1964; Riley et al., 1966). This conclusion is confirmed by comparing the enthalpic values for such triplexes (Ohms and Ackermann, 1990; Pilch et al., 1990b; see also "Thermodynamics of Triplexes" below).

Studies of mixed PyPuPy DNA/RNA complexes of mixed sequence using thermal denaturation and band-shift assay (Escudé et al., 1992, 1993; Roberts and Crothers, 1992; Wang and Kool, 1994a, 1995) or quantitative affinity cleavage (Han and Dervan, 1993) have shown that the RR*D and DR*D (D = deoxyribo, R = ribo) triplexes are much less stable than the other six variants. The exact orders of stability for six hybrid DNA–RNA triplexes were somewhat different in these studies (Figure 3.14). These differences might reflect variations in sequences and in experimental conditions, such as pH and salt concentration. However,

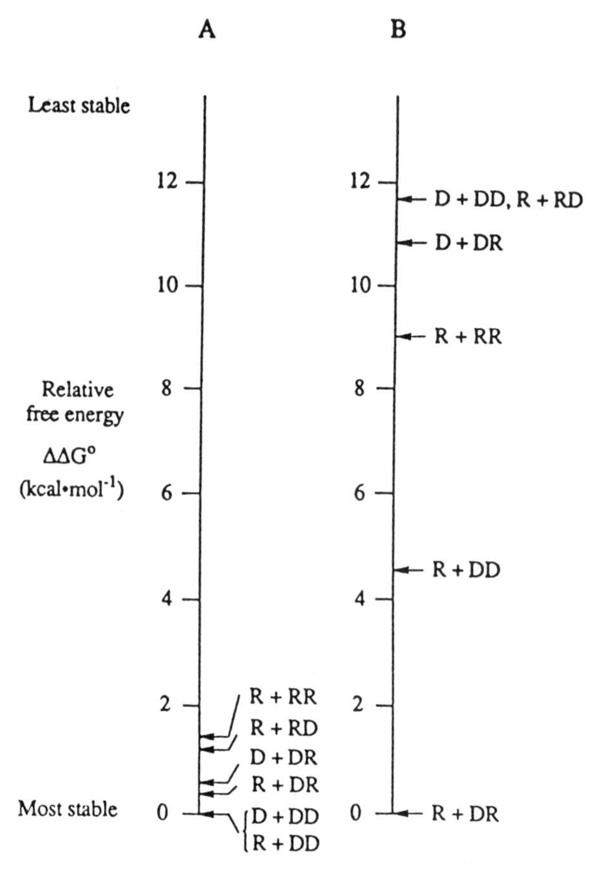

FIGURE 3.14. Comparison of relative stabilities [$\Delta\Delta G^O$, $kcal \cdot mol^{-1}$] for six experimentally observed mixed DNA-RNA triplexes. (A) Data from Han and Dervan (1993); (B) data from Roberts and Crothers (1992). Reproduced from Han and Dervan (1993).

a simple guiding principle has been suggested: if the duplex contains the D purine strand, the R or D third strand will bind; if the duplex contains the R purine strand, only the R third strand will bind (Han and Dervan, 1993). When the R third strand is targeted to the DD duplex, the resulting triplex is as stable as DD*D (Han and Dervan, 1993) or even stronger (Escude et al., 1992; Roberts and Crothers, 1992).

Wang and Kool (1995b) studied separate effects of the 2′-hydroxyl and C5-methyl groups, which distinguish RNA from DNA, on the PyPuPy triplex by synthesizing nucleic acid strands containing all possible combinations with and without such groups. Both types of substitutions are stabilizing and independent of one another; however, their relative contributions depend on the nature of the purine duplex strand. In the case of a purine DNA strand, the stabilizing effect of the thymine methyl group is relatively larger than the effect of the hydroxyl group. In the case of the purine RNA strand, the results are different: stabilization by the hydroxyl group is larger than the stabilization by methyl group.

According to the affinity cleavage patterns for different combinations of ribo- and deoxyribonucleotide strands (Han and Dervan, 1994), stable complexes may be separated into two groups: (1) DD*D, DD*R, RD*D, and RD*R where the purine strand of the duplex is of the deoxyribonucleotide type, and (2) DR*R and RR*R with the purine strand of the ribonucleotide type. Model building showed that the differences in strand cleavage for two groups may result from the B-like conformations of the targeted duplexes in the first group and the A-like conformations of the duplexes in the second group. A conclusion that the type of purine strand is a major determinant of resulting triplex conformations (Han and Dervan, 1994) is in general agreement with the CD data of Hung et al. (1994), which found that the D(purine) · R(pyrimidine) duplexes had CD spectra equally different from those of the B and the A forms, whereas the R(purine) · D(pyrimidine) duplexes had CD spectra closer to those of the A than the B form. More detailed knowledge of the geometries of hybrid DNA/RNA triplexes awaits NMR and X-ray investigations.

The data on the mixed DNA/RNA triplexes of the PyPuPu type are limited. Skoog and Maher (1993a) were unable to bind the Pu-type RNA third strand to a targeted DNA duplex. In a subsequent paper, Semerad and Maher (1994) came to the conclusion that the PyPuPu triplex is stable "only when all three of the substituent strands are DNA." However, some nucleic acid-binding ligands may significantly change this trend. Recently, Pilch and Breslauer (1994) found that the poly(dT) · poly(rA) · poly(rA) triplex does not form under ordinary conditions but may be induced by the nucleic acid-binding drugs berenil and 4′,6-diamino-2-phenylindole. Moreover, they were successful in inducing the poly(dT) · poly(rA) · poly(dT) triplex with berenil, 4′,6-diamino-2-phenylindole, ethidium, and netropsin. Thus, ligand binding seems to significantly change the energetics of strand interactions in triple helices.

Kinetics of Triplex Formation

The early kinetic experiments on poly(rU) · poly(rA) · poly(rU) and oligo(rU) · oligo(rA) · oligo(rU) using UV spectrophotometry showed that the association–dissociation rates of triple helices are more than 100 times lower than those of double helices (Blake and Fresco, 1966; Blake et al., 1967, 1968; Porschke and Eigen, 1971). The association rates depend on the cation concentrations present in solution. For poly(rU) · poly(rA) · poly(rU), these rates increase upon increasing the concentration of both monovalent and divalent metal cations (Felsenfeld and Rich,

1957). However, concentrations of Na^+ about a hundred times greater than those of Mg^{2+} or Mn^{2+} are necessary to produce an equivalent effect. In the case of triplexes containing protonated cytosines in the third strand, divalent cations increase the association rate, whereas increasing the Na^+ concentration either decreases the association rate or generally destabilizes the triplex (Lipsett, 1964; Maher et al., 1990; Lyamichev et al., 1991). An increased association rate at increased Na^+ concentration has also been reported (Rougée et al., 1992). As shown by Wilson et al. (1994), the type of buffer is of significant importance for cytosine-containing triplexes. Thus, the above discrepancy might originate from the differences in buffers employed.

For a 21-mer oligonucleotide of irregular sequence, the association rate constant for the PyPuPy triplex, measured by the restriction nuclease protection technique, was of the order of $10^3\,M^{-1}\,s^{-1}$, which is 10^3 times lower than the corresponding constant for duplex formation. The triplex dissociation rate constant was in the 10^{-5}–$10^{-4}\,s^{-1}$ range, resulting in a triplex lifetime of 1–10 h at 37°C (Maher et al., 1990). Similar values for the association and dissociation rate constants were obtained in UV-melting experiments for a 22-mer pyrimidine oligonucleotide/target DNA complex (Rougée et al., 1992) and in DNase I footprinting experiments for acridine-linked 10-mer oligonucleotide/target DNA complexes (Fox, 1995). Rougée et al. (1992) also studied the effect of mismatches on the kinetic properties of triplexes. It was shown that the mismatch TA*X ($X \neq T$) in the center of the triple helix had no effect on the association rate, but increased the dissociation rate up to 10^3-fold (Rougée et al., 1992). The presence of Mg^{2+} ions increased the association rate, presumably due to a more effective (compared to monovalent ions) reduction in the repulsion of negatively charged phosphate backbones (Maher et al., 1990; Rougée et al., 1992).

Triplex formation by the association of a third strand with a host duplex should be a bimolecular process. Consequently, the position of this equilibrium and the melting temperature, T_m, should depend on the total concentration of strands, C_T. This dependence is assessed from a slope of a plot of ${T_m}^{-1}$ versus $\ln C_T$ (Cantor and Schimmel, 1980a). Dependence of ${T_m}^{-1}$ versus $\ln C_T$ for the 15-mer PyPuPy triplex was determined spectrophotometrically (Plum et al., 1990) and showed the slope to be 30 times lower than expected from calorimetrically measured triplex melting enthalpy, and suggested a bimolecular process (Marky and Breslauer, 1987). Therefore, the triplex formation approached the limit of pseudo first-order kinetics, similar to association–dissociation equilibria in polymer DNA.

Thus, triplex formation from a duplex and a third strand appears to proceed similarly to the well-studied duplex formation from complementary single strands, albeit at a slower rate. Based on the data of Porschke and Eigen (1971), Craig et al. (1971) developed a nucleation-zipping model of polynucleotide strand association (see also Cantor and Schimmel, 1980b). A different theoretical approach by Anshelevich et al. (1984) yields the same results as the model of Craig et al. (1971) in the case of short oligomers (which is applicable to triplex formation). The nucleation-zipping model requires the formation of a three- or four-base-pair intermediate, which then zips more rapidly than it dissociates into separate components. In accordance with this model, the association constant is independent, while the dissociation constant depends on the presence of mismatches in the third strand (Rougée et al., 1992).

The changes in binding due to the cooperativity of oligodeoxyribonucleotides bound to adjacent sites by triple helix formation were studied by affinity cleavage titrations (Strobel and Dervan, 1989; Distefano et al., 1991; Distefano and Dervan, 1992, 1993; Colocci et al., 1993; Colocci and Dervan, 1995) and sequence-specific alkylation of a target duplex (Froehler et al., 1992a). The data showed that a 3.5- to 20-fold enhancement in equilibrium association constant is realized for 9- to 11-mers binding in the presence of an occupied abutting site. The cooperativity was not observed for two binding sites separated by one base pair. The priming of a second oligonucleotide at a duplex template is probably favored by a stacking interaction between the terminal bases of this oligomer and the oligomer already bound at the adjacent site (Colocci et al., 1993).

The study of association–dissociation kinetics forms the basis for one important application of triplex methodology. Triplex-forming oligonucleotides have relatively high affinities for double-helical DNA. For example, dissociation constants for the complex of 21–22-mer pyrimidine oligonucleotides with their DNA target may be as low as 1 to 10 n*M* (Maher et al., 1990; Durland et al., 1991; Rougée et al., 1992). These values approach those observed for many sequence-specific DNA-binding proteins. If there is a PyPu tract in the promoter or gene sequence, the protein factors which take part in transcription cannot easily displace the oligomer bound to this tract. Furthermore, when associated with double-helical DNA near physiological pH and ionic conditions, the bound oligonucleotide forms a triplex with a half-life on the order of 10 hours. The initiation of transcription or chain elongation can be inhibited in such a way when the bound oligonucleotide hampers RNA polymerase binding to its promoter site (see Chapter 7). It should be kept in mind, however, that the rate of triplex formation is slow. Therefore, for oligonucleotides competing with protein factors for binding to a specific sequence on DNA, kinetic phenomena might become the limiting factor. The desired effect caused by oligonucleotide binding can be increased by increasing its concentration, as demonstrated by Maher et al. (1989) in experiments on *Ava*I restriction nuclease protection assay.

Kinetic studies of intramolecular triplex formation in supercoiled DNA are scarce. The conversion of the duplex into an H DNA in a negatively supercoiled plasmid containing the $(GAA)_4TTC(GAA)_4$ insert is completed in less than 2 minutes at 25°C when the pH is shifted from 8.0 to 5.2 (Hanvey et al., 1989; Kang et al., 1992b). However, the triplex extrusion appears to depend on the GC content of the PyPu tract. Triplex formation at room temperature is extremely slow in plasmid DNA containing the $(G)_n \cdot (C)_n$ and $(GA)_n \cdot (TC)_n$ tracts (Panyutin et al., 1989; Panyutin and Wells, 1992). However, incubation at a higher temperature (50°C) for at least several minutes promotes triplex formation (Panyutin and Wells, 1992).

The reverse transition, from triplex to duplex, is relatively fast. After the removal of negative supercoiling with an excess of topoisomerase, all triplexes relax into duplexes in about 3 minutes, as confirmed by single-strand-specific nuclease P1 probing (Hanvey et al., 1989).

Thermodynamics of Triplexes

There has been a limited number of thermodynamic studies of triplexes, because of the more stringent conditions (as compared to duplexes) required to attain stability. The complexity of the nearest-neighbor pattern makes it difficult to distinguish the contributions of hydrogen-bond interactions of bases in triplexes and base stacking. Therefore, all the available data indicate a total interaction between the third strand and the target duplex. Some examples of thermodynamic data for the triplexes of various sequences have been collected in Table 3.3. The limited set of data and varied conditions (pH, ionic strength, strand concentrations, etc.) do not allow a direct comparison of the data for different third strand sequences. A comparison is additionally difficult because the available thermodynamic data come either from DSC (and are model-independent) or from UV melting (and are model-dependent and calculated under different assumptions). Therefore, only a preliminary description of the thermodynamic characteristics can be provided at the moment.

Generally, triplexes are less stable than their parent duplexes (Manzini et al., 1990; Pilch et al., 1990a; Plum et al., 1990; Rougée et al., 1992). Enthalpic values for third-strand dissociation from the triplex are several-fold smaller than for duplex melting. The only exception is the example of the PyPuPu triplex (entry 7), where these values are approximately equal.

As discussed by Cheng and Pettitt (1992a), under equal conditions, enthalpic values for hairpin or loop triplexes (entries 1–3, 11, 14) are consistently higher than for intermolecular triplexes (entries 4, 6, 16) with similar sequence compositions.

Cytosine protonation at acidic pH stabilizes CG^*C^+ triads (entries 8 and 13). Increasing ionic strength stabilizes triplexes that do not contain protonated bases (entry 12). The RNA UA*U triad is enthalpically more stable than the DNA TA*T triad, as can be seen in a comparison of entries 9, 5 and 12. The studies of the relative stabilities of PyPuPy triplexes composed of identical sequences but different RNA or DNA strand combinations (Figure 3.14) gave a qualitative agreement with respect to stabilities (Roberts and Crothers, 1992; Han and Dervan, 1993; Wang and Kool, 1995a). However, the differences in the order of the stabilities of RNA/DNA combinations and the values of the stabilizing effect should be pointed out.

Figure 3.14 shows a comparison of the results obtained by Han and Dervan (1993) on a 35-base-pair duplex and by Roberts and Crothers (1992) on a 12-base-pair hairpin duplex (left and right panels, respectively). The differences in the scale of variation of absolute free energies (1.2 kcal/mol vs 12 kcal/mol) and the order of stability may be due to different methods of analysis or other experimental differences. Among them are the differences in length, base composition, cation concentration, and stability of the target duplex (duplex fragment vs hairpin duplex). As in the case of DNA and RNA triplexes, some additional work is necessary in order to establish more reliable thermodynamic correlations for DNA/RNA hybrid triplexes.

Recent studies have shown that the free energy penalty for introducing a single mismatch ranges from 2.5 to 6 kcal/mol (Belotserkovskii et al., 1990; Roberts and Crothers, 1991; Rougée et al., 1992; Xodo et al., 1993), which is close to the corresponding values for DNA and RNA (Gralla and Crothers, 1973; Tibanyenda et al., 1984). Theoretical studies of the influence of mismatches on triplex stability (van Vlijmen et al., 1990; Mergny et al., 1991b) are consistent with experimental results. The enthalpy of triplex formation in sequences containing one or two bulges is the same as in a perfect triplex (Roberts and Crothers, 1991). The lack of effect of bulges on the enthalpy suggests that extra bases do not significantly disturb stacking in the helix and are probably extruded. The exact orientation of a mismatched base is not known, but the decrease in enthalpy for triplexes with mismatches could be explained by the disruption of one or two stacking interactions. The dependence of triplex structures on stacking interactions was also revealed in studies of abasic sites in the third strand (Horne and Dervan, 1991). A single abasic site eliminated two stacking interactions, and the resulting decrease in affinity was similar to that observed for imperfectly matched natural triads. A clear understanding of the thermodynamic properties of imperfect triads in various sequence contexts allows one to use the triplex-mediated methodology of physical separation of duplexes (Roberts and Crothers, 1991) (see Chapter 7). The thermodynamic data on triplex structure continues to accumulate, and the reader is referred to a review of Plum et al. (1995) for an extensive study on the subject.

TABLE 3.3. Thermodynamic data for third-strand dissociation from the triplex.*

Entry	Sequence of the third strand	Length (bp)	Media conditions	ΔH (kcal/mol)	T_m (°C)	ΔG (kcal/mol)	Reference
1[a]	5′-CTTCCTCCTCT(e)	11	50 mM Na^+ 10 mM Mg^{2+} pH 5.0	−6.6	52	—	Manzini et al. (1990)
2[a,b]	5′-CTCTTCTTTC(e)	10	50 mM Na^+ 10 mM Mg^{2+} pH 5.0	−5.8	—	—	Xodo et al. (1990)
3[a]	5′-CTTCCTCCTCT(f)	11	50 mM Na^+ 10 mM Mg^{2+} pH 6.0	−5.0	62	ΔS = 16.6 eu	Xodo et al. (1991)
4[a]	5′-CCCTTTTCCC	10	2.0 M Na^+ 50 mM Mg^{2+} pH 5.5	−3.4 −4.2	29.9 20.6	−0.75 −0.64	Pilch et al. (1990a)
5[a]	5′-TTTTTTTTTT	10	50 mM Mg^{2+} pH 7.0	−2.3	20.6	−0.66	Pilch et al. (1990a)
6[b]	5′-TTTTTCTCTCTCTCT	15	210 mM Na^+ pH 6.5	−2.0	30.0	−0.09	Plum et al. (1990)
7[a]	3′-GGGAAAAGGG	10	10 mM Na^+ 50 mM Mg^{2+}	−8.1	54.0	−1.34	Pilch et al. (1991)
8[c]	5′-CTCTCTCTCTCTCTC(g)	15	0.2 M Na^+, pH 4.34 0.2 M Na^+, pH 4.42 0.2 M Na^+, pH 4.63 0.2 M Na^+, pH 4.88			−0.39 −0.36 −0.29 −0.23	Lyamichev et al. (1989)
9[b]	rA_xU_y ($x = 5$ or 7; $y = 3–11$)		0.1 M Na^+ pH 7.0	−5.1 (average for increasing y)	—	—	Ohms and Ackermann (1990)
10[a]	5′-TTTCCTCCTCTTCTTCTTTTTT	22	0.1 M Na^+ pH 6.8	−5.0	—	—	Rougée et al. (1992)

11[a]	5′-CCTCTCCTCCCT	12	0.1 *M* Na^+ pH 5.0	−4.7	56.0	—	Roberts and Crothers (1991)
12[a]	5′-TTTTTTTTTTTT(h)	12	60m*M* Na^+	−3.6	22.2	—	Durand et al. (1992a)
			0.11 *M* Na^+	−4.2	31.4	—	
			0.26 *M* Na^+ pH 7.0	−4.9	44.4	—	
13[d]	5′-TCCTCTTCTTTTCTTTCTT	19	50 m*M* Na^+, pH 5.0	—	—	−0.54	Shindo et al. (1993)
			pH 5.6	—	—	−0.48	
			pH 6.2	—	—	−0.41	
			pH 6.8	—	—	−0.37	
14[a]	5′-CTCTCTCTTT(i)	10	50 m*M* Na^+	−5.9	41.0	—	Völker et al. (1993)
14[b]	5′-CTCTCTCTTT(i)	10	pH 6.7	−4.0	41.0	—	
15[a]	5’-CTTCTCC(i)	7	1 *M*, Na^+	−5.3	30.3		Plum and Breslauer (1995)
15[b]	5’-CTTCTCC(i)	7	pH 6.34	−2.9	30.3		
16[k]	5’-TTTTTCTCTCTCTCT	15	100 m*M* Na^+ 1 m*M* spermine, pH 7.0	−2.1		−0.59	Singleton and Dervan (1994)

* Entries 1–9 taken from Cheng and Pettitt (1992a).
[a] van't Hoff two-state model (UV or calorimetry).
[b] Direct calorimetry.
[c] ΔG for intramolecular triplex derived from statistical mechanical calculations.
[d] van't Hoff two-state model (filter-binding assay).
[e] Third strand + duplex hairpin.
[f] 5-methylated cytosines.
[g] H DNA.
[h] Three strands bridged by hexaethylene glycol chains.
[i] Twice-folded oligonucleotide.
[k] Affinity cleavage titrations.

New Variants of Triplex Structure

Some special variants of triplex structures have been discovered in natural DNA sequences or have been designed for various purposes.

Structures formed in sequences that have functional significance for the cell have been the subject of extensive examination. There is hope that conformations different from the usual B form can participate in some key biological processes, such as DNA replication and recombination, gene expression, chromosome packaging, etc. (see Chapter 6). Appropriate sequences have been cloned in plasmid DNA and the possibility of adopting altered structure has been investigated. Triple-stranded H DNA has been found inside some sufficiently complex intramolecular structures.

Belotserkovskii et al. (1992) formed a complex under acid conditions composed of an intramolecular protonated PyPuPy triplex and a duplex made up of the looped part of DNA with a complementary oligonucleotide (Figure 3.15). This bound oligonucleotide stabilized the triplex-containing structure so as to permit its observation at neutral pH. Oligonucleotide trapping seemed to stabilize the triple-stranded complexes in two other studies (Michel et al., 1992; Ulrich et al., 1992) in which the authors did not study the structures of the complexes in detail. The real importance of such structures is in the possibility of stabilizing intramolecular triplexes by single-strand fixation. It was suggested that the nascent mRNA or other endogenic single-stranded nucleic acids can trap and stabilize H DNA (Reaban and Griffin, 1990; Belotserkovskii et al., 1992). A number of proteins specific to single pyrimidine strands in PyPu tracts (Davis et al., 1989; Yee et al., 1991; Muraiso et al., 1992) could bind the unpaired strand in H DNA and stabilize the whole structure under physiological conditions in the absence of the stabilizing effect of low pH. However, it is not clear enough if these proteins have sufficiently short binding sites, since the well-known single-stranded binding protein from *E. coli*, which has a binding site approximately 70 nucleotides long, actually destabilizes the triplex (Klysik and Shimizu, 1993) (see also Chapter 6).

A Z-triplex structure can be formed in a sequence containing adjacent blocks capable of forming left-handed and triplex conformations (Bianchi et al., 1990; Weinreb et al., 1990; Brinton et al., 1991; Pestov et al., 1991). H DNA is formed in corresponding sequences under mildly acidic conditions ($pH < 6$), whereas the Z form can be observed at a more common physiological pH. This structural motif seems to be biologically important, as it has been found in the origin of replication of the Chinese hamster *dhfr* amplicon (Bianchi et al., 1990), the site of unequal chromatid exchange in the mouse myeloma cell line (Weinreb et al., 1990), and the mouse c-Ki-*ras* promoter (Pestov et al., 1991).

Using 2-D gel electrophoresis and chemical and enzymatic probing techniques, Voloshin et al. (1992) found that, when cloned within a plasmid double-stranded human telomeric repeat, $(T_2AG_3)_{12} \cdot (C_3TA_2)_{12}$

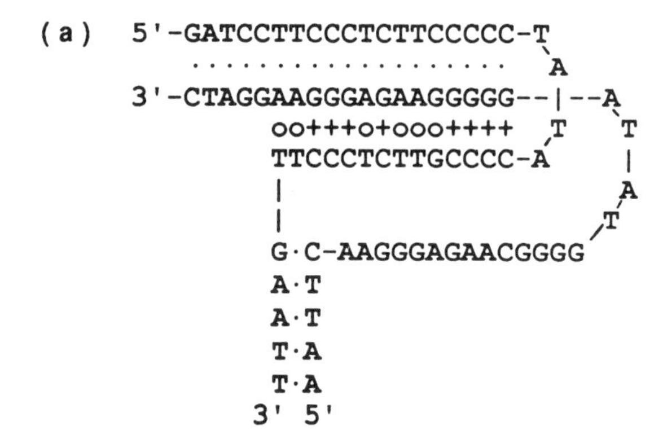

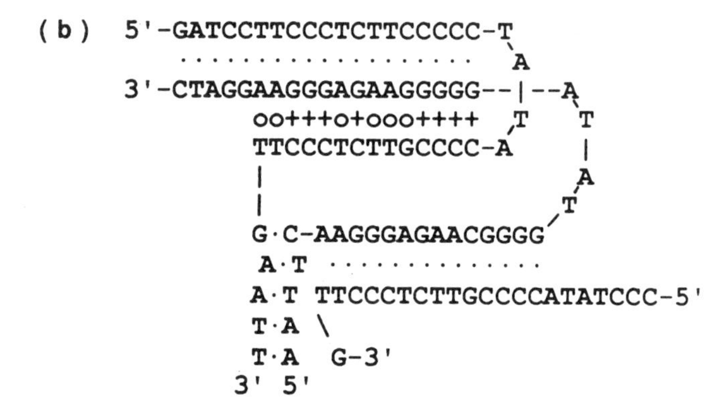

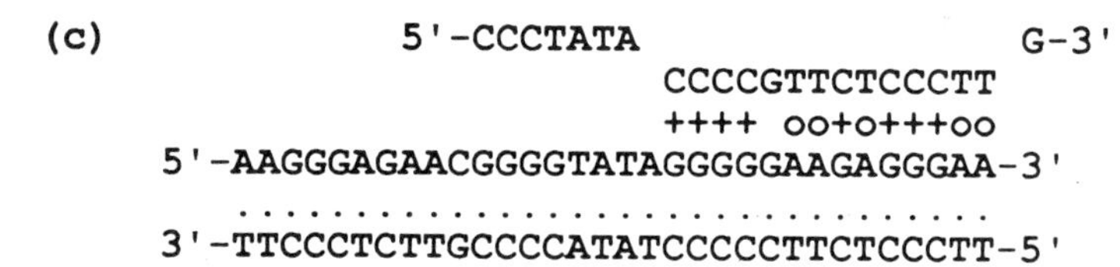

FIGURE 3.15. Possible structures formed between the PyPu tract in a supercoiled plasmid and an oligonucleotide partially complementary to the Pu strand. (a) H DNA; (b) H DNA where a purine single strand additionally forms a duplex with an oligonucleotide, which results in stabilization of the entire protonated structure and allows it to exist at higher pH values (up to pH 7); (c) intermolecular triplex between the PyPu tract and an oligonucleotide which forms when the superhelical density of the plasmid is not high enough to extrude the H form. Reproduced from Belotserkovskii et al. (1992) *Nucleic Acids Res* by permission of Oxford University Press.

forms a protonated superhelicity-induced structure. They proposed an eclectic model for this structure in which four different elements coexist: a nonorthodox intramolecular triplex stabilized by the canonical protonated CG^*C^+ base triads and highly enriched by noncanonical base triads; an intramolecular quadruplex formed by a portion of the G-rich strand; a single-stranded region encompassing a portion of the G-rich strand; and, probably, the (C,A)-hairpin formed by a portion of the C-rich strand.

In addressing the question of the appropriate length of the spacer between the two halves of the PyPu sequence, it was found that a long (40–50 base pairs) spacer inhibits triplex formation (Klysik, 1992). However, when such a spacer contains an inverted A-T repeat, which can form the cruciform, the interaction of remote PyPu tracts and the formation of a triple-stranded structure becomes possible (Figure 3.16). The significance of this finding may lie in the possibility of providing the interaction of remote PyPu tracts, which act as an enhancer–promoter pair (Guarente, 1988; Collado-Vides et al., 1991) and determine in concert the transcriptional efficiency of the gene. Such tracts can be found, for example, in the 5′-flanking sequence of the *Drosophila hsp26* gene, where the two sequences are separated by approximately 200 base pairs (Lu et al., 1993). The persistence length of DNA does not allow such tracts to be brought together by smooth duplex bending. Their interaction could be provided by some structural distortion in the intermediate sequence (either by the formation of a second unusual structure or by some protein that kinks the double helix).

The parallel orientation of certain duplex-forming DNA sequences is well established (see Rippe and Jovin, 1992, for review). Although the biological significance of such a structure is not understood, it was tested in supercoiled DNA under superhelical stress (Klysik et al., 1991). The parallel-stranded $(dA)_{15} \cdot (dT)_{15}$ insert in DNA adopts a right-handed helical form and, upon increasing the negative superhelical density to >0.03, undergoes a major transition which results in a short triplex (approximately five triads) and a relaxation of approximately 2.5 helical turns.

Although the presence in plasmids of several combinations of unusual structures, including triplexes, has been well documented using 2-D gel electrophoresis and chemical probing, the biological significance of these structural motifs in vivo has not been confirmed yet.

One specific kind of intermolecular triplex could be significant in the cell cycle. It can be formed when the DNA molecule donates one strand of the PyPu tract to form a triplex with another PyPu tract remote along the sequence. This mechanism has been suggested to play a role in chromosome condensation (Hampel et al., 1993). Such a possibility was demonstrated for two plasmid molecules (Hampel et al., 1993) and for a single yeast chromosome (Hampel and Lee, 1993). Earlier, such a

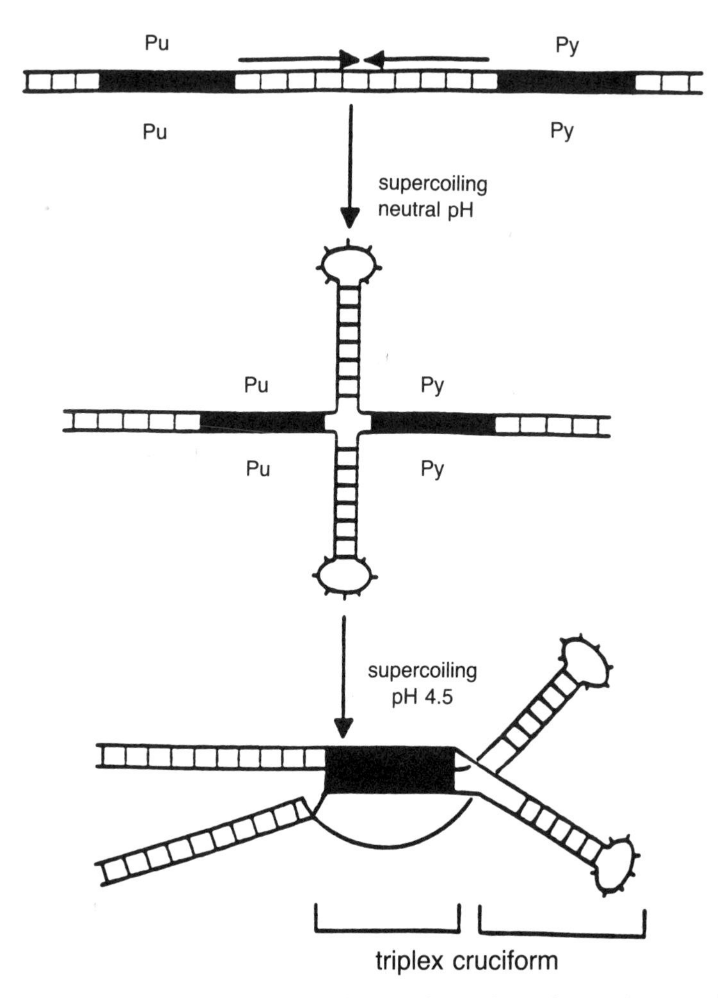

FIGURE 3.16. A proposed mechanism for the formation of a tandem cruciform/triplex structure in a supercoiled plasmid. Two mirror-repeated PyPu tracts cannot form a triplex when they are spatially separated by a sufficiently long duplex. If this duplex contains a sequence capable of forming the cruciform, then PyPu tracts can be brought into physical proximity by cruciform extrusion. Under acidic pH, both structures can coexist. From Klysik (1992). Reprinted with permission.

condensation was suggested to be mediated by a single-stranded RNA molecule (Lee and Morgan, 1982).

Several kinds of model triplexes were designed for structural and thermodynamic studies.

An intermolecular triplex can be formed between a single-stranded oligonucleotide and an oligonucleotide folded into a hairpin duplex

(Manzini et al., 1990; Xodo et al., 1990, 1991, 1993; Roberts and Crothers, 1991, 1992; Völker et al., 1993). Triplex formation between a hairpin-duplex binding site and a corresponding Hoogsteen complementary oligonucleotide results in a change in melting temperatures sufficient for separation of the temperatures for double-helix melting and third-strand dissociation (Roberts and Crothers, 1991, 1992).

Certain circular oligonucleotides can carry two regions with strong binding affinities for single-stranded DNA and RNA. In this case, a triplex structure can be formed after binding these short DNA or RNA molecules to both regions (Kool, 1991; Prakash and Kool, 1992; Paner et al., 1993; Wang and Kool, 1994a,b, 1995a,b). This new kind of molecular recognition has been made more versatile by designing a conformationally switching structure (Figure 3.17) in which a relatively long circular pyrimidine oligonucleotide (macrocycle) can bind either of the two shorter purine oligonucleotides (Rubin et al., 1993). Antisense targeting to RNA is a simple example of potential application of this structure. A thermodynamic comparison of complexes showed that a considerably larger loss of free energy results from the binding of a mismatch-containing oligonucleotide to a circular target than to a single-strand target (Kool, 1991).

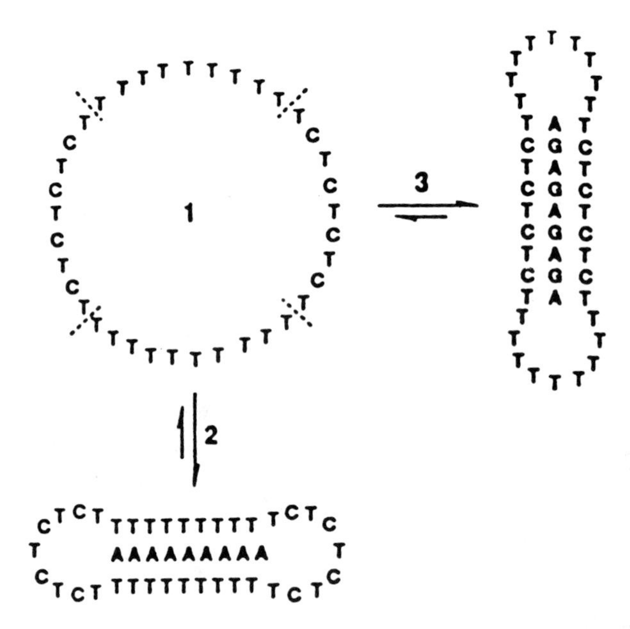

FIGURE 3.17. Circular oligonucleotides can be complexed to a target single strand on two sides via a triplex mechanism. This new kind of molecular recognition has been made more versatile by designing a conformationally switching structure in which a pyrimidine oligonucleotide macrocycle can bind either of the two shorter purine fragments. Reprinted with permission from Rubin et al. (1993). © 1993 American Chemical Society.

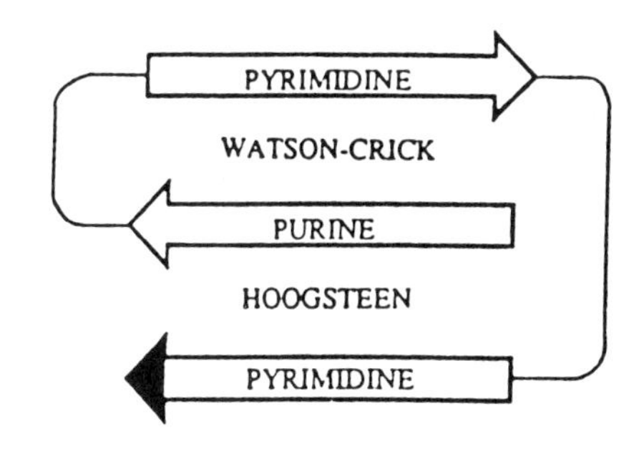

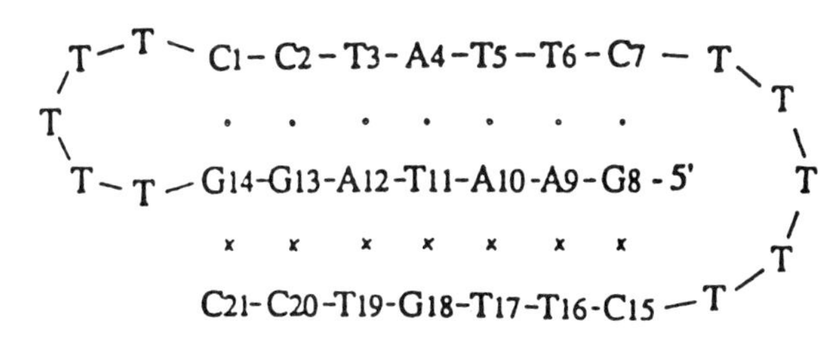

FIGURE 3.18. An intramolecular triplex formed from a relatively long oligonucleotide (~20–30 bases) which folds back twice on itself. This structure is useful for a number of physicochemical studies, since it is more stable than model triplexes composed of separate strands. Reprinted with permission from Radhakrishnan et al. (1992a). © 1992 American Chemical Society.

An intramolecular triplex formed from a relatively long oligonucleotide (~20–30 bases) which folds back twice on itself is often used in structural studies (Sklenar and Feigon, 1990; Chen, 1991; Radhakrishnan et al., 1991a,b,c, 1992a,b; Durand et al., 1992a,b; Lin and Patel, 1992; Macaya et al., 1992a,b; E. Wang et al., 1992; Radhakrishnan and Patel, 1993a,b; Völker et al., 1993; Plum and Breslauer, 1995). Compared to a triplex formed from single oligonucleotide strands, such an intramolecular triplex (Figure 3.18) is more stable and does not require excessive concentrations of stabilizing agents, such as salts, polyamines, or low temperatures. This makes it more convenient in NMR and thermodynamic studies. In some cases, the folding parts of the oligonucleotide in such a triplex can be bridged by flexible linkers, e.g., oligoethylene glycol chains (Durand et al., 1992a,b; Dittrich et al., 1994; Shchyolkina et al., 1994) or a terephthalamide linker group (Salunkhe et al., 1992).

The potential for applications of triplex methodology in the control of gene expression, in targeted mutagenesis, as artificial nucleases, and so on (see Chapter 7) have resulted in several other appropriate model triplexes. Antisense oligonucleotides complementary to a part of an mRNA are sometimes ineffective because RNA hairpins weaken or prevent their binding (Verspieren et al., 1990). In an attempt to bind an oligonucleotide to a hairpin, a new variant of triplex has been designed (Figure 3.19). It consists of a single-stranded RNA or DNA and an oligonucleotide capable of folding into a hairpin so as to donate one half of its structure for duplex formation and the second half to serve as the third strand (Giovannangéli et al., 1991; Brossalina and Toulmé, 1993; Brossalina et al., 1993; Kandimalla and Agrawal, 1994; Azhaeva et al., 1995; François and Hélène, 1995). A variant of this structure with a methylphosphonate oligonucleotide analog has been used to inhibit protein synthesis on a single-stranded mRNA template (Reynolds et al., 1994).

As an approach to increase the oligonucleotide affinity to its duplex target, Dieter-Wurm et al. (1992) created a model of platinated DNA triplex. In this model, a metal entity with a linear coordination geometry (e.g., *trans*-$(NH_3)_2Pt(II)$ or Ag(I)) may formally replace a Hoogsteen hydrogen bond between guanine and cytosine, keeping the two bases in a nearly planar arrangement. Cocrystallization of the platinated Hoogsteen G*C pair with 1-methyl-cytosine allowed the formation of the desired base triad, providing evidence that covalent binding of pyrimidine oligonu-

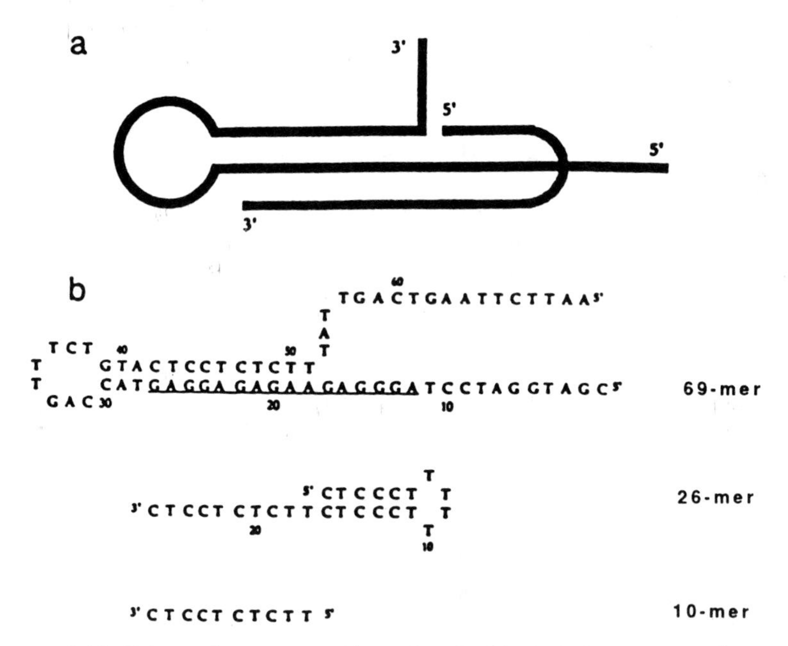

FIGURE 3.19. Schematic representation of a double hairpin structure (panel a), which may be relevant to an antisense targeting mRNA which is capable of partial folding and sequences of experimentally developed target and incoming oligonucleotide (panel b). Reprinted with permission from Brossalina and Toulmé (1993). © 1993 American Chemical Society.

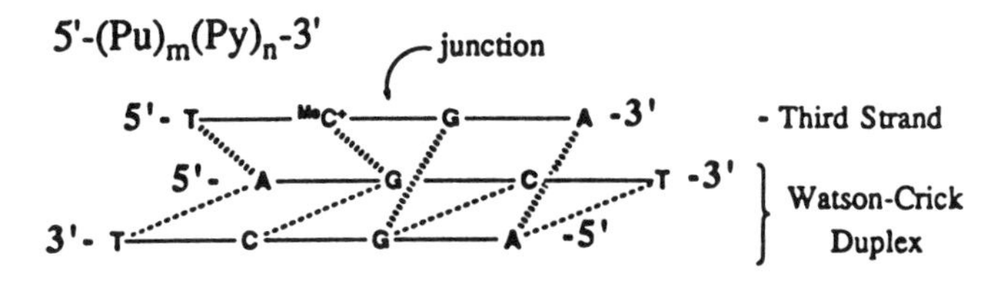

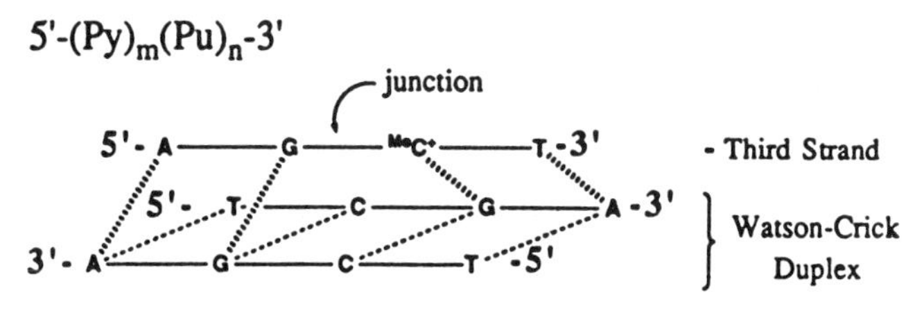

FIGURE 3.20. Schematic representation of alternate-strand triplex formation, where oligonucleotides bind in the major groove and recognize sequential purine tracts in the duplex by forming Hoogsteen hydrogen bonds first with purines in one strand and then with purines in another strand. Two possible target sites, 5′-$(Pu)_m(Py)_n$-3′ and 5′-$(Py)_m(Pu)_n$-3′, are shown. The junction between purines and pyrimidines on one strand of the duplex is also shown. Reprinted with permission from Beal and Dervan (1992b). © 1992 American Chemical Society.

cleotide to double-stranded DNA via a linear *trans*-$(NH_3)_2Pt(II)$ entity is sterically feasible.

In order to increase the repertoire of triplex-forming sequences, the triplex structures at adjacent Pu and Py tracts were designed using a third strand which can switch from Hoogsteen bonding with one target strand in the duplex to that with another strand. In one variant of alternate-strand triple-helix formation, the recognition of purine blocks on opposite strands was accomplished by a third-strand oligomer where homopyrimidine blocks have either 3′–3′ or 5′–5′ linkages (Horne and Dervan, 1990; McCurdy et al., 1991; Ono et al., 1991a; Froehler et al., 1992a). In a different approach, a third strand consisting of both purine and pyrimidine blocks and having normal phosphodiester linkages switches strands at the junction between the oligopurine and oligopyrimidine tracts but maintains the required strand polarity (Figure 3.20) (Haner and Dervan, 1990; Sun et al., 1991a; Beal and Dervan, 1992b; Jayasena and Johnston, 1992a,b, 1993; see also Chapter 4 for more details).

If a pair of target sites is separated along the sequence, an oligonucleotide hybrid possessing two triplex-forming oligomers connected by a flexible polymeric linker chain can form a stable complex. This was tested using a segment of the herpes simplex virus-1 D-glycoprotein promoter with a pair of 12-base-pair target sites separated by a turn of double helix and linkers of 20 to 25 rotatable bonds (Kessler et al., 1993).

An unusual triple-stranded complex can be formed between the duplex DNA and a peptide nucleic acid (PNA) (Figure 3.21) (Egholm et al., 1992a,b). A 2:1 complex of PNA and DNA forms when two homopyrimidine PNA strands bind to the duplex and displace one DNA strand (Nielsen et al., 1991; Egholm et al., 1992a,b; Cherny et al., 1993a). Such complexes have an unprecedented high stability (Egholm et al., 1992a,b), are right-handed, and have base conformations very similar to those of a conventional DNA triplex composed of analogous bases (Kim et al., 1993). They are of great potential importance, for example, in the inhibition of transcription, since the PNA oligomer is extremely stable in physiological media (Demidov et al., 1994b).

A complication arises with the orientation of the third strand in the triplex structure when the triplex-forming oligonucleotide contains only guanines and thymines. Thymines in the purine-rich third strand can form traids with TA pairs of the duplex, allowing a PyPuPu triplex to form. The third strand is antiparallel to the duplex purine strand and parallel to the duplex pyrimidine strand. When the number of thymines gradually

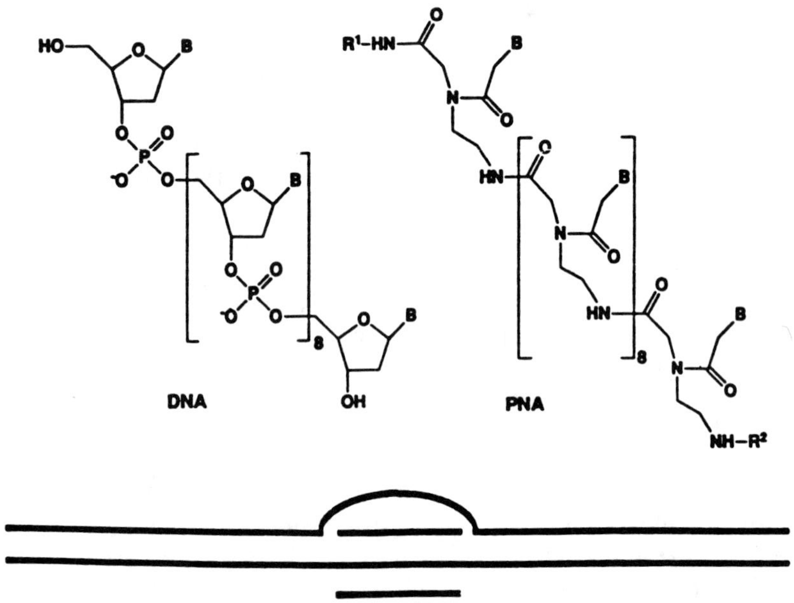

FIGURE 3.21. Peptide nucleic acid (PNA) (shown in the upper panel) was designed to mimic the sugar–phosphate backbone of DNA. Due to the lack of repulsion between the uncharged PNA backbone and the charged DNA backbone, very stable complexes form in which two strands of PNA and one strand of DNA form a triplex and the other DNA strand is displaced. Since PNA has neither natural peptide backbone nor sugar–phosphate backbone, it is resistant to the usual proteolytic and nucleolytic enzymes. Reproduced from Demidov et al. (1993).

```
YP1  5'------------->3'
       GGTTTGTTTTGTTT
YA1  3'<-------------5'

                               TARGET II
                             --------------
5' GCTAGTCAGGAAAGAAAAGAAACCCCTTTTTTTTTTCAGCTCGA 3'
3' CGATCAGTCCTTTCTTTTCTTTGGGGAAAAAAAAAAGTCGAGCT 5'
         --------------
             TARGET I
                           3'<-------------5'  YP2
                             GGGGTTTTTTTTTT
                           5'------------->3'  YA2
```

FIGURE 3.22. Guanine and thymine that compose three strands hybridize to their target duplex depending on the number of TpG and GpT steps (or ApG and GpA steps in the duplex). A 44-base pair DNA fragment was designed to bind targeted oligonucleotides that were synthesized to provide parallel (P) and antiparallel (A) orientations relative to the purine strand in the target duplex. When the guanines and thymines were clustered in the sequence, the oligonucleotide YP2, but not YA2, formed a triplex and protected the duplex from DNase digestion. This indicates an antiparallel orientation of longer pyrimidine tracts and a parallel orientation of shorter purine tracts. When guanines were dispersed in the oligonucleotide sequence, YA1 protected the duplex from DNase digestion more effectively than YP1. This indicates a parallel orientation of short thymine tracts dispersed in the oligonucleotide relative to the pyrimidine strand in the duplex. From Sun et al. (1991a). Reprinted with permission.

increases, the third strand may contain an equal number of thymines and guanines or a lesser amount of guanines. Thus, at some critical number of thymines, the third strand should change the orientation of hybridization. This case was studied by Hélène's group (Sun et al., 1991; Giovannangéli et al., 1992a). Figure 3.22 shows an example of tested third-strand sequences and their target (Sun et al., 1991a). Both oligonucleotide sequences contain four guanines and ten thymines. When the guanines and thymines were clustered in the sequence, an antiparallel orientation of longer pyrimidine tracts and a parallel orientation of shorter purine tracts was found in the DNase I digestion experiment. A similar result was obtained for oligonucleotides containing one 5-methylcytosine interrupting the thymine tract (Giovannangéli et al., 1992a). When guanines were dispersed in the sequence, an oligonucleotide was oriented in a parallel fashion to the pyrimidine strand in the duplex. The authors explained these results by referring to their energetic minimizations of triplexes containing distortions in the third-strand backbone due to the presence of smaller pyrimidine and larger purine bases. Theoretical considerations predict the dependence of third-strand orientation on the number of ApG and GpA steps in the homopurine sequence (Sun et al.,

1991a–c; see also Chapter 4). It should be noted that the triplexes formed were not very stable, as seen in DNase or copper–phenanthroline footprinting patterns (Sun et al., 1991; Giovannangéli et al., 1992a). To some extent, the situation seems similar to the orientation of the third strand on the *c-myc* promoter sequence containing an imperfect PyPu tract (Cooney et al., 1988; Durland et al., 1991).

Conclusion

Triple-stranded nucleic acid structures are formed on a core homopurine–homopyrimidine duplex using the third strand of an appropriate sequence that lies in the major groove and binds to the duplex via Hoogsteen hydrogen bonds. Either a mirror repeat of the same duplex (in the case of intramolecular H or H* DNA in supercoiled molecule) or an external homopyrimidine or purine-rich oligomer (in the case of an intermolecular triplex) can be used. Some variants of these structures have been designed to better understand the biological role of H (H*) DNA, to study the molecular details of triplexes, and to be used in numerous medical and biotechnological applications. A family of triple-stranded nucleic acids contains various structures that differ with respect to backbone conformations and, provided that the strand sequences are irregular, are heterogeneous in helical parameters along the strands. The triplexes can exist for a sufficiently long time, and this offers many opportunities for their application; however, their thermodynamic stability is lower than that for core duplexes. Among the triplex-stabilizing factors are cationic neutralization of phosphate charges, acidic pH for protonated triplexes, multivalent cations for the PyPuPu triplexes, supercoiling and stabilization of single-stranded loops for H (H*) DNA, and the use of some modified bases. The search for appropriate base analogs is also promising for the extension of triplex-recognition schemes necessary for defining the biological roles of triplexes and the further development of applications based on triplex methodology. This search will be discussed in Chapter 4.

4
Triplex Recognition

Extension of Triplex Recognition Schemes

The homopyrimidine strand of duplexes cannot bind the strand with high affinity, because its interaction with the forthcoming third strand would involve only one Hoogsteen-like bonding; in contrast, in the homopurine strand, the bases are able to form two bonds with each third-strand base.

Significant limitations imposed by the classic triplex structure, which imply targeting the incoming third strand at the homopurine tract of the existing duplex, have stimulated the search for new triplex recognition schemes. Another stimulus stems from the necessity of stabilizing the triplex under physiological conditions. At least two major reasons justify such a search.

First, several biological roles of triplex structures have been suggested (see Chapter 6 for discussion). Due to the relatively rare occurrence of perfect PyPu mirror repeats, these biological roles could be significantly limited. However, if some nonstandard types of triads could be used, the biological significance of the triplexes would have a much more solid basis.

Second, several biotechnological and genetic methodologies have been proposed which are based on intermolecular triplex formation (see Chapter 7 for discussion). Again, a discovery of new structural motifs capable of triplex forming may significantly extend the field of triplex applications. Some of these applications use oligonucleotide delivery to its target in the living cell. As a response to this demand, metabolically stable oligonucleotide analogs that retain their triplex-forming ability are designed.

Several approaches are being developed to understand base recognition in triplex formation. Among them are (1) testing new unusual base triads composed of natural bases, (2) synthesizing novel base analogs to extend the triad code, (3) excluding the recognition of certain base pairs in the PyPu tract by synthesizing oligonucleotides with abasic sites, and (4) designing third strands capable of binding to alternate strands of the duplex target.

Natural Bases in Unusual Triads

A number of possible nonstandard base triads have been tested in intermolecular triplexes. The first unusual base combinations were considered by Johnson and Morgan (1978) who suggested TA*G and CG*A^+ triads with G and A^+ in the third strand of the PyPuPy triplex (Figure 4.1). However, their own hydrodynamic, fluorometric, and UV-melting data for complexes composed of homopurine and homopyrimidine polynucleotides with repeating triplets were not consistent with the existence of such triads. More recent studies were devoted to determining the stability of PyPuPy triplexes containing various point substitutions (Griffin and Devan, 1989; Belotserkovskii et al., 1990; Macaya et al., 1991; Mergny et al., 1991b; Pei et al., 1991; Roberts and Crothers, 1991; Kiessling et al., 1992; Rougée et al., 1992; Yoon et al., 1992; Chandler and Fox, 1993; Fossella et al., 1993; Miller and Cushman, 1993; Best and Dervan, 1995). Because of the different methods employed, somewhat different sequences, different buffer media conditions, and the absence of quantitative treatment of the results in some cases, only a preliminary qualitative comparison can be made (Table 4.1). The relative stabilities of isolated triads in regular PyPuPy triplexes are indicated. It is clear that the well-known CG*C^+ and TA*T triads in all cases are the most stable ones. The triads AT*G and GC*T, in which duplex purine and pyrimidine bases are interchanged, also consistently demonstrated sufficient stability. A slightly lower stability was found for the CG*T triad. All these triads can be drawn as containing one Hoogsteen bond each (Figure 4.2).

The isolated AT*G and GC*T triads inside regular PyPuPy triplexes were studied by the NMR technique, which provided detailed structural information (Radhakrishnan et al., 1991b, 1992a,b; E. Wang et al., 1992; Radhakrishnan and Patel, 1994a,c,d). It was found that in order to avoid gross structural deformation of the triple helix, the orientation of the glycoside bond relative to the sugar-phosphate backbone and the conformation of the sugar ring of the third strand guanosine significantly differ from similar parameters of other third-strand nucleosides (E. Wang et al., 1992; Radhakrishnan et al., 1992a; Radhakrishnan and Patel, 1994a,c).

The molecular modeling studies (van Vlijmen et al., 1990; Mergny et al., 1991b) show that many unusual triads lack Hoogsteen hydrogen bonds or that third-strand bases have unfavorable contacts with the bases of the duplex and are therefore bulged out of the triple helix. For example, if the Watson–Crick base pair is GC with C in the homopurine strand, then a pyrimidine third-strand base is preferable. In spite of similar unfavorable contacts between the duplex cytosine and mismatched pyrimidines, the latter smaller bases can be well stacked within the third strand. On the contrary, due to steric hindrance of the bulky amino group at position 4 of cytosine in the duplex and the mismatched purine base in

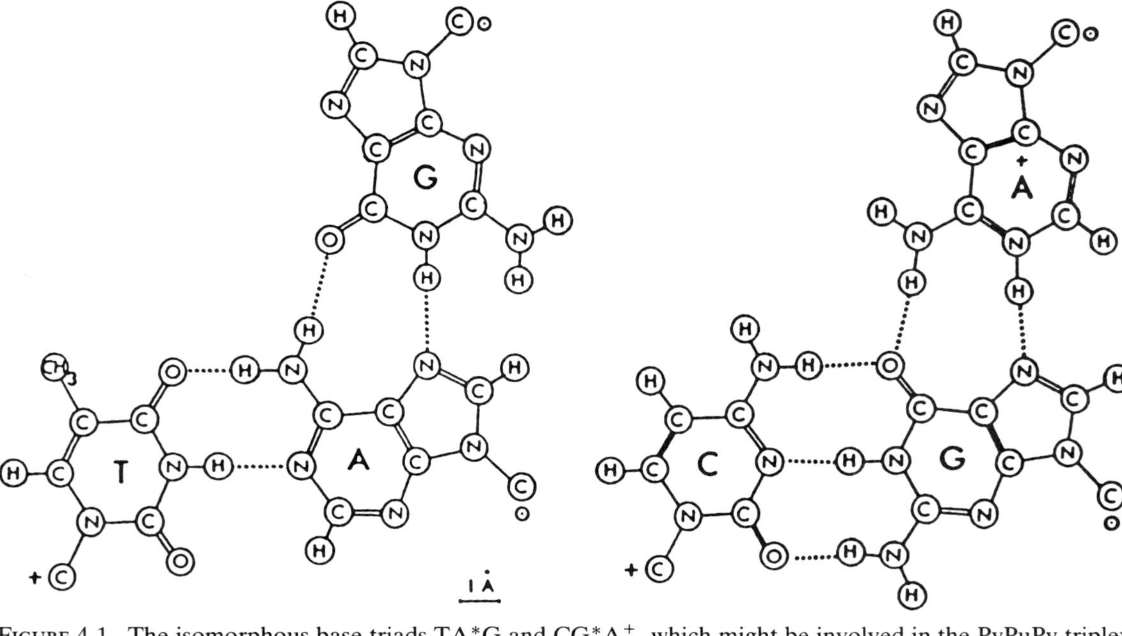

FIGURE 4.1. The isomorphous base triads TA*G and CG*A^+, which might be involved in the PyPuPy triplex. They destabilize the triplex, since the significant deviations of positions and orientations of glycoside bonds for the third-strand bases from those characteristic for CG*C^+ and TA*T triads do not allow their smooth incorporation into the PyPuPy triplex. Reproduced from Johnson and Morgan (1978).

TABLE 4.1. Relative stabilities of 16 possible isolated triads inside the regular PyPuPy triplex.

Source	CG*C⁺	CG*A⁺	CG*G	CG*T	TA*C	TA*A	TA*G	TA*T	GC*C	GC*A	GC*G	GC*T	AT*C	AT*A	AT*G	AT*T
Griffin and Dervan (1989)	++++	−−	−−	++	−−	−−	−−	++++	+	−−	−−	−−	−−	−−	++++	−−
Belotserkovskii et al. (1990)	++++	++	+	++	+	+	+	++++	+	−−	+	++	++	+	+	+
Mergny et al. (1991b)	++++	+++	+++	++	+	++	++	++++	++	+	++	++	−−	+	+++	+
Yoon et al. (1992)	++++	++	++	++	++	+	++	++++	+	+	+	+++	+	−−	++++	+
Macaya et al. (1991)	++++	++	++	+												
Pei et al. (1991)				+++	+++											
Roberts and Crothers (1991)			+	+++		+										
Rougée et al. (1992)					+	++	+	++++								
Fossella et al. (1993)	++++	++	+	+++	+	+	+	++++	++	−−	+	++	−−	−−	+++	+
Miller and Cushman (1993)								++++	−−	−−	−−	++	−−	−−	++	−
													+	++	++	++
													(AU*C)			(AU*U)
Chandler and Fox (1993)	++++	−	+	+					+	−	−	++	−	−	++++	+
Best and Dervan (1995)	++++	−	+	+	++	+	+	++++	++	−	+	++	−	+	++	+

The first and second bases in triads are those in the pyrimidine and purine strands of the target duplex, respectively, the third bases after * sign are those in the triplex-forming pyrimidine strands.

Data of UV melting (Macaya et al., 1991; Fossella et al., 1993; Rougée et al., 1992; Mergny et al., 1991b; Roberts and Crothers, 1991; Miller and Cushman, 1993); supercoil energy necessary for H-form extrusion (Belotserkovskii et al., 1990); affinity cleaving (Griffin and Dervan, 1989; Best and Dervan, 1995); band-shift assay (Yoon et al., 1992), affinity chromatography (Pei et al., 1991), and DNase digestion (Chandler and Fox, 1993) have been used. Stability was qualitatively evaluated by comparing melting temperature of complexes, amounts of oligonucleotides in duplex and triplex electrophoretic bands, or amounts of Fe–EDTA cleaved products in different complexes.

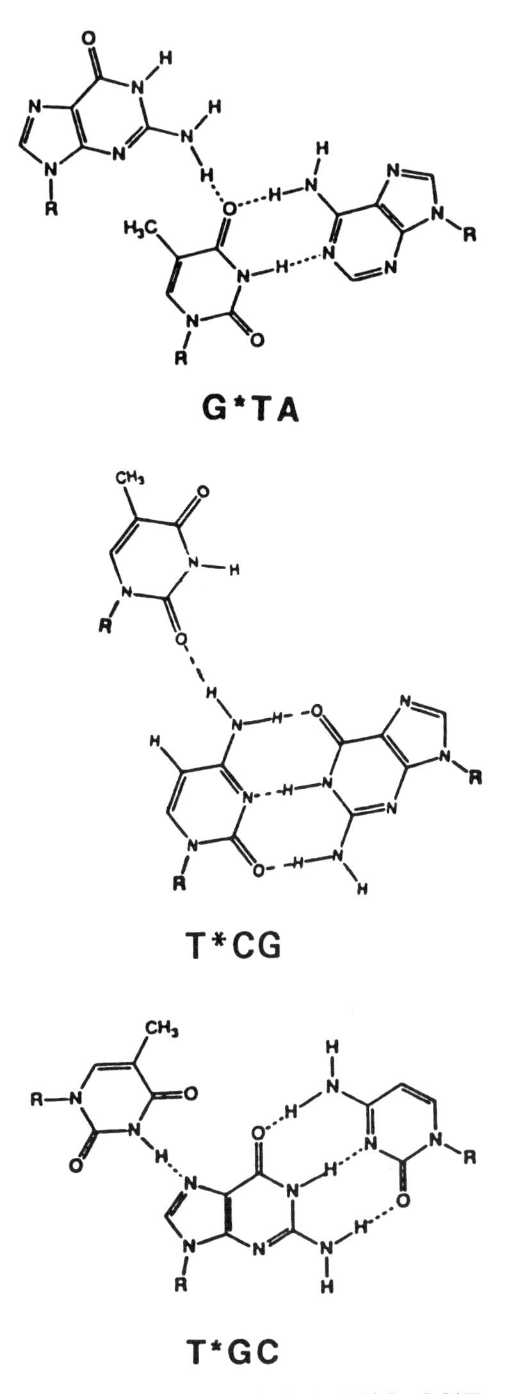

FIGURE 4.2. The structures of the unusual triads-AT*G, GC*T and CG*T which are fairly easily accommodated in the regular PyPuPy triplex. They can form single Hoogsteen hydrogen bonds. In order to avoid structural distortion of the triple helix, they have unusual configurations of glycoside bonds and sugar rings for the third strand bases.

the third strand, the latter base is expelled into the major groove at the expense of the stacking energy. Almost all mismatched bases are expelled out of imperfect AT*Z (where Z is A, G, or C) triads due to unfavorable interactions with the 5-methyl group in thymine. Guanine in such a mismatched triad is less disadvantageous, because it can make a hydrogen bond between its amino group and O4 of thymine.

There are less data on acceptable variations of the PyPuPu-type triads. Beal and Dervan (1991; 1992a) tested various base combinations (Figure 4.3) using affinity cleaving. As expected, TA*A, TA*T, and CG*G triads are the most stable (Table 4.2). The base triads with one Hoogsteen bond each (CG*A, TA*C, and GC*T) were also sufficiently stable to be detected. These data are in perfect agreement with those of Greenberg and Dervan (1995), who used DNase I footprinting assay to determine association constants and free energies of binding the purine-rich oligonucletides to the PyPu target. The triad CG*A can be made more stable in acidic media where N1 of adenine becomes protonated (Malkov et al., 1993a). Incorporation of protonated adenine into the third strand triplex allows the formation of the PyPuPu triplex at acidic pH without any multivalent cations. However, the addition of divalent metal cations stabilizes the triplex at higher pH (Malkov et al., 1993a). Similar stabilization may be expected in the case of the TA*C triad (protonation of N3 of cytosine). Durland et al. (1994), who measured the binding of oligonucleotides to their target duplex with GC inversion in the PyPu tract, found the GC*T imperfect triad the least destabilizing. The NMR data show that the GC*T triad may exist in equilibrium between a nonhydrogen-bonded form and the $T(O4)\cdots C(NH_2)$ bonded form. The structural perturbation around the GC*T inversion site extends to at least two adjacent triads on either side (Dittrich et al., 1994). There is no obvious way to further stabilize the GC*T triad.

Thus, one can find some unusual triads that distort the triple helix relatively slightly. The magnitude of the effect depends on the type of specific triad and the flanking sequence (Griffin and Dervan, 1989; Kiessling et al., 1992). Being isolated, these mismatches do not significantly destabilize the structure of the complex, providing the basis for triplex formation on imperfect PyPu tracts. However, little is known about the stability of complexes containing several such interruptions.

→

FIGURE 4.3. Positions of the bases in 16 possible natural base triads for the PyPuPu triplex. The third-strand bases were fitted to the Watson–Crick pairs so as to maximize hydrogen bond complementarities and maintain the third-base glycoside bonds similar to the stable CG*G, TA*A, and TA*T triads. All bases have *anti* conformations about the glycosidic bonds. Reproduced from Beal and Dervan (1992a) *Nucleic Acids Res* by permission of Oxford University Press.

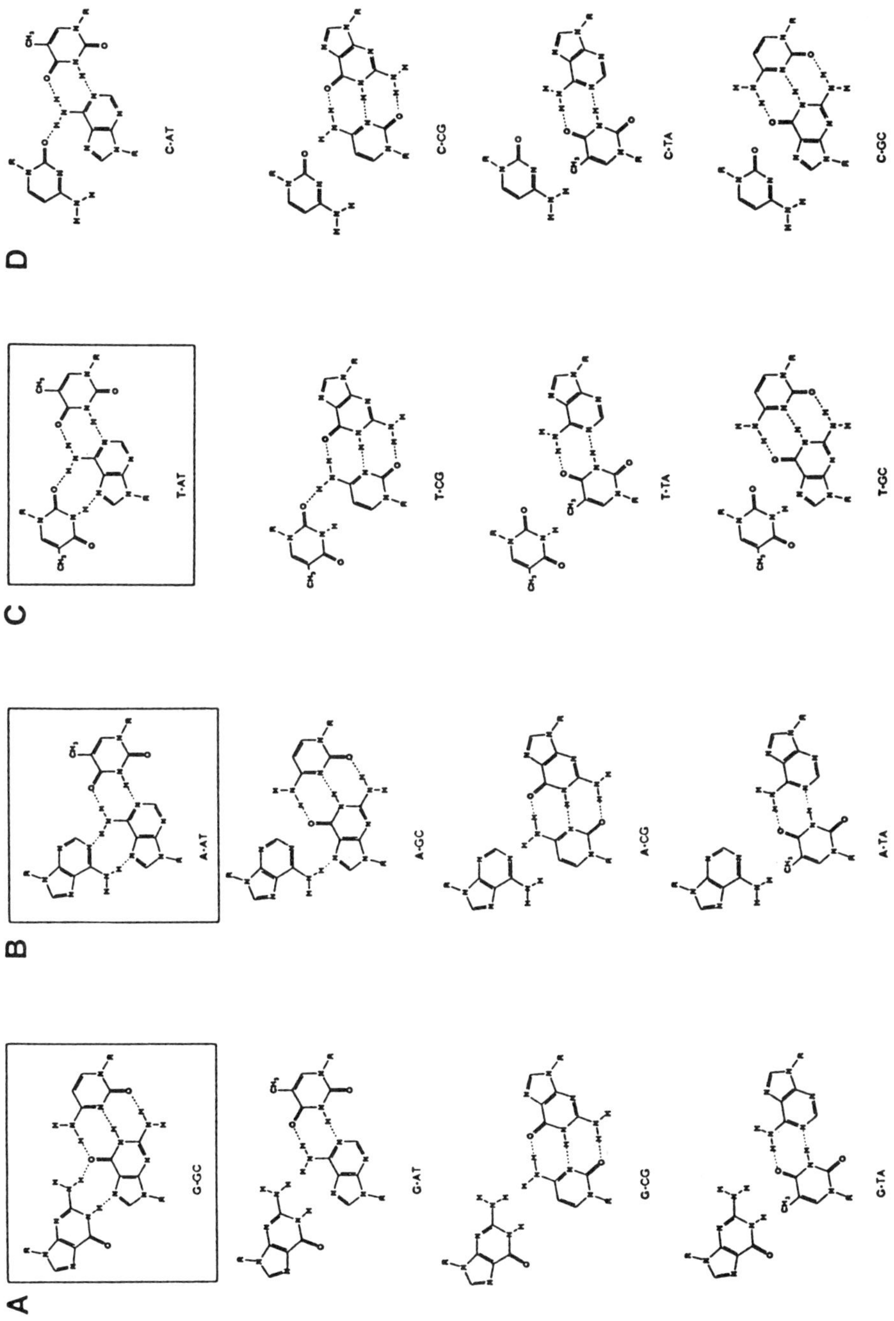

A
G·GC
G·AT
G·CG
G·TA
B
A·AT
A·GC
A·CG
A·TA
C
T·AT
T·CG
T·TA
T·GC
D
C·AT
C·CG
C·TA
C·GC

TABLE 4.2. Relative stabilities of 16 possible triads inside the PyPuPu triplex.

Duplex pair	Third-strand bases			
	A	G	C	T
TA	+++	−	+	+++
CG	+	+++	−	−
GC	−	−	−	+
AT	−	−	−	−

Relative stabilities based on quantitative analysis of duplex cleavage.
(+++) >80%; (++) 60–80%; (+) 40–60%; (−) <40%.
Reproduced from Beal and Dervan (1992a).

Triplexes with Base and Nucleoside Analogs in the Third Strand

The data on regular polymer complexes in Chapter 1 show that some modifications of pyrimidine bases at position 5 led to stabilized homopolymer triplexes. In line with these data, a more recent study showed that the third-strand oligonucleotide containing 5-bromouracil binds the same mixed homopurine DNA target as its thymine-containing analog, but with greater affinity and over an extended pH range (Table 4.3) (Povsic and Dervan, 1989).

In order to overcome the requirement of acidic pH for majority of PyPuPy triplexes, several base analogs of cytosine have been designed. According to earlier studies, 5-methylcytosine (m^{5}C) had a stabilizing effect on the homopolymer triplex, as judged from an increase in the

TABLE 4.3. Comparison of stabilities for triplexes containing natural bases and brominated base analog in the third strand.

Oligonucleotide	Cleavage efficiency at pH		
	6.6	7.0	7.4
5′-TUUUUCUCUCUCUCT-3′	+	+	−
5′-TTTTTTCTCTCTCTCT-3′	++	++	−
5′-TUUUUCUCUCUCUCT-3′	+++	+++	+

Stability of triple-stranded complexes was evaluated using the efficiencies of cleavage of the duplex by EDTA–Fe(II) attached at oligonucleotide 5′ end as foolows: + (2–4%); ++ (5–7%); +++ (8–10%). 5-Brominated uracils are underlined. Reproduced from Povsic and Dervan (1989).

melting temperature. Introduction of m^5C into a simple repeating pyrimidine polynucleotide resulted in the stabilization of the triple helix at neutral pH (Lee et al., 1979, 1984). The stabilizing effect of m^5C was also found for triplexes with mixed (irregular) sequences (Povsic and Dervan, 1989; Xodo et al., 1991). Figure 4.4 shows that the pH stability curve for the oligonucleotide triplex with m^5C in the third strand is shifted to higher pH as compared to triplexes with C in the third strand (Xodo et al., 1991). The stabilizing influence of the 5-methyl group may result from favorable changes in hydrogen bonding, solvation, and base-staking interactions (Xodo et al., 1991; Hausheer et al., 1992; Singleton and Dervan, 1992b). A recent study shows an interesting effect of oligonucleotides containing 5-methylcytosines on triplex formation in a model system. When oligonucleotides with methylated cytosines are added to a model oligonucleotide duplex, they displace a pyrimidine strand from the duplex and form a triplex with two strands containing 5-methylcytosines (Xodo et al., 1994a). Such a displacement phenomenon might be used for a depletion of frequently occurring RNA hairpins in antisense targeting of oligonucleotides.

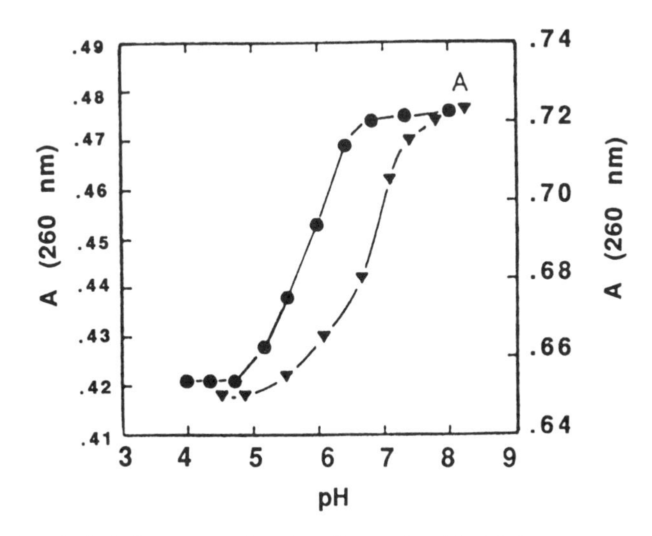

FIGURE 4.4. pH-Stability curves for the oligonucleotide triplex

```
+         + +     + +     +
5′ C-T-T-C-C-T-C-C-T-C-T
5′ G-A-A-G-G-A-G-G-A-G-A-T-T
3′ C-T-T-C-C-T-C-C-T-C-T-T-T
```

with third strand cytosines (●) and 5-methylcytosines (▼) in 50 m*M* NaCl, 10 m*M* $MgCl_2$ at room temperature. The ordinate shows the increase of UV absorption during dissociation of the third strand. Reproduced from Xodo et al. (1991) *Nucleic Acids Res.* by permission of Oxford University Press.

A pyrimidine analog, $m^{5ox}C$, which is the 6-keto derivative of 5-methylcytosine, was used for the pH-independent recognition of GC base pairs (Xiang et al., 1994). Of the possible tautomers of this analog, the one which contains a proton at N3 turned out to be preferable in solution. Thus, the hydrogen bond between N3 of the cytosine analog and N7 of guanine may be formed at neutral pH. At slightly acidic pH, the stability of the PyPuPy triplex containing $m^{5ox}C$ was lower than that of the triplexes containing C and m^5C; however, the stability was higher at pH 8.0 and 8.5.

The third strand may contain some modified bases, such as 8-oxo-derivatives of adenine in place of cytosine. Substitution at position 8 favors a *syn* conformation of nucleoside (Howard et al., 1974, 1975; Limn et al., 1983), making it possible to form hydrogen bonds between 6-amino hydrogen and N7 of the adenine derivative and O6 and N7 of guanine in the duplex. Using oligonucleotides with N6-methyl-8-oxoadenine, Krawczyk et al. (1992) formed stable triplexes at pH 7.2. The stability of triplexes containing up to eight 8-oxoadenine-modified triads (Figure 4.5) at neutral pH was shown in several studies (Miller et al., 1992; Davisson and Johnsson, 1993; Jetter and Hobbs, 1993; Q. Wang et al., 1994). Note that the orientation of the adenine derivative in the CG*M triad is about 90° relative to that in the triad suggested by Johnson and Morgan (1978) (Figue 4.1).

In order to stabilize the PyPuPy triplex at neutral pH, a specific nucleoside analog of cytidine (2′-O-methylpseudoisocytidine) was tested (Ono et al., 1991b). This analog contains a proton at the N3 position for hydrogen bonding with guanine in the Hoogsteen pair of the triad (Figure 4.6). A third pyrimidine strand containing a cytidine analog was shown to

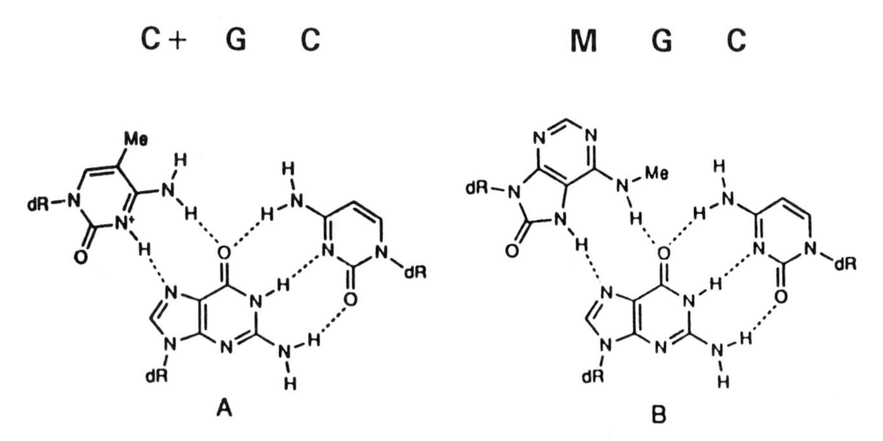

FIGURE 4.5. Suggested analogy in binding schemes for the CG*C$^+$ triad (A) and the CG*M triad (B). C$^+$ = protonated 5-methylcytosine; M = N6-methyl-8-oxo-2′-deoxyadenine. Reproduced from Krawczyk et al. (1992).

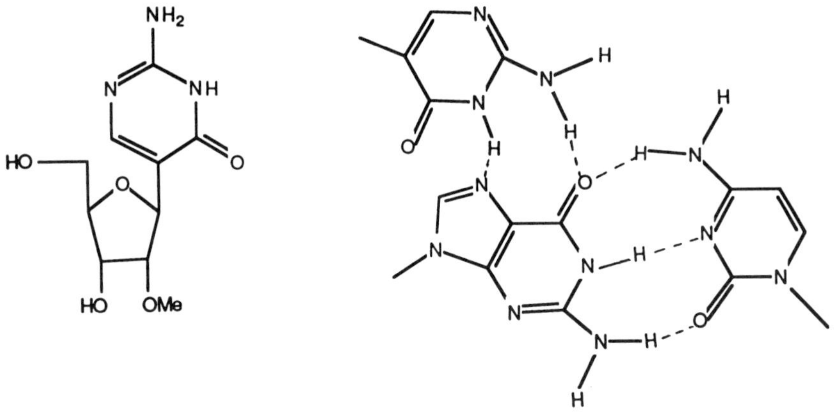

FIGURE 4.6. Structure of 2′-O-methylpseudoisocytidine and its pairing scheme with the Watson–Crick CG pair. Due to the presence of a proton at the N3 position of 2′-O-methylpseudoisocytidine, a stable triad formation does not require acidic conditions. Hydrophobic methoxy substitution at the 2′ position of ribose also contributes to triad stabilization. Reprinted with permission from Ono et al. (1991b). © 1992 American Chemical Society.

make the PyPuPy triplex more stable than a third strand with 5-methyl-2′-deoxycytidine. Two contributions are responsible for the stabilizing effect: the introduction into the base of a proton-donating group for hydrogen bond formation and the introduction of hydrophobic methoxy substitution at the 2′ position of ribose. The stabilizing effect of the latter contribution was confirmed by the enhanced stability of triplexes containing the 2′-O-methyl RNA oligomer compared to normal DNA and RNA oligomers (Escudé et al., 1992; Shimizu et al., 1992, 1994). The differences in hybridization properties between RNA and 2′-O-methyl RNA may be due to steric and conformational differences which result from the presence or absence of 2′-O-methyl groups (Wang and Kool, 1995a).

Based on the data that 5-methyl substitution in third-strand pyrimidines exerts a stabilizing effect on the PyPuPy triplex, one can expect that other hydrophobic substitutions can have a similar effect. A comparison of the thermal stability of triplexes containing 5-methylated and 5-propynylated analogs of 2′-deoxyuridine and 2′-deoxycytosine (Figure 4.7) showed that 5-(1-propynyl)-2′-deoxyuridine significantly stabilizes the triple helix (+2.4°C per substitution), whereas 5-(1-propynyl)-2′-deoxycytosine destabilizes it (−3.4°C per substitution) (Froehler et al., 1992b). In the latter case, the instability of the triplex is due to a decrease in pK_a for N3 (3.30 vs 4.35 for the 5-methyl derivative).

In an attempt to increase the pK_a of the 2′-deoxycytidine and thus stabilize the triplex at neutral pH, oligonucleotides containing deoxyribose analogs in which the ring oxygen was replaced by a methyl group (Figure

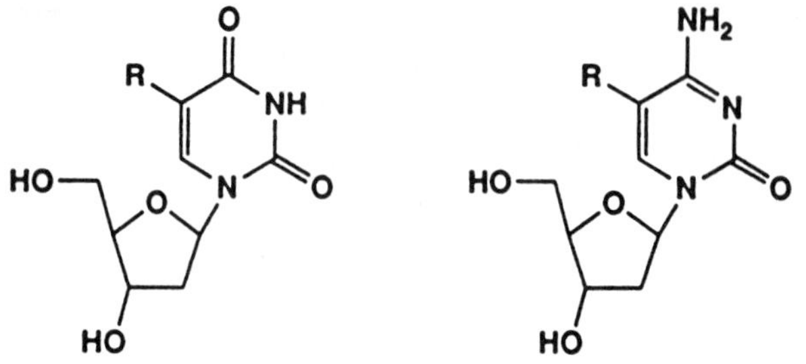

1) R = $-C{\equiv}C-CH_3$ 3) R = $-C{\equiv}C-CH_3$

2) R = $-CH_3$ 4) R = $-CH_3$

FIGURE 4.7. Structures of 5-substituted analogs of deoxyuridine (left) and deoxycytidine (right) tested for triplex stabilization. Methyl substitution has been known to stabilize triplexes compared with those containing unsubstituted nucleosides. A further increase in hydrophobicity of analogs by incorporation of propynyl groups (R represents $-C{\equiv}C-CH_3$) results in stabilization for the deoxyuridine analog and destabilization for the deoxycytidine analog. In the latter case, destabilization is a result of a significant decrease in pK_a for N3. Reprinted from *Tetrahedron Lett.* 33, Froehler et al. (1992b). Oligonucleotides containing C-5 propyne analogs of 2′-deoxyuridine and 2′-deoxycytidine, pp. 5307–5310. © 1992, with kind permission from Elsevier Science Ltd, The Boulevard, Langford Lane, Kidlington OX5 1GB, UK.

4.8) were tested for triplex formation (Froehler and Ricca, 1992). The carbocyclic analog of 5-methyl-2′-deoxycytidine stabilizes the triple-helical complex relative to the furanosyl nucleoside (by about 2°C per substitution). This stabilization is at least partly explained by the increased pK_a of the base. The carbocyclic analog of thymidine decreases triplex stability relative to thymidine (by about 4°C per substitution), presumably due to the unfavorable configuration of the cyclopentane ring.

The nonnatural deoxyribonucleoside 1-(2-deoxy-β-D-ribofuranosyl)-4-(3-benzamido)phenylimidazole (D_3) (Figures 4.8 and 4.10), was designed to sterically match the edges of PyPu base pairs in the major groove and facilitate favorable stacking with the bases in the third strand (Griffin et al., 1992; Kiessling et al., 1992). When included in a pyrimidine oligonucleotide, D_3 has greater affinity for purine–pyrimidine base pairs than for pyrimidine–purine base pairs (Figure 4.9). According to affinity cleavage data, the stabilities of base triads decrease in the order $AT{*}D_3 \sim GC{*}D_3 > TA{*}D_3 > CG{*}D_3$ (Griffin et al., 1992). It is not obvious which hydrogen bonds are potentially formed. The study of an intramolecular oligonucleotide triplex with the $AT{*}D_3$ triad showed that the D_3 was positioned between its target base pair and the 3′ triad (Figure 4.10) (Koshlap et al., 1993). This indicates a sequence-specific intercalation of D_3, since it recognizes AT and GC pairs in a nearest-neighbor-dependent

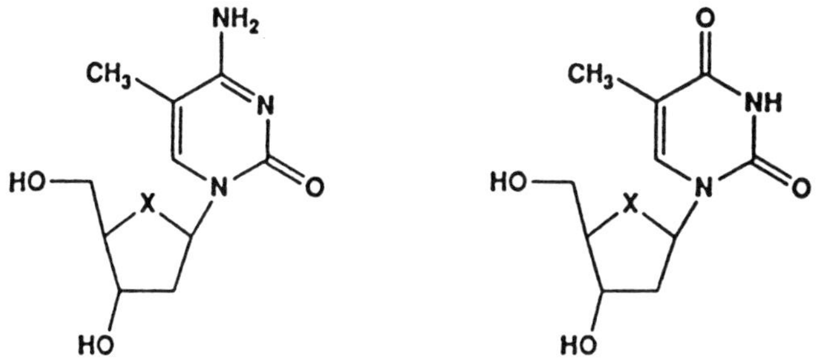

1; X = CH_2
2; X = O

3; X = CH_2
4; X = O

FIGURE 4.8. Structures of carbocyclic ribose analogs of 5-methylcytidine (left) and thymidine (right). Increasing pK_a for the 5-methylcytidine analog results in enhancement of triplex stability. Triplex destabilization by the thymidine analog is probably caused by the unfavorable configuration of the cyclopentane ring. Reprinted with permission from Froehler and Ricca (1992). © 1992 American Chemical Society.

way (Griffin et al., 1992). Thus, some base analogs can be accommodated in the triplex structure in a completely new way.

The nonnatural pyrido[2,3-d]pyrimidine nucleoside F probably exists in different tautomeric forms within double-helical and triple-helical complexes (Staubli and Dervan, 1994). The hydrogen bond patterns for triads including F remain to be established.

The deoxyribonucleoside 1-(2-deoxy-β-D-ribofuranosyl)-3-methyl-5-amino-1*H*-pyrazolo[4,3-*d*]pyrimidin-7-one (P1) was designed so that one its edge mimics N3-protonated cytosine (Figure 4.11) (Koh and Dervan, 1992). When incorporated into a pyrimidine oligonucleotide, it was expected to substitute for cytosine in forming the Hoogsteen hydrogen bonds with guanine. This suggestion was confirmed by NMR data (Radhakrishnan et al., 1993b). The affinity-cleaving data showed that the stabilities of base triads decreased in the order CG*P1 ≫ GC*P1 ≫ TA*P1 ~ AT*P1. Thus, the selectivity of P1 is similar to C, but the triplex is stable over an extended pH range, as it was shown that up to six contiguous GC base pairs in a 16-nucleotide-long site are recognized by P1 analogs via triplex binding at pH 7.4 (Koh and Dervan, 1992). Sequence composition effects on triplex formation by oligonucleotides containing multiple P1 substitutions have been determined using quantitative DNase I titration (Priestley and Dervan, 1995). According to the equilibrium association constants, when six P1 in the third strand alternate with thymines, oligonucleotide binding is 4 orders of magnitude weaker than for a similar oligonucleotide containing m^5C in place of P1. However, when six P1 are clustered, oligonucleotide binding is 5 orders of magnitude stronger than for m^5C-containing oligomer. The observed phenomenon

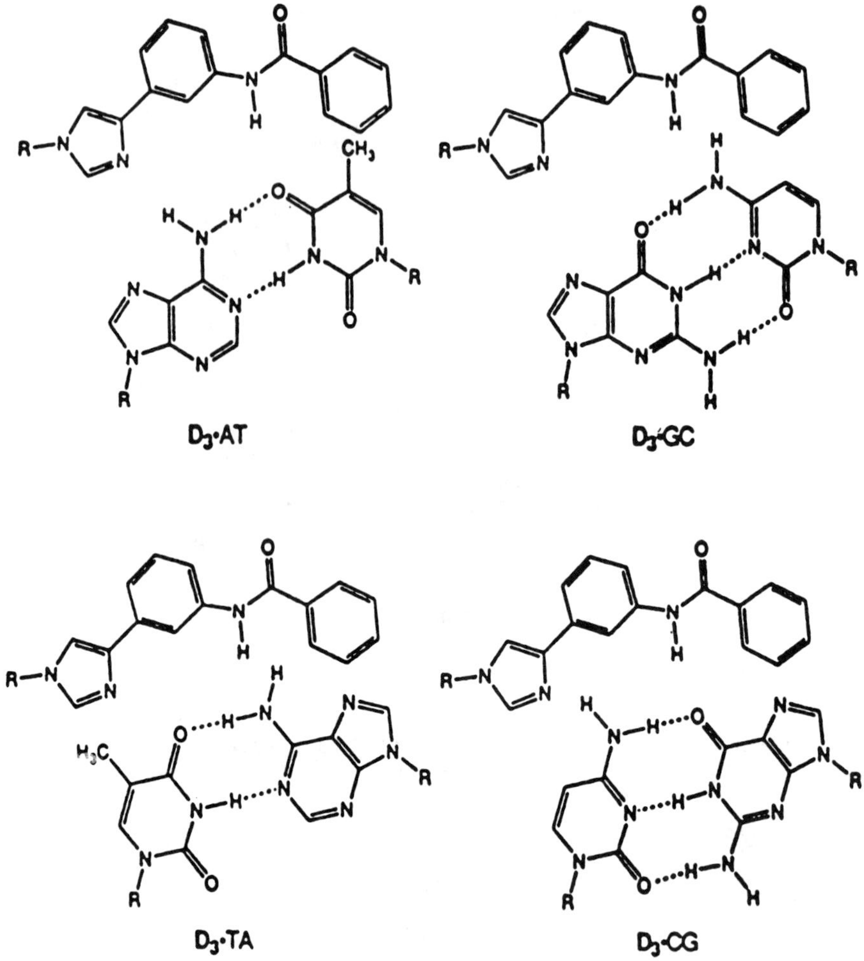

FIGURE 4.9. The deoxyribonucleoside analog 1-(2-deoxy-β-D-ribofuranosyl)-4-(3-benzamido)phenylimidazole (D_3) has more favorable interactions with the AT and GC pairs in the major groove of the PyPuPy triplex (bottom) compared to the TA and CG pairs (top). No obvious hydrogen bonds could be drawn, and it was suggested that D_3 recognizes the overall shape of the purine–pyrimidine base pair. Reprinted with permission from Griffin et al. (1992). © 1992 American Chemical Society.

might be explained by the lack of structural isomorphism of the TA*T and CG*P1 triads which results in a gross structural deformation and strong triplex destabilization in the case of alternating triads. This structural deformation is significant only at the boundaries of the cluster of CG*P1 triad; therefore, triplex destabilization is less pronounced

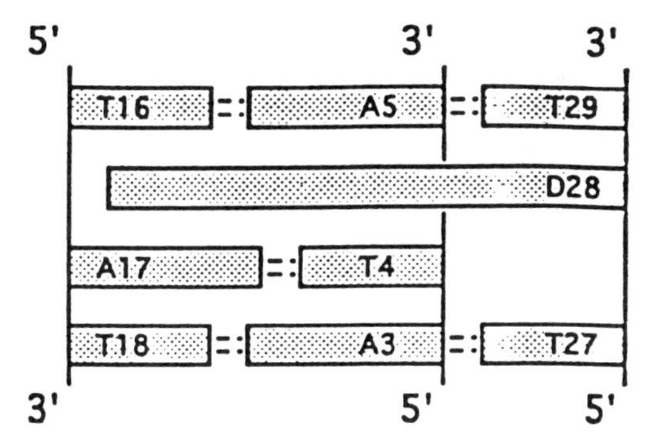

FIGURE 4.10. According to NMR data, the nonnatural deoxyribonucleoside analog 1-(2-deoxy-β-*D*-ribofuranosyl)-4-(3-benzamido)phenylimidazole (D_3) is accommodated in the structure of the PyPuPy triplex by intercalation. Reprinted with permission from Koshlap et al. (1993). © 1993 American Chemical Society.

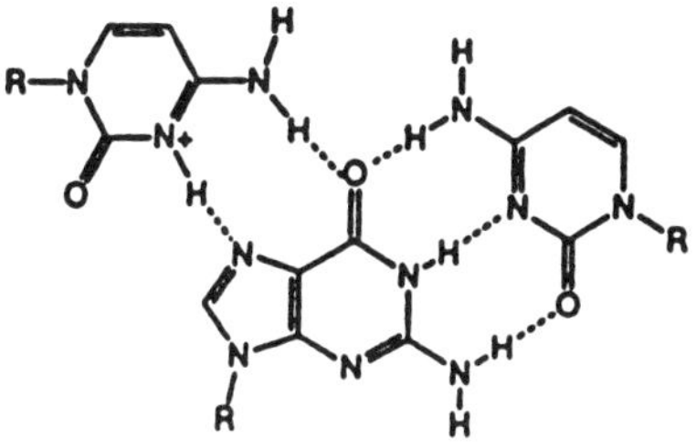

C+GC

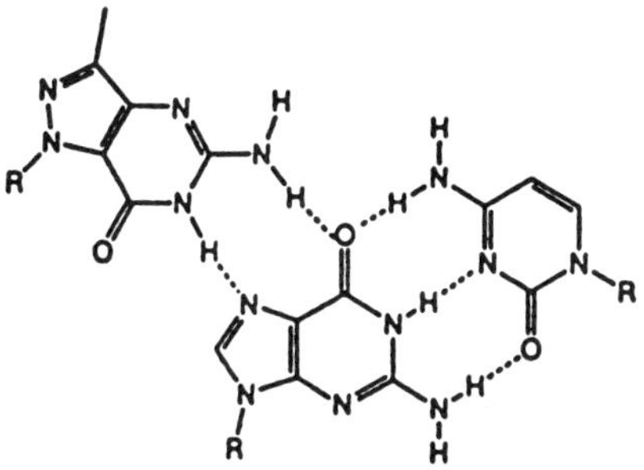

P1·GC

FIGURE 4.11. A synthetic deoxyribonucleoside 1-(2-deoxy-β-*D*-ribofuranosyl)-3-methyl-5-amino-1*H*-pyrazolo[4,3-*d*]pyrimidin-7-one analog (P1) mimicking the hydrogen-bonding edge of N3-protonated cytosine forms a stable triad with the Watson–Crick CG pair. The triads are stable over a broad range of pH values and up to six of them may be incorporated into a 16-mer triplex. Reprinted with permission from Koh and Dervan (1992). © 1992 American Chemical Society.

(Priestley and Dervan, 1995). The stabilization advantage over the triplex containing contiguous m^5C might be explained by the strongly unfavorable clustering of protonated bases in the third strand of the latter triplex (Jayasena and Johnston, 1992b; Kiessling et al., 1992; Völker and Klump, 1994; see also section "Factors That Destabilize Triplexes" in Chapter 3).

Modification of the hydrophobicity and hydrogen-bonding capability of a base analog was exploited in a design of N4-(6-aminopyridinyl)-2′-deoxycytidine (Huang and Miller, 1993). When incorporated into a pyrimidine third strand, this analog recognizes TA and GC base pairs in the duplex target. The hypothetical hydrogen-bonding schemes suggested on the basis of stereochemical considerations do not explain the experimentally determined stability of triads. Although it is possible to draw three hydrogen bonds between this analog and the TA pair, the resulting triplex is more stable for the case of two bonds between the GC pair and the analog. In addition to hydrogen bonding, there might be some other mechanism of interaction between the analog and the base pairs in the duplex.

Milligan et al. (1993) studied the consequences of replacing thymine by 7-deazaxanthine and guanine by 7-deazaguanine in the purine-rich third strand (Figure 4.12). At physiological pH, 7-deazaxanthine is a purine complementary base of thymine in its hydrogen-bonding capability and introduces less distortion into the sugar–phosphate backbone of the third strand than does thymine. The expected better binding of the 7-deazaxanthine-containing oligonucleotide was confirmed by an experiment that showed a 100-fold increase in affinity compared to the thymine-containing oligonucleotide. Since G-rich oligonucleotides have a propensity to form tetrastranded complexes in the presence of K^+ and Na^+, the key hydrogen bond donor relevant to this process was eliminated by using 7-deazaguanine in the third strand. This replacement prevented oligomers from forming a tetrastranded structure but also significantly decreased their triplex-forming properties. This effect can be tentatively explained by the need of N7 in guanine to coordinate with divalent metal cations with a concomitant triplex-stabilizing effect (see Chapter 5 for discussion). Another way to prevent guanine from participating in tetrastranded complexes is to substitute O6 atoms that are engaged in specific hydrogen bonding in tetrads and coordination of monovalent cations. Two 6-thioguanine substitutions for guanine in 15-mer (Gee et al., 1995) and four similar substitutions in 14-mer (Olivas and Maher, 1995b) oligonucleotides efficiently prevented the formation of tetrastranded complexes and resulted in the formation of the desired triplexes.

Xanthine was introduced to the third strand of pyrimidine-rich 15-mer oligodeoxynucleotides to recognize cytosine in the purine strand (G-C inversion) in a homopyrimidine–homopurine duplex target (Shimizu et al., 1991). Although the third-strand xanthine may form one hydrogen bond with cytosine in the purine strand (GC*X triad), the experimentally

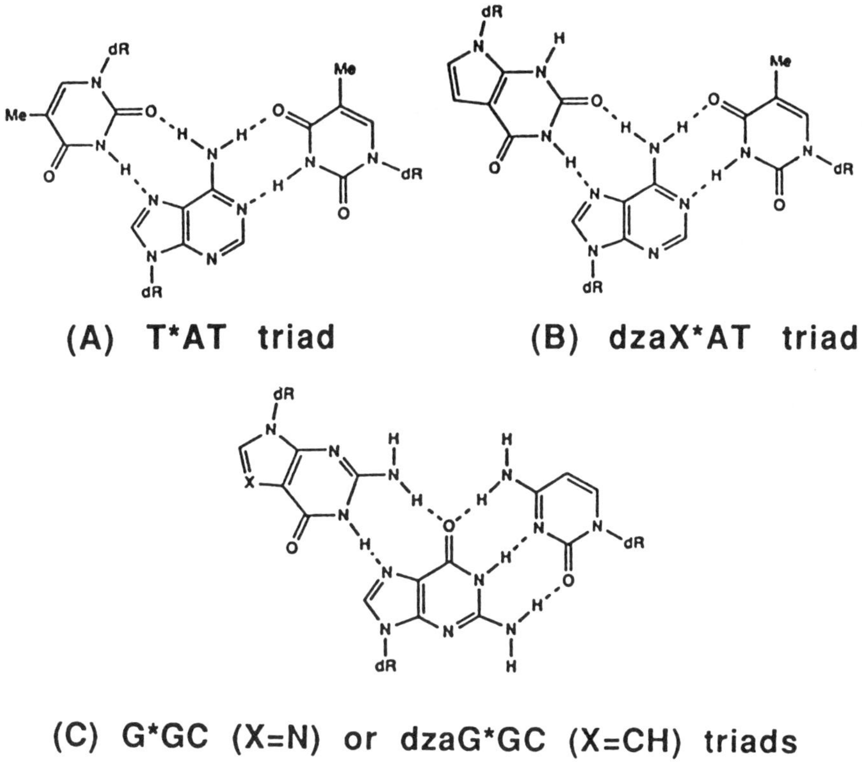

(A) T*AT triad **(B) dzaX*AT triad**

(C) G*GC (X=N) or dzaG*GC (X=CH) triads

FIGURE 4.12. Some hydrogen-bonding schemes for the PyPuPu-type triplex. (A) TA*T triad, where the smaller-sized thymine introduces some irregularity in the structure of the purine strand sugar-phosphate backbone; (B) TA*dzaX (dzaX = 7-deazaxanthine), which is smoothly incorporated into the purine third strand; (C) CG*G (X = N) or CG*dzaG (dzaG = 7-deazaguanine, X = CH). Reproduced from Milligan et al. (1993) *Nucleic Acids Res* by permission of Oxford University Press.

determined stability of the triplex was lower than that for triplexes containing imperfect GC*C and GC*T triads (see, e.g., appropriate columns in Table 4.1).

The triad-forming properties of hypoxanthine were examined in some early studies of polyribonucleotide complexes where the IA*I, CI*I, and CI*C triads were identified (Rich, 1958a,b; Chamberlin, 1965; Thiele and Guschlbauer, 1969). A more recent study also showed the formation of other triads: TA*I and CG*I (Letai et al., 1988). This finding is relevant to mixed DNA–RNA triplexes, which may be used in some biotechnological methodologies. It should be also noted that an attempt to form the DNA triplex $(dT)_{17} \cdot (dA)_{17} \cdot (dI)_{17}$ was not successful (Hogeland and Weller, 1993).

According to model considerations, the nucleoside analog 2-deoxynebularine (N) can form one Hoogsteen hydrogen bond with cytosine and adenine (Figure 4.13) (Stilz and Dervan, 1993). The affinity cleavage data show that the stabilities of base triads in the PyPuPu triplex decrease in the order GC*N ~ TA*N ≫ CG*N ~ AT*N. If a regular PyPu tract is interrupted by a GC inversion, the GC*N triad is more stable than any of the triads of natural bases. Therefore, 2-deoxynebularine could be used for a third-strand recognition of imperfect PyPu tracts.

The ability of certain azole-substituted oligodeoxyribonucleotides to form the PyPuPu type triplex with isolated GC and AT inversions was studied by Durland et al. (1995). According to the band-shift assay data, the oligonucleotides containing pyrazole, imidazole, 1,2,4-triazole and 1,2,3,4-tetrazole formed more stable triplexes than oligonucleotides with unmodified bases. The selectivity in triplex formation exhibited by certain azoles implies some base pair-specific interactions, although their nature has not been studied.

Adenine bases in the purine third strand can often be replaced by thymine bases. When the number of thymines in the third purine-rich strand increases at some proportion, a question arises about the orientation of this strand in the triplex structure: is it antiparallel to the duplex purine strand (as in the PyPuPu triplex) or parallel (as in the PyPuPy triplex)? (See also the section New Variants of Triplex Structure in Chapter 3.) This orientation was shown to depend on the number of GpA and ApG steps in the homopurine targeted sequence (Sun et al., 1991a–c; Giovannangeli et al., 1992a). A similar problem arises in the case of α-anomeric oligonucleotides, which are expected to behave opposite to natural β-anomers (see Hélène, 1991, for review). Thus, not only may thymine base introduce some irregularity in a largely PyPuPu triplex, but also guanine may distort the configuration of a mostly PyPuPy triplex. Figure 4.14 shows the CG*G triad in the third strand oriented parallel to the homopurine tract. To diminish the structural distortion of the sugar–phosphate backbone due to the presence of larger guanine and smaller thymine bases, a new triad involving the 2-aminopurine was proposed (Roig et al., 1993). Both oligo-β- and -α-deoxyribonucleotides composed of 2-aminopurine and guanine were synthesized. If the triplex with the third strand containing guanines and 2-aminopurines turns out to be stable, then this will be the first example of a PyPuPy triplex with a completely purine third strand.

Two different types of pyrimidine analogs have been designed to modify electrostatic interactions in DNA. Oligonucleotides bearing pyrimidine 5-ω-aminohexyl substituents are worth discussing here, because these zwitterionic analogs represent an example of internal neutralization of phosphate charges (Hashimoto et al., 1993). Modified bases distinguish matched from mismatched nucleotides in complementary DNA strands with a specificity comparable to that of natural bases and form a duplex

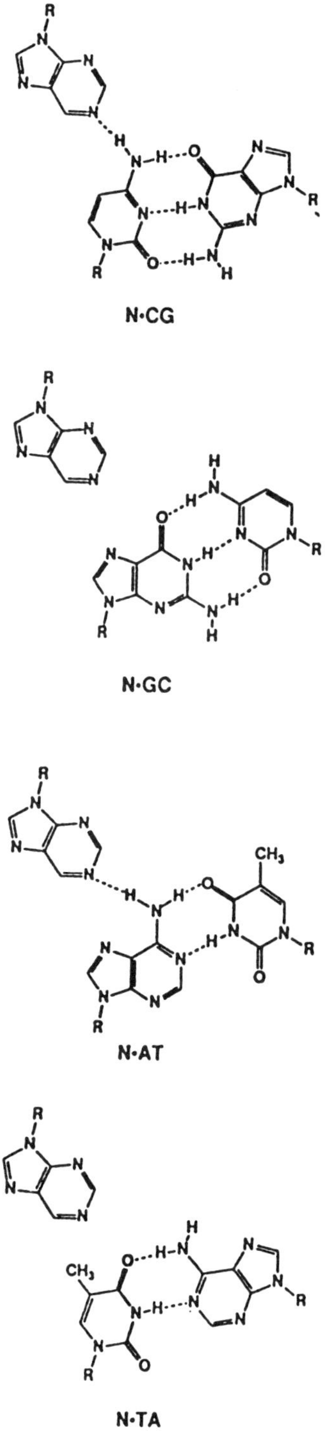

N•CG

N•GC

N•AT

N•TA

FIGURE 4.13. A hypothetical fit of a nucleoside analog, 2-deoxynebularine, to Watson–Crick base pairs within the PyPuPu triplex. According to the experimental data, the upper triad GC*N is the most stable of the four and can possibly be used for third strand recognition of a PyPu tract containing cytosine in the purine sequence. Reprinted with permission from Stilz and Dervan (1993). © 1993 American Chemical Society.

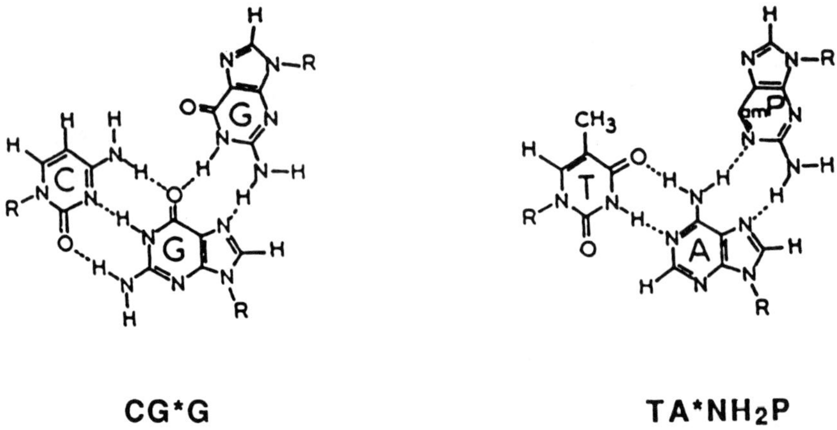

CG*G

TA*NH2P

FIGURE 4.14. The CG*G triad in the PyPuPy triplex (left). To diminish the structural distortion of the sugar-phosphate backbone due to the presence of larger guanine and smaller thymine bases, a new triad involving 2-aminopurine in place of thymine was proposed (right). The triplex-forming properties of guanine and 2-aminopurine containing third strands await detailed investigation. Reprinted from *Tetrahedron Lett.* 34, Roig et al. Oligo-β-deoxyribonucleotides and oligo-α-deoxyribonucleotides involving 2-aminopurine and guanine for triple-helix formation, pp. 1601–1604. © 1993, with kind permission from Elsevier Science Ltd, The Boulevard, Langford Lane, Kidlington OX5 1GB, UK.

which is more stable than one composed of only natural bases. However, oligonucleotides with 5-ω-aminohexyl substituted pyrimidines have not been tested for triplex formation. Based on the well-known triplex stabilization by polyamines, Barawkar et al. (1994) created a cytosine analog by conjugating 5-methyl-cytosine with spermine. The triplexes with cytosine-spermine conjugates are most stable at pH 7.1, which is desirable for many applications. However, nothing is known about the specificity of triplex formation by such oligonucleotides bearing multiple positive charges on spermine that may potentially overstabilize the triplexes with mismatches.

Abasic Sites in the Third Strands

In an attempt to overcome the problem of mismatches in an otherwise perfect triplex, Horne and Dervan (1991) studied the effect of an abasic site (ϕ) on triplex formation. In the PyPuPy triplex, all abasic triads TA*ϕ, CG*ϕ, AT*ϕ, and GC*ϕ are significantly less stable than TA*T, CG*C^+, and AT*G triads. Imperfectly matched base triads and abasic sites in the third strand destabilize triplexes to a similar extent, with TA*ϕ and CG*ϕ being less stable than AT*ϕ and GC*ϕ.

Base Analog in the Duplex Part of the Triplex

An attractive option for targeting single-stranded nucleic acids (e.g., RNA) by the addition of two oligonucleotides as the second and the third strand has been described by Trapane et al. (1994). To recognize adenosine and guanosine in a pyrimidine-rich first strand, a second strand contains deoxypseudouridine and deoxypseudoisocytidine, respectively, which are able to form Watson–Crick hydrogen bonds with the mentioned nucleosides in the first strand and two Hoogsteen-like hydrogen bonds with cytidines in the third strand. Thus, new triads AΨ*C and GΨi*C (where Ψ = deoxypseudouridine; Ψi = deoxypseudoisocytidine) give a definite hope for the recognition of any naturally occurring single-stranded nucleic acid which may serve as the first strand in a triplex.

Alternate Strand Triplex Formation

In order to increase the complexity of the target sequence recognized by the triplex mechanism, a new, elegant approach was proposed (Horne and Dervan, 1990). Sufficiently long tracts with purines in one strand providing stable triplex formation are not always found within biologically important sequences. However, two shorter adjacent PyPu tracts, with purines in different strands, are conceivable. Therefore, targeting the third strand at one duplex strand with a subsequent switch to the other strand would result in a sufficiently long structure made up of two abutting triplexes. Such a structure would be stabilized by a fairly large number of Hoogsteen hydrogen bonds, but would be destabilized by the absence of stacking interactions between third-strand bases at the point of strand switch and probably by the structural strain in a polynucleotide chain crossing the major groove from one purine stretch to the other.

The development of alternate-strand triple-helix formation increased the number of sites capable of being recognized by a pyrimidine strand made of two 3′–3′ or 5′–5′ coupled blocks (Horne and Dervan, 1990; McCurdy et al., 1991; Ono et al., 1991a; Froehler et al., 1992a). Pyrimidine oligonucleotides coupled in such a way have the necessary orientations for alternate-strand purine targeting in 5′-$(Pu)_m(Py)_n$-3′ and 5′-$(Py)_m(Pu)_n$-3′ tracts, respectively. An essential step in the design of such bidirectional oligomers is the choice of covalent linker. Horne and Dervan (1990), based on their model-building studies, introduced an abasic deoxyribose analog (ϕ) which should maintain structural continuity of the sugar–phosphate backbone in the 3′–3′ internucleotide junction of the oligomer as it switches from recognizing one duplex strand to another (Figure 4.15). In this work, the hybrid oligomer contained two 9-mer PyPu tracts separated by two interrupting base pairs. According to the affinity-cleavage data, a half-long 9-mer was not able to bind stably to its duplex target. As the number of Hoogsteen bonds in the hybrid oligomer was

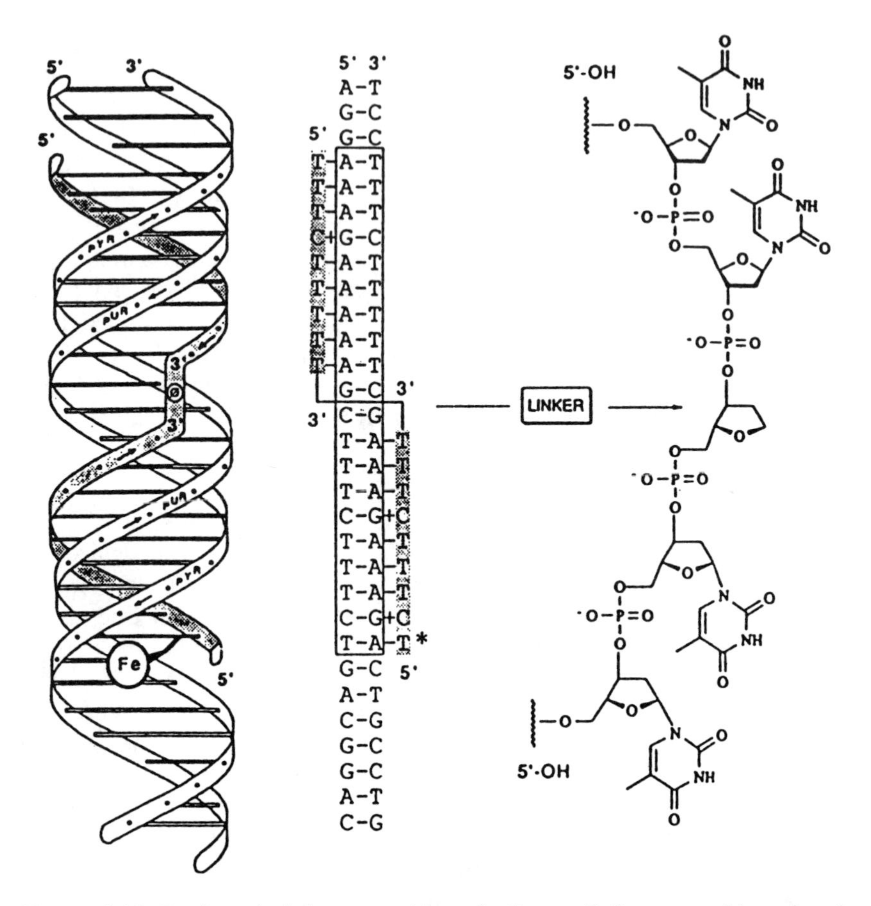

FIGURE 4.15. Design of triplex recognition of adjacent PyPu tracts with purines in different strands. A third pyrimidine strand is composed of two oligomers covalently coupled head-to-head or tail-to-tail by a relatively flexible linker. Shown is a triplex recognition with the third strand oligopyrimidines coupled $3'-3'$ via an abasic deoxyribose analog ϕ. Using the affinity cleavage technique, this hybrid third strand was shown to form a triplex, whereas neither of its halves did. Reprinted with permission from Horne and Dervan (1990). © 1990 American Chemical Society.

twice as large, its specific binding to the target with the double-strand cleavage was observed.

The nature of internucleotide linkage has received considerable attention in the study of alternate-strand triplexes. Ono et al. (1991a) generated three-dimensional A conformations of triplexes by computer graphics. From these computer models, the minimum possible distances between the $5'$ ends of the third strand at the $5'-5'$ junction are much longer than those between the $3'$ ends at the $3'-3'$ junction. The mutual orientations of triads near the junction point are such that the two hydroxyl groups of the 5′ ends to be linked are pointing in opposite directions (an estimated distance of 17–23 Å). At the 3′–3′ linking site, the two 3′ ends are closest to each other (5–8 Å) when the binding third strand skips two base pairs of the duplex between the 3′ ends similar to the model of Horne and Dervan (1990). Therefore, it was suggested to use the two phosphates and one 1,3 propane-diol in a spacer at the 3′–3′ linkage (P2) and four phosphates and three 1,3 propane-diols in a spacer at the 5′–5′ linkage (P4) (Figure 4.16) (Ono et al., 1991a). These authors indeed observed the binding of a designed hybrid oligopyrimidine to the target duplexes as judged from UV-melting experiments. Probably, the choice of the two phosphates and one 1,3 propane-diol spacer was not optimal. First, it was proposed on the basis of a triplex model in the A conformation, which is hardly correct. Second, the melting temperature of the hybrid triplex in 0.75 *M* NaCl is too far from that for the duplex (one should expect close thermal transitions for the duplex and triplex in such a high salt concentration (Blake et al., 1967; Durand et al., 1992a). Third, the data of Horne and Dervan (1990) show that under otherwise equivalent conditions, 1,3 propane-diol linkage results in less stable hybrid triplexes than abasic site linkage.

One more type of linkage was proposed by Froehler et al. (1992a). They used a xylose dinucleoside linker which minimizes the entropy of

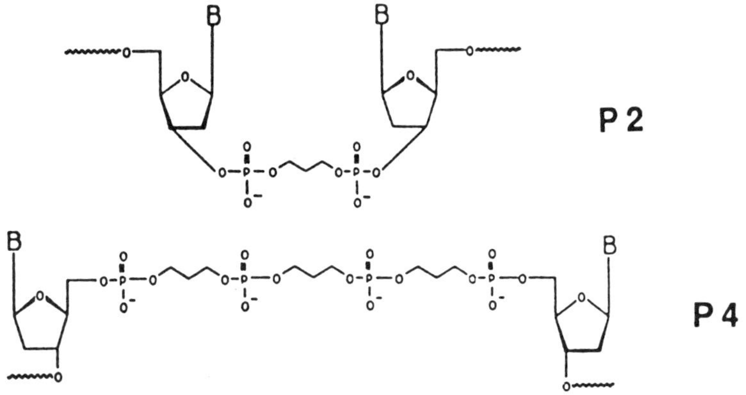

FIGURE 4.16. Structures of linker groups designed to couple oligopyrimidines oriented in opposite directions while hybridizing with sequential purine tracts in different strands. On the basis of computer modeling, a linker with two phosphate groups and one propanediol (P2) was suggested for 3′–3′ coupling, and a linker with four phosphate groups and three propanediols (P4) for 5′–5′ coupling. The formation of triplexes for such hybrid third strands was confirmed by UV-melting experiments. Reprinted with permission from Ono et al. (1991a). © 1991 American Chemical Society.

oligomer binding due to junction rigidity. The 3′–3′ linked pyrimidine oligonucleotides with nine nucleotides on each side of the linker had a higher melting temperature than a 21-mer that was bound to a single polypurine tract. Although this finding points to the feasibility of a stable hybrid triplex, the reason for its excessive stability is not clear. It well may be that the xylose linker "is superior to the phosphate diester linkage of a 3′–5′ oligodeoxynucleotide for triple-helix binding in terms of steric and enthropic considerations" (Froehler et al., 1992a). However, the hybrid triplex should be destabilized by the lack of stacking interaction between the bases at the linkage point as compared with the single-tract triplex, which is also stabilized by three extra triads.

Recently, a new type of oligonucleotide linkage for the 3′–3′ junction was proposed (Asselin and Thuong, 1993). Two pyrimidine oligomers are tethered via a tetramethylene linker, bridging the 5 positions of 5-methylcytosines. The hybridizing properties of such alternate strands are still awaiting a detailed investigation.

Further progress in studying recognition schemes was based on using natural 3′–5′ bonded oligonucleotides made up of both purine and pyrimidine blocks and alternate-strand triplex formation. The feasibility of such triplexes was first shown for oligonucleotides which can fold twice on themselves (Haner and Dervan, 1990; Jayasena and Johnston, 1992a). Such intramolecular triplexes are more stable than those composed of separate strands (Chen, 1991; Durand et al., 1992a).

The alternate-strand triplexes between the duplex DNA fragments and appropriate oligonucleotides, which might be relevant to biotechnological applications, were studied in more detail (Sun et al., 1991a; Beal and Dervan, 1992b; Jayasena and Johnston, 1992b, 1993; Olivas and Maher, 1994b; Washbrook and Fox, 1994a,b). The order of consecutive Pu and Py blocks in targeted sequences imposes some limitations. The triplex formation at a 5′-$(Py)_m(Pu)_n$-3′ motif is not equivalent to that at a 5′-$(Pu)_m(Py)_n$-3′ motif, presumably due to unequal crossover distances between the adjacent binding sites in two types of triplex-forming sequences (Beal and Dervan, 1992b; Jayasena and Johnston, 1992b; Washbrook and Fox, 1994a). Structurally, the triplex at the 5′-$(Pu)_m(Py)_n$-3′ target is similar to the above-discussed hybrid triplex with a 3′–3′ junction, whereas the triplex at 5′-$(Py)_m(Pu)_n$-3′ target is similar to the triplex with a 5′–5′ junction. Figure 4.17 shows the schematic of third-strand oligonucleotide targeting at adjacent PyPu tracts. It is clear that for the sequence 5′-$(Pu)_m(Py)_n$-3′, the site of strand switching is located where abutting purine strands in the duplex overlap (left panel). Exept for a phosphodiester, no special linker is required to make a crossover between strands. Probably, the two bases at the junction do not form specific triads and are expelled into the major groove. This provides the third-strand backbone with more freedom to adjust its conformation so as to maintain helical continuity (Beal and Dervan, 1992b). For the sequence

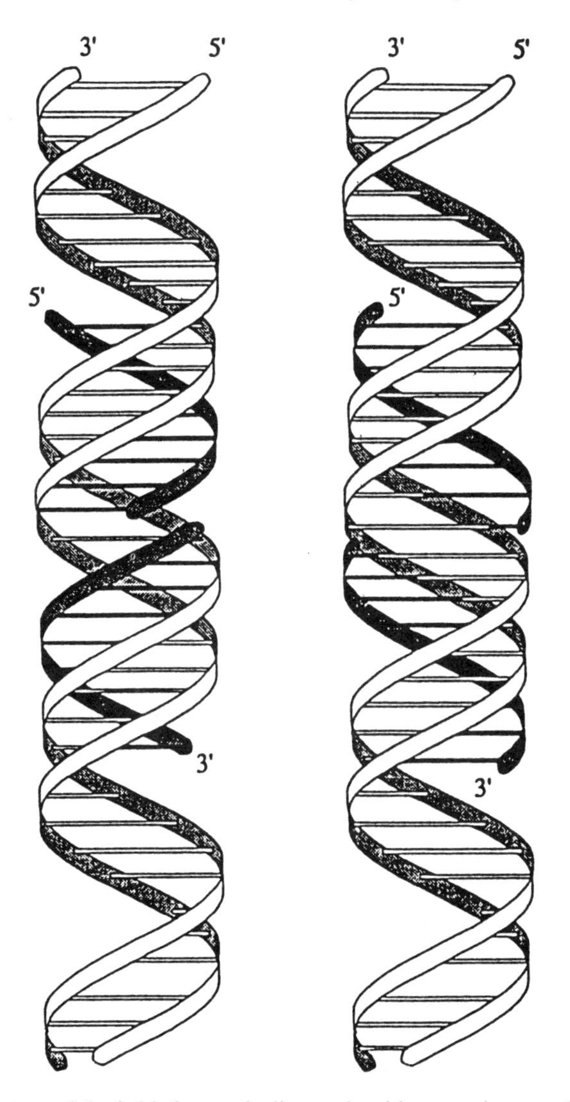

FIGURE 4.17. A model of third-strand oligonucleotide targeting at adjacent PyPu tracts. For the sequence 5′-$(Pu)_n(Py)_m$-3′, the site of strand switching is located where the abutting purine strands in duplex overlap (left panel). No special linker, except for a phosphodiester, is required to make a crossover between strands. For the sequence 5′-$(Py)_n(Pu)_m$-3′, there is no overlap of adjacent binding sites (right panel). Reprinted with permission from Beal and Dervan (1992b). © 1992 American Chemical Society.

$5'$-$(Py)_m(Pu)_n$-$3'$, there is no overlap of adjacent binding sites. This necessitates some kind of a linker, which was determined to be two nucleoside residues (Beal and Dervan, 1992b).

It is possible that the formation of triplexes by natural oligonucleotides at the $(Py)_m(Pu)_n$ targets is sequence-dependent. Jayasena and Johnston (1992b) described the formation of alternate-strand triplexes by oligonucleotides without any special spacers between the pyrimidine and purine parts. Triplex formation at a $5'$-$(Py)_m(Pu)_n$-$3'$ target occurs with a lower yield than that at a $5'$-$(Pu)_m(Py)_n$-$3'$ target. These authors found further limitations in alternate-strand recognition by oligonucleotides with only phosphodiester linkages (Jayasena adn Johnston, 1993). The triplex forms readily a the $5'$-A_8C_8-$3'$ sequence. However, it is very weak or does not even exist at the $5'$-C_8A_8-$3'$ sequence, in spite of the opportunity of third-strand sequence slippage to form the necessary spacer for a strand switch. Perhaps the poor ability of the $5'$-C_8A_8-$3'$ sequence to form a triplex is connected with the unusual B′ form structure of the 3′-end adenine tract (Alexeev et al., 1987; Nelson et al., 1987; Crothers et al., 1990; Hagerman, 1990).

Sequence effects in the alternate strand recognition were studied using substitution of adenines by thymines in part of the third strand (Washbrook and Fox, 1994b). Within the hybrid triplex with adjacent PyPuPy and PyPuPu helices, the relative stability of the reverse Hoogsteen triads TA*A and TA*T depends on the order of purine and pyrimidine blocks in the sequence and on the type of divalent cation (Mg^{2+} or Mn^{2+}).

In contrast to artificial DNA sequences optimized for alternate-strand recognition, a substantial enhancement of affinity was not observed using this strategy for a natural target sequence of the human p53 gene containing three adjacent purine-rich tracts on opposite DNA strands (Olivas and Maher, 1994b). Compared with the oligomer targeted to a single 12-base-pair tract, only a modest (1.3- to 1.4-fold) increase in affinity was found for an oligonucleotide that simultaneously recognizes all three purine tracts. Different variations in the junctions between oligonucleotide parts targeted to consecutive purine tracts did not result in a significant increase in affinity.

Thus, it is clear that in many genetic and biotechnological applications, intermolecular triplexes can be formed in relatively short adjacent PyPu tracts using an alternate-strand recognition strategy. The accumulated data provide certain guidelines for design of the third-strand oligonucleotides.

Modified Oligomer Backbones

Backbone modification is one of the developing approaches to increase the metabolic stability of oligonucleotides and improve their complex-forming properties in antisense and triplex-based antigene methodologies.

A number of phosphorus-modified internucleotide linkages have been designed as substituents of phosphodiester bonds (see Uhlmann and Peyman 1990; Nielsen, 1991; Crooke, 1992; Marshall and Caruthers, 1993; Milligan et al., 1993; Varma, 1993; Wagner et al., 1993; for reviews). Perhaps the most extensively studied modifications are the methylphosphonate and phosphorothioate internucleotide linkages (Figure 4.18). Modification of pentose (2′-methyl-ribose) and replacement of the entire deoxyribose–phosphate backbone by a structurally homomorphous polyamide (peptide) backbone add to the series of oligonucleotide analogs that were tested for triplex formation. All of these modifications generate certain desirable oligomer characteristics, such as resistance toward nucleases and retention of the ability to form stable hydrogen bonds with natural DNA or RNA.

Modifications in the Sugar–Phosphate Backbone

The lack of repulsion between negative phosphate groups of natural DNA molecules and nonionic methylphosphonate oligonucleotide analogs was expected to stabilize the duplexes and triplexes. However, the data obtained suggest more complex properties of methylphosphonate-containing multistranded structures. Duplexes of dT_{19} or dU_{19} with dA^*_{19} (* denotes a methylphosphonate analog) have increased melting temperatures (T_m), while the duplexes of dA_{19} with either dT^*_{19} or dU^*_{19}

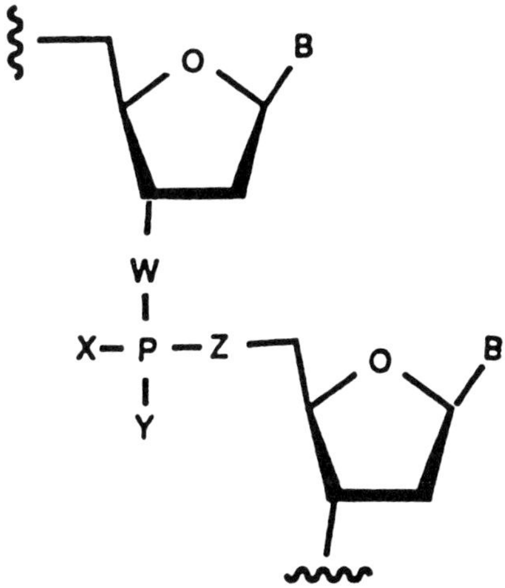

FIGURE 4.18. Phosphorus-modified derivatives of the DNA internucleotide linkage. Phosphodiester (W, X, Y, and Z = O); methylphosphonate (W, X, and Z = O, and Y = CH_3); phosphorothioate (W, X, and Z = O, and Y = S); phosphoramidate (W, X, and Z = O, and Y = NR_2, where R is alkyl or aryl substituent). Reprinted with permission from Marshall & Caruthers (1993) *Science* 259, 1564–1570. © 1993 AAAS.

have reduced T_m values. It appears that reduced phosphate charge repulsion is offset by unfavorable steric and other substituent effects (Kibler-Herzog et al., 1990). No triplex was found when a methylphosphonate strand replaced any phosphodiester-linked strand. More recent data show that 1:1 and 1:2 complexes of oligoadenines and oligothymines can be stabilized by designing alternating phosphodiester and methylphosphonate linkages in one or both oligomers (Kibler-Herzog et al., 1993). The weakness of such modified triplexes is illustrated by the finding that the pyrimidine methylphosphonate oligonucleotide analog required a 10-fold greater concentration (compared to that of normal phosphodiester oligonucleotide) to protect duplex DNA from DNase I digestion due to triplex formation (Matteucci et al., 1991). The above data are in conflict with the findings of Callahan et al. (1991), who reported that a PyPuPy triplex involving the completely methylphosphonated analog $d(C^*T^*)_8$ and $d(AG)_8$ was formed as efficiently as a triplex composed of normal strands. Compared to their natural prototypes, more stable triplexes $d(CT)_8 \cdot d(A^*G^*)_8 \cdot d(A^*G^*)_8$ and $d(T)_{16} \cdot d(A^*)_{16} \cdot d(A^*)_{16}$ were obtained (Ts'o et al., 1992). Molecular dynamics simulations predict that the oligonucleotide third strand with methylphosphonate backbone is more favorable to triplex formation than the normal phosphodiester backbone strand (Hausheer et al., 1990).

The stability of complexes with phosphorothioate analogs has not been yet understood. The stability of duplexes depends on the position of phosphorothioate linkages relative to purines or pyrimidines. Triplex formation is also dependent on the presence of a phosphorothioate group in a purine or pyrimidine strand: poly[d($G^{\#}A^{\#}$)] ($^{\#}$ denotes phosphorothioate analog) and poly[d(TC)] spontaneously dismutate into a triplex at neutral pH, whereas triplex formation between poly[d($T^{\#}C^{\#}$)] and poly[d(GA)] is inhibited (Latimer et al., 1989). The 34-mer duplex containing 15 central adenines in one strand and 15 complementary thymines in the other strand yielded no detectable triple helix upon combination with $dT_{15}(S_{14})$ with all phosphorothioate interbase linkages (S_{14}). But 34-mer duplex formed triple helices with different stereoisomers of $dT_{15}(S_1)$, with only one phosphorothioate interbase linkage (Kim et al., 1992).

A complication with these two types of analogs arises because the phosphorus center is rendered chiral by phosphorothioate or methylphosphonate substitutions. For a deoxyoligonucleotide with n singly modified internucleotide linkage, there are 2^n stereoisomers with variable binding properties.

New developments in stereoselective chemistry could hopefully overcome the complexities in triplex-targeting these analogs to duplex DNA. Yet there are many problems to understand. For example, Hacia et al. (1994b) enzymatically produced pure (Rp)-phosphorothioate oligonucleotides and studied their triplex-forming properties. Purine-rich oligonucleotides containing a diastereomeric mixture of phosphorothioate linkages or stereoregular phosphorothioate linkages formed triple helices with affinities similar to those of the corresponding natural phosphodiester oligomers, whereas diastereomeric or regular pyrimidine phosphorothioate oligonucleotides did not have measurable affinity to double-stranded DNA (Hacia et al., 1994b). These results were explained by the relatively small difference between phosphodiester and phosphorothioate backbones in the purine strand and a corresponding significant difference for pyrimidine oligonucleotides (Hacia et al., 1994b).

The formation of achiral internucleotide bonds does not have difficulties intrinsic to stereoselective synthesis. Matteucci et al. (1991) compared the triplex-binding properties of oligonucleotides with natural phosphodiesters, chiral R,S-methylphosphonate and R,S-methoxyethylamidate, and achiral formacetal and 5′-thioformacetal internucleotide linkages (Figure 4.19). All modified oligomers contained four unusual bonds each. The formacetal analog was bound to the 14-nucleotide target as efficiently as the phosphodiester oligomer. The affinity of the methylphosphonate analog was an order of magnitude lower. The methoxyethylamidate and 5′-thioformacetal analogs had the lowest affinities.

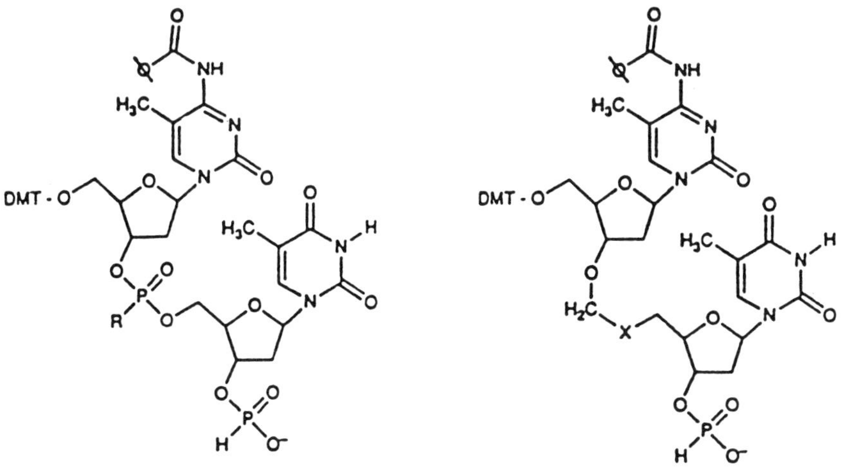

FIGURE 4.19. Dimer synthons containing various types of internucleotide linkages.
Left: R = O^-, phosphodiester
R = CH_3, methylphosphonate
R = $NH\text{-}CH_2\text{-}CH_2OCH_3$, methoxyethylamidate
Right: X = O, formacetal
X = S, 5′-thioformacetal.
Reprinted with permission from Jones et al. (1993a). © 1993 American Chemical Society.

Subsequent studies showed that achiral 3′-thioformacetal linkage can result in comparable binding properties of phosphodiester and modified backbone oligomers (Jones et al., 1993a). However, an increasing number of formacetal linkages reduced the triplex-binding affinity. The difference in binding properties of oligomers can be understood in terms of subtle differences between the length of the backbone linkage, the conformations of deoxyribose rings, and the rigidity of the backbone. The 3′-thioformacetal bond is approximately 0.5 Å longer than the formacetal bond and comparable with the natural phosphodiester bond. At a low number of formacetal bonds, the shortening of the oligomer is not significant, and it has high affinity to the duplex target. A further increase in the number of formacetal bonds results in a less perfect fit of the duplex and the third-strand oligomer, with a decrease in affinity. Poor binding of the 5′-thioformacetal analog results from unfavorable interactions of sulfur in the 5′ position and either ribosyl oxygen or H6 of the pyrimidine bases. Thus, of the analogs studied, the 3′-thioformacetal analog is the most promising, as it has the highest binding affinity.

In order to restrict the conformational freedom of unbound oligonucleotide to resemble its DNA-bound conformation, Jones et al. (1993b) replaced the phosphodiester internucleotide linkage with a structurally more rigid "riboacetal" linkage (Figure 4.20). The thermal denaturation studies showed that compared with the phosphodiester oligomers, the temperature of modified third-strand dissociation increases at a rate of approximately 4°C per riboacetal substitution. DNase I digestion analysis showed that the riboacetal 15-mer analog protected the duplex DNA at a 100-fold lower concentration as compared with its phosphodiester prototype. This significantly increased affinity does not deteriorate the selectivity of binding. Unspecific binding of the riboacetal analog to the

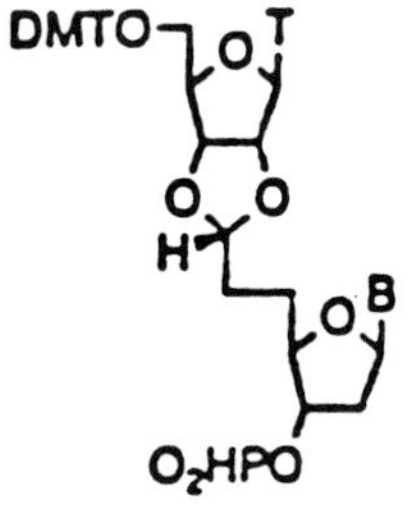

FIGURE 4.20. "Riboacetal" internucleotide linkage, which restricts the conformational freedom of the third-strand oligonucleotide analog. The oligonucleotide backbone seems to be restricted to the conformational range close to that required for triplex formation, since every riboacetal substitution increases the triplex melting temperature by 4°C. Reprinted with permission from Jones et al. (1993b). © 1993 American Chemical Society.

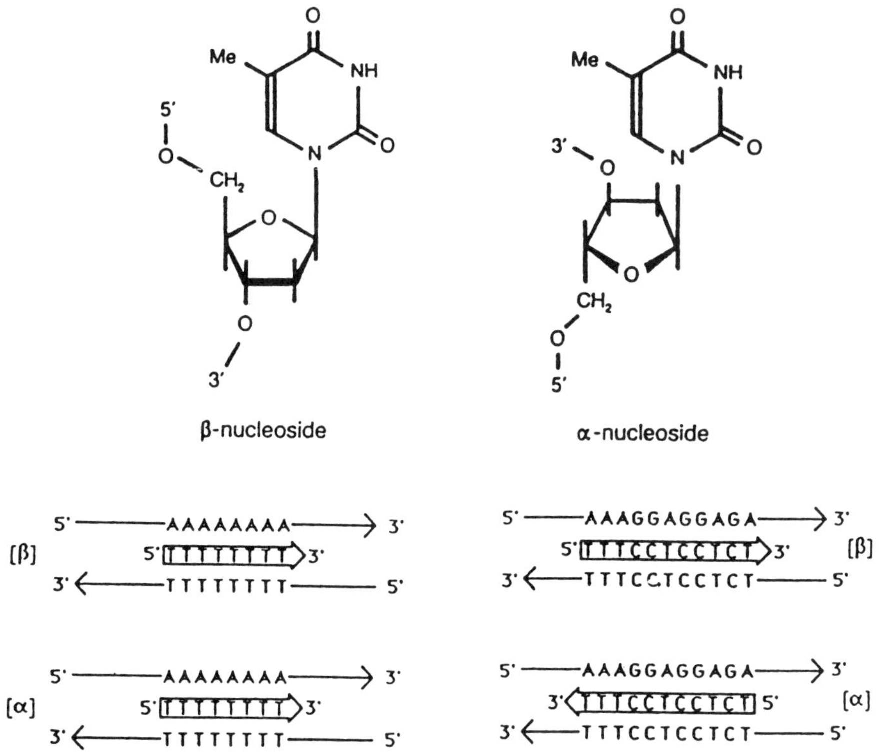

FIGURE 4.21. Structures of α-nucleoside and natural β-nucleoside. The orientation of pyrimidine α and β-oligomers in the triple helix is sequence-dependent. Oligo(dT) binds parallel to the purine strand irrespective of whether it is synthesized with α- or β-anomers. In contrast, both cytosine- and thymine-containing oligomers bind parallel (β) or antiparallel (α) to the purine strand. From Hélène (1991). Reprinted with permission.

single-base mismatched duplex was found only at a 100-fold higher concentration than that required to bind to the exact target.

The triplex-forming properties of pyrimidine oligonucleotides with the α-anomers of nucleotides in place of natural β-anomers (Figure 4.21) have been tested (Le Doan et al., 1987; Praseuth et al., 1988a,b; Sun et al., 1991b; Debart et al., 1992). The base relative to the deoxyribose is positioned on the 3′-OH side in the α-anomeric configuration and on the 5′-OH side in the β-anomers. Due to the change in the anomeric configuration, the third-strand oligo-α-deoxynucleotides are expected to adopt the opposite orientation to the natural anomers. However, oligo-α-T binds the duplex in the same orientation as the corresponding oligo-β-T (Le Doan et al., 1987). This implies Hoogsteen hydrogen bonds for TA*T triads in the β-anomers and reverse Hoogsteen bonds for α-anomers. When cytosines are introduced into the oligo-α-deoxynucleotide, reverse Hoogsteen bonds would result in nonisomorphous TA*T and CG*C$^+$ triads, therefore, triads with Hoogsteen bonds are preferable, which means that the pyrimidine third strand is antiparallel to purine and parallel to pyrimidine strands in the duplex (Sun et al., 1991b; Sun and Lavery, 1992). The formation of a triplex involving two α-RNA and one β-DNA strands was demonstrated (Debart et al., 1992).

An unusual triple-stranded complex can be formed between the duplex DNA and a peptide nucleic acid (Figure 3.23) (Egholm et al., 1992a,b). The 2:1 complex of PNA and DNA forms when two homopyrimidine PNA strands bind to the duplex and displace one DNA strand (Nielsen et al., 1991; Egholm et al., 1992a,b; Cherny et al., 1993b). Such complexes have unprecedentedly high stability (Egholm et al., 1992a,b), they are right-handed, and their base conformation is very similar to that of the conventional DNA triplex composed of analogous bases (Kim et al., 1993). In addition to the formation of stable triplexes, PNA is very promising in genetic and biotechnological application because of its high metabolic stability. Almost no degradation was detected during PNA incubation in several metabolic enzyme-rich media such as nuclear extracts, sera, etc. (Demidov et al., 1994b).

Conjugated Oligonucleotides

A duplex target site may consist of a series of isolated PyPu domains linked by spacer DNA elements with a random sequence inappropriate for triplex formation. For example, in the gene of the herpes simplex virus D glycoprotein, a pair of 12-nucleotide-long PyPu sites are separated by a 10-nucleotide spacer. Thus, two triplexes are positioned on the same face of the duplex and are separated by one turn of the helix. Kessler et al. (1993) tested a series of hybrid oligomers that had two triplex-forming oligonucleotides linked by simple flexible spacers. According to the band shift and DNase I footprinted assays, such hybrids bind to both target sites simultaneously with an affinity significantly higher than either of the separate parts of the hybrid, provided that the length of flexible spacers is longer than 20–25 rotatable bonds. Two models for short and long polymeric linkers are conceivable: one in which the short linker element stretches across the minor groove parallel to the helix axis, and one in which the long linker element (50–60 bonds) continues in the major groove path, making a full helix turn.

Attachment of intercalating molecules to the oligonucleotides, or chemical or photochemical cross-linking of oligonucleotides, stabilizes triple-helical complexes. The triplex–duplex junction on the 5′ side of the third strand has been shown to be a strong intercalation site for several drugs (Collier et al., 1991a; Sun et al., 1991c). An acridine moiety coupled to the 5′ end of a homopyrimidine oligonucleotide strongly stabilized the triplex upon intercalation into the duplex (Sun et al., 1989) or upon stacking interaction at the end of a short model duplex (Stonehouse and Fox, 1994). Triple-stranded structures were also significantly stabilized by an intercalating benzo[e]pyridoindole derivative (Mergny et al., 1992). The use of oligonucleotide–intercalator conjugates allows one to use short triplex-forming oligomers. For example, a 7-mer purine oligomer–oxazolopyridocarbazolo conjugate formed a triplex with the U3 long terminal repeat sequence of HIV-1, which was thermostable up to 40°C (Mouscadet et al., 1994b). Stabilization of the triplex structure might be accomplished through covalent bonding of the third strand to its target duplex. For example, a psoralen–oligonucleotide conjugate under long-wavelength UV irradiation is cross-linked to the target PyPu site. The hybrids of oligonucleotides with intercalating or cross-linking terminal moieties are promising for prolongation of a transcriptional repression by triplex-forming oligonucleotides (see Chapter 7).

Oligonucleotide conjugates with cholesterol were proposed to facilitate their transport into the cell (Letsinger et al., 1989). A bulky cholesteryl group also serves as a protecting group from exonuclease digestion in biological media. The studies of duplex- and triplex-forming properties showed that the presence of a cholesteryl group generally increases the stability of model oligonucleotide duplexes and triplexes (Gryaznov and Lloyd, 1993). In model oligonucleotide triplexes (Figure 4.22), a single terminal cholesteryl group stabilizes the triplex structure, probably due to an additional stacking interaction with the terminal base or base pair. In the case of hydrophobic intercholesteryl contacts via proper spatial arrangement, stabilization becomes more significant. For example, coupling cholesteryl groups to the ends of oligonucleotides significantly stabilized "clamp" triplexes (Figure 4.22). They consist of a single-stranded RNA or DNA and a hairpin-folded oligonucleotide (Giovannangéli et al., 1991; Brossalina and Toulmé, 1993; Brossalina et al., 1993; François and Hélène, 1995) and are prospective agents in antisense strategies (see Chapter 7). Synthesis of oligonucleoticles conjugated to cholesterol via a triglycyl linker has been recently described (Vu et al., 1994).

Another kind of terminal oligonucleotide modification was proposed by Tung et al. (1993). Based on the stabilization of triplex structures by soluble polyamines, they suggested linking covalently various polyamines that could bind to DNA in the same major groove as the main nucleotide oligomer and thereby stabilize the triplex. The UV- and CD-melting data showed an enhanced thermal stability of such modified triplexes. Moreover, the protonated PyPuPy triplex did not require free polyamine or divalent cations to be formed.

For many modified triplexes, a similar question arises. The terminal stabilizing moieties can provide excessive stabilization, which can result in significant oligomer binding to a mismatched, nontarget DNA. Therefore,

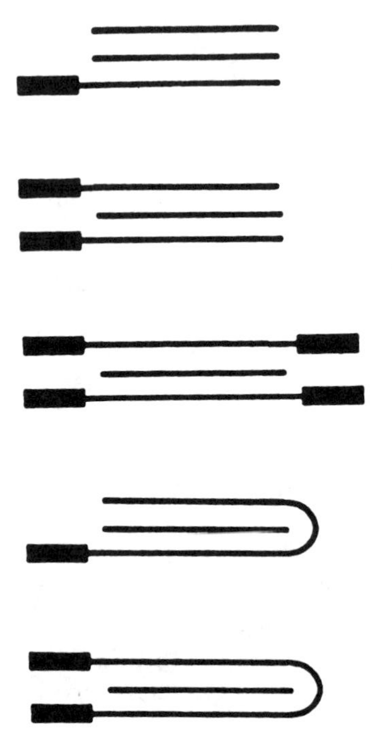

FIGURE 4.22. A few examples of triple-stranded complexes stabilized by cholesterol conjugation. ■ denotes cholesterol moiety. Hydrophobic interaction of the cholesteryl moiety with nucleobases or another cholesteryl is a likely source of triplex stabilization. Reproduced from Gryaznov and Lloyd (1993) *Nucleic Acids Res* by permission of Oxford University Press.

the specificity of triplex formation must be carefully examined. In the case of cholesterol-conjugated oligonucleotides, the triplex is destabilized by 11°C for one mismatch and is not formed in the presence of two mismatches in a 16-nucleotide-long target. A similar analysis for polyamine-conjugated third strands is in progress.

The data accumulated show that some modification in the thirdstrand oligomers can enhance the thermodynamic and metabolic stability of triplexes. However, a number of problems await resolution. Among them are insufficient knowledge of the structural compatibility of the modified third strands and target duplexes, difficulties in stereoselective synthesis of potentially racemized oligomers, the insufficiently studied relation between the enhancement in stability and the loss in selectively of triplexes, etc. However, the rapidly growing experimental and computational data provide optimistic hope for the feasibility of forming triplexes with predictable properties.

Protein–DNA Interactions and Triplex Formation

Protein–DNA interactions at the sites of potential triplex formation have several important aspects.

First, interactions of specific proteins with their recognition sites on DNA can be inhibited by third-strand binding. Such triplex-mediated prevention of restriction nuclease action (François et al., 1989c; Maher et al., 1989; Hanvey et al., 1990; Q. Wang et al., 1994) is used for temporary restriction site protection and subsequent rare DNA cleavage in a variant of the "Achilles heel" methodology of rare chromosomal DNA cutting (Strobel and Dervan, 1991, 1992; see Chapter 7 for more detail). Binding and further progression along the template of DNA polymerase (Hacia et al., 1994a) and of RNA polymerase and transcription factors is inhibited by triplex-forming oligonucleotides which are used for the modulation of gene expression (see Table 7.1 for references). It was shown, however, that the change in DNA conformation caused by triplex formation is highly localized and does not interfere with the formation of a complex between the origin of replication and its binding protein at an immediately adjacent site (Huang et al., 1992). Therefore, inhibition of protein binding is expected to be effective only when the third-strand target site overlaps the protein-binding site.

Second, the structural aspect of protein binding near the triplex-forming site is important in considering the biological role of triplexes (see Chapter 6). This is relevant, in particular, to the suggested triplex assistance in protein–DNA interactions (including promoter and enhancer regions, and RNA polymerase and possible DNA looping proteins) (Lee et al., 1984; Kinniburgh, 1989; Pestov et al., 1991). Triplex formation is also important in DNA replication (Lapidot et al., 1989; Baran et al., 1991; Dayn et al., 1992b). A number of proteins that bind to homopurine or homopyrimidine single strands have been described (Biggin and Tjian, 1988; Davis et al., 1989; Gilmour et al., 1989; Hoffman et al., 1990; O'Neill et al., 1991; Yee et al., 1991; Kolluri et al., 1992; Muraiso et al., 1992; Aharoni et al., 1993). They could potentially stabilize the H DNA in the cell by binding the unpaired strand (Lee et al., 1984). Unfortunately, nothing is known about the interactions of these PyPu-specific proteins with the looped part of H DNA.

Interactions of other single-strand-binding (SSB) proteins (e.g., from *E. coli*) with locally unwound sites in superhelical DNA result not only in the stabilization of such unwound regions, but also in a further increase in the size of the denaturation bubble with concomitant supercoil relaxation (Glikin et al., 1983; Langowski et al., 1985). *E. coli* SSB was shown to remove single-strand folds or other secondary structures from DNA, facilitating recombination (Muniyappa et al., 1984) and transcription (Baran et al., 1991). Interaction with the looped part of the triplex in supercoiled plasmid DNA results in extended unpairing in the H DNA-forming region (Klysik and Shimizu, 1993). Hence, it well may be that the homopurine or homopyrimidine single-strand-binding proteins destabilize rather than stabilize triplexes. The biological significance of such protein–DNA interactions will be discussed in Chapter 6.

Finally, a protein that preferentially binds to a triple-stranded DNA structure has been found (Kiyama and Camerini-Otero, 1991). However, its affinity for triplexes was only two- to fivefold greater than for duplex DNA, and its biological role remains to be elucidated.

Triplex DNA–Drug Interactions

Triplex formation can change the binding properties of various DNA-interacting ligands, including heterocyclic compounds. The substances in question can be divided into two major groups: groove-binders and intercalators. Interactions of both ligand groups with the triplex DNA have dual importance. First, some of them are antibiotics (e.g., actinomycin D, duocarmycin A), and the formation of a triple-stranded structure can significantly change their interactions with the DNA target. Second, a number of such compounds are known to stabilize the helical structure of DNA upon binding. There have been a few studies of triplex–drug interactions.

Groove-Binders

As we saw earlier in this chapter, the attachment of cholesterol or polyamines to the end of triplex-forming oligonucleotides increases the number of stabilizing contacts between the duplex target and the third strand. The plausible mechanism of triplex stabilization by cholesterol consists in its interaction with hydrophobic surfaces of accessible nucleobases (Gryaznov and Lloyd, 1993). The underlying basis for triplex stabilization by polyamines may be in the electrostatic interactions of positively charged amino and imino groups with negatively charged phosphates of DNA and in hydrogen bonding with phosphate oxygens and corresponding groups of bases (Suwalsky et al., 1969; Feuerstein et al., 1986; Jain et al., 1989; Schmid and Behr, 1991). No matter how the DNA target interacts with cholesterol and polyamines conjugated to third-strand oligonucleotides, this results in overall triplex stabilization. Cholesterol and polyamines have no sequence specificity against DNA, and we described them separately. Unlike the unspecific hydrophobic and electrostatic interactions of cholesterol and polyamines, a number of agents interact with nucleic acids more specifically.

Among the other (specific) externally binding agents (Figure 4.23) whose interactions with triplex nucleic acids have been studied are netropsin, Hoechst 33258, berenil, 4′,6-diamidino-2-phenylindole (DAPI), distamycin A and its analog, mitramycin, chromomycin A_3, duocarmycin A, and methyl green.

A minor groove-specific drug, netropsin (Figure 4.23), binds to the PyPuPy DNA triplex containing TA*T triads without displacing the

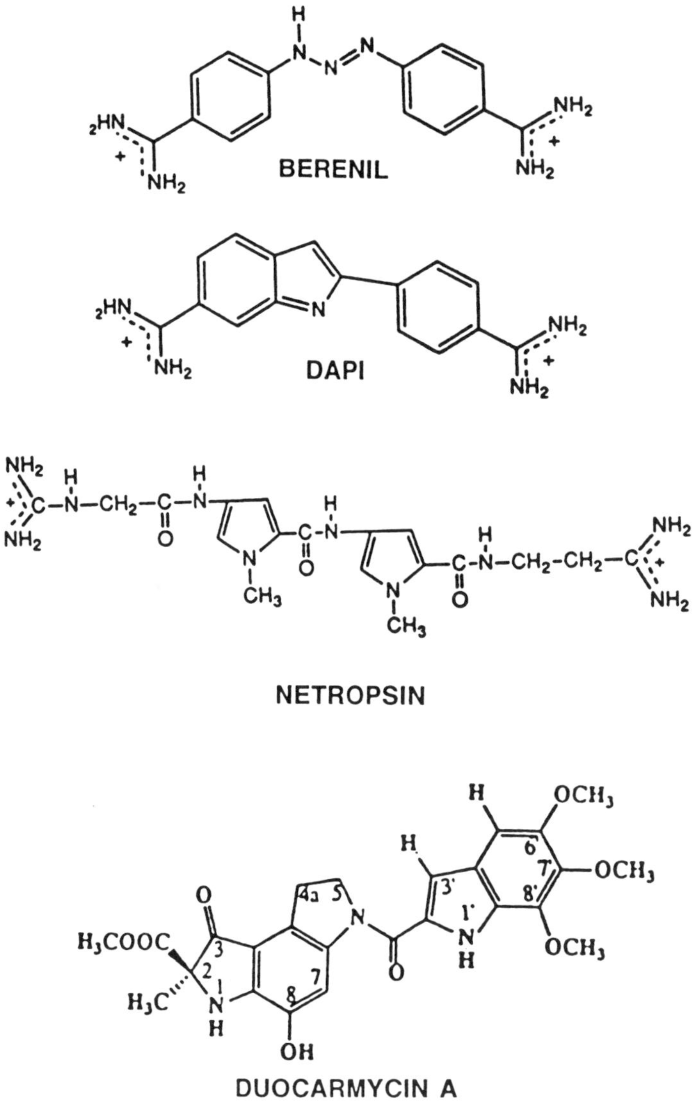

BERENIL

DAPI

NETROPSIN

DUOCARMYCIN A

FIGURE 4.23. Structures of some of the groove-binding drugs that were tested for their interaction with the triple-helical nucleic acids.

major groove-bound third strand (Durand et al., 1992b; Park and Breslauer, 1992; Wilson et al., 1993). The affinity of netropsin for the triplex was found to be lower than that for the duplex. Thermal denaturation experiments show that netropsin destabilizes the DNA triplex but stabilizes the duplex. Netropsin binding to the DNA triplex also decreases the cooperativity of the triplex–duplex transition. Similar to netropsin, but to a somewhat lesser extent, another minor-groove binder, Hoechst 33258, also destabilizes the DNA triple helix (Durand et al., 1994a). The interaction of berenil with triplex DNA depends on ionic strength. It stabilized the triplex form of $(dA)_{12}$-x-$(dT)_{12}$-x-$(dT)_{12}$ in the absence of NaCl and slightly destabilized the triplex at 1 *M* NaCl concentration (Durand et al., 1994b).

Unlike the above-described studies of netropsin interaction with preformed DNA triplexes, experiments with hybrid DNA–RNA triplexes, where netropsin was included at the stage of triplex formation, have shown that netropsin may induce the formation of triplexes that would otherwise not form (Pilch and Breslauer, 1994). The poly(dT) · poly(rA) · poly(dT) triplex induced by netropsin includes a ribonucleotide strand in the duplex, which is known to result in the least stable triplex among the DNA–RNA hybrids (Roberts and Crothers, 1992; Han and Dervan, 1993, 1994). The poly(dT) · poly(rA) · poly(dT) triplex was also induced by DAPI and berenil (Figure 4.23). The exact mechanism by which these compounds stimulate triplex formation has not been studied; yet it is worth elucidating, because one of the drugs, berenil, is such a strong triplex-promoter that it allows the formation of the poly(dT) · poly(rA) · poly(rA) triplex, which has never been observed before (Pilch and Breslauer, 1994).

Upon interaction with the PyPuPu triplexes, the minor-groove-binder chromomycin A_3 has a binding constant at least an order of magnitude lower than that for the duplex (Chen, 1991). This weak activity binding may be explained by the need to accommodate one of the carbohydrate side chains of chromomycin A_3 in the major groove, which is already occupied by the triplex third strand (Chen, 1991).

Distamycin A, which binds in the B form but not in the A form minor groove, forms a complex with the triple helix and partly destabilizes it, but does not displace the third strand (Howard et al., 1992). The distamycin analog, Dst 2, also destabilizes the triple helix (Umemoto et al., 1990). Its interactions with the host triplex alter the nature of the thermal transition from direct melting of $(dT)_6 \cdot (dA)_6 \cdot (dT)_6$ into single strands to two-step melting, when the transition from triplexes to single-stranded molecules proceeds through a double-stranded intermediate (Umemoto et al., 1990).

The DNA alkylating agent duocarmycin A (Figure 4.23), related to antibiotic CC-1065, covalently binds in the minor groove and affects the PyPuPy triplex–duplex equilibrium by reducing the transition midpoint to a lower pH (Lin and Patel, 1992).

A DNA major-groove-binding drug, methyl green, was found to be excluded from binding to the triple helical poly(dT) · poly(dA) · poly(dT) in which the major groove was occupied by the third strand (Kim and Norden, 1993).

Intercalators

The well-known intercalating agent ethidium bromide (EtBr) (Figure 4.24) was the first agent used in the study of triplex nucleic acids–intercalator interactions. The interactions studied were those of EtBr with poly(rU) · poly(rA) · poly(rU) and poly(rU) · poly(dA) · poly(rU) (Le Pecq and Paoletti, 1965; Le Pecq, 1971; Waring, 1974; Lehrman and Crothers, 1977), poly(dT) · poly(dA) · poly(dT) (Morgan et al., 1979; Scaria and Shafer, 1991), poly[d(TC)] · poly[d(AG)] · poly[d($TC^{\dagger}$)] (Lee et al., 1979, 1984), and mixed-sequence triplex DNA oligomers (Xodo et al., 1990; Callahan et al., 1991; Chen, 1991; Mergny et al., 1991a). There is still no clear understanding of EtBr interactions with triplex structures.

Upon interaction with polyadenylate and polyuridilate, EtBr destabilized a triple-stranded complex, promoting the dissociation of the third strand (Le Pecq and Paoletti, 1965; Le Pecq, 1971; Waring, 1974; Lehrman and Crothers, 1977). This destabilization could be relevant to the A-type conformation of complexes containing polyribonucleotides, since in other studies EtBr was shown to intercalate the triplex structures of polydeoxyribonucleotides, which are closer to the B-type conformation (Morgan et al., 1979; Mergny et al., 1991a; Scaria and Shafer, 1991).

In DNA triplexes, EtBr clearly intercalates between the TA*T triads. The EtBr binding site spans 2.8 base triads, compared to 2.4 base pairs (Scaria and Shafer, 1991). Binding constants, determined from absorbance measurements, indicate that EtBr interaction with the poly(dT) · poly(dA) · poly(dT) triplex is substantially stronger than with the duplex poly(dA) · poly(dT) (Scaria and Shafer, 1991). The latter duplex adopts an unusual B′ conformation (Alexeev et al., 1987; Nelson et al., 1987), and the difference in affinity could be due to a poor EtBr binding to the B′ form rather than to a very strong EtBr binding to the triplex (Sun and Hélène, 1993). Clearly, high cooperativity of EtBr binding to poly(dA) · poly(dT) as a result of conformational change induced by intercalation (Scaria and Shafer, 1991) supports this explanation. EtBr fluorescence, as an indicator of intercalation, increases upon binding to the poly(dA) · poly(dT) duplex and becomes even greater upon binding to the poly(dT) · poly(dA) · poly(dT) triplex (Morgan et al., 1979; Scaria and Shafer, 1991). There are two conceivable reasons. First, more EtBr molecules may bind to the triplex than to the duplex structure. Second, compared to the duplex, in the triplex structure EtBr may be less accessible to solvent and therefore to fluorescence quenchers, resulting in a higher fluorescence yield (Olmsted and Kearns, 1977). For this reason, triplex-bound EtBr can produce stronger fluorescence than a greater amount of duplex-bound EtBr. In a recent study, EtBr was shown to promote the formation of a poly(dT) · poly(rA) · poly(dT) triplex, supporting the idea of intercalative interaction with triple-helical nucleic acids (Pilch and Breslauer, 1994).

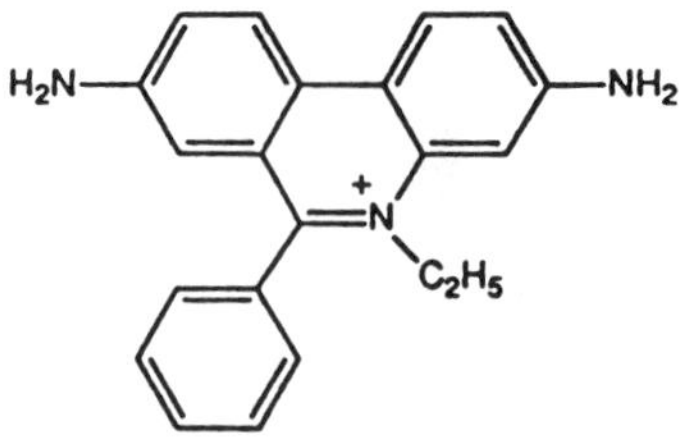

H_2N NH_2 N^+ C_2H_5

ETHIDIUM BROMIDE

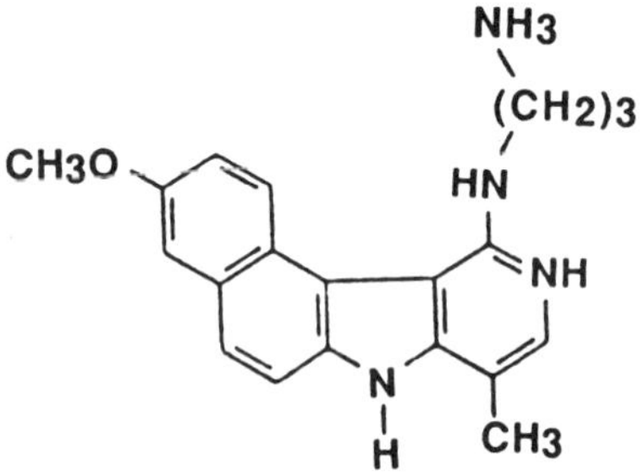

NH3 (CH2)3 HN NH CH3O N H CH3

BEPI

$NHCH_2CH_2\overset{+}{N}H(CH_3)_2$ N + H

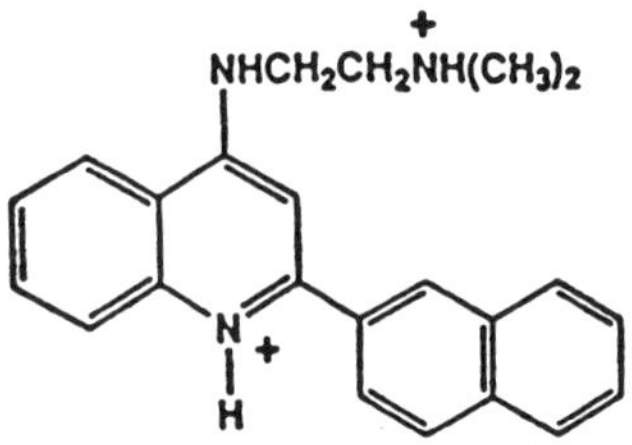

NAPHTHYLQUINOLINE

FIGURE 4.24. Some of the intercalating drugs employed in studies of triplex–drug interactions.

EtBr binds poorly to the CG*C^+-containing regions, presumably because of electrostatic repulsion between the cationic drug and the protonated cytosines (Lee et al., 1979). In the special case of TA*T triads clustered at the ends of the oligonucleotide complex, binding of EtBr to the mixed sequence PyPuPy triplex was only 40% lower than to the duplex (Mergny et al., 1991a). Since the contacts between EtBr and

CG^*C^+ triads are not favorable, and pairs of adjacent TA*T triads appropriate for EtBr intercalation do not often occur in triple helices, the number of binding sites in the mixed sequence triplex is expected to be lower than in the duplex. Therefore, the interaction of EtBr with the triplexes of mixed sequence should generally result in severalfold lower fluorescence enhancement than in the case of duplexes (Lee et al., 1979, 1984; Xodo et al., 1990; Callahan et al., 1991; Mergny et al., 1991a). This feature is used to detect triplex formation (Lee et al., 1979, 1984; Morgan et al., 1979; Xodo et al., 1990; Callahan et al., 1991). Another intercalative ligand, actinomycin D, also has lower affinity for the triplex than for the duplex (Chen, 1991), presumably due to increased rigidity of the triplex structure.

A family of drugs, the benzopyridoindoles, shows promise for stabilizing the triplexes (Mergny et al., 1992). Like EtBr, these agents are cationic, avoid the CG^*C^+ triads, and intercalate between the TA*T triads. Yet, the intercalation of one of these drugs, 3-methoxy-7H-8-methyl-11-[(3′-amino)propylamino]-benzo[e]pyrido[4,3-b]indole (BePI) (Figure 4.24), was shown to result in stronger stacking interactions in the triplex than in the duplex, which may account for its triplex-stabilizing effect (Mergny et al., 1992; Pilch et al., 1993; Duval-Valentin et al., 1995). A series of quinoline derivatives and propidium resulted in a general increase in the triplex melting temperature (Wilson et al., 1993, 1994). According to molecular modeling, one of the intercalators studied, naphthylquinoline (Figure 4.24), has very good stacking interactions with the bases at triplex intercalation sites. This favorable stacking allows the observation of a triple helix even with racemic methylphosphonate and phosphorothioate oligonucleotide analogs, by outweighing the unfavorable conformational distortions in the modified backbones which do not permit the formation of triplexes in the absence of stabilizers (Wilson et al., 1993, 1994). An anti-tumor benzo[a]phenazine derivative NC-182 was shown to intercalate into B-form DNA with no apparent sequence specificity (Tarui et al., 1994). The intercalation of NC-182 into a poly(dT) · poly(dA) · poly(dT) triplex stabilizes it against thermal denaturation. At a molar ratio of NC-182/nucleotide > 0.06, the melting curves of triplex and duplex structures coincide.

Several studies have shown that triplex–duplex junctions contain some kind of structural discontinuity in the underlying duplex. According to molecular modeling, this discontinuity is most likely a result of a bend in duplex DNA (Chomilier et al., 1992). Besides being recognized by structure-sensitive chemical probing agents (Hartman et al., 1992), it is the site of strong intercalation for several agents, such as acridine, ellipticine, BePI (Sun et al., 1989; Perrouault et al., 1990; Collier et al., 1991a; Ono et al., 1991a). This finding resulted in the development of two different applications: (1) ligand-mediated stabilization of the triplex-stranded complex, and (2) localized photomodification of a DNA target.

The intercalation binding mode of a ligand coupled to a triplex-forming oligonucleotide was applied to enhance triplex stability using acridine and ellipticine (Sun et al., 1989; Collier et al., 1991b). This stabilization is due to the increased number of stabilizing contacts between duplex DNA and the oligonucleotide–intercalator conjugate. It was shown that the highest affinity is obtained by intercalator attachment to the 5′ end of the triplex-forming oligonucleotide (Sun et al., 1989). Recent experiments show that coupling BePI, which is a triplex-stabilizer itself, to an oligonucleotide results in a much higher stabilization than from oligonucleotide–acridine (or ellipticine) conjugates, presumably due to BePI intercalation into a stretch of TA*T triads at the end of the triple helix (Sun and Hélène, 1993). An investigation of a joint influence of acridine (covalently bound to oligonucleotide) and a naphthylquinoline derivative (which intercalates the triplex) on triplex stability showed that the stabilizing contributions of two compounds are not additive (Cassidy et al., 1994). Perhaps, binding of one compound makes the structure of the DNA template less appropriate for subsequent binding of the other compound.

Localized intercalation of a photochemically active agent forms the basis for a site-specific photomodification of a DNA target (e.g., cross-link or strand break). In order to make the triple-stranded complex permanent, Hélène and coworkers used a strategy of photochemical cross-linking of an intercalated agent. They irradiated the target-bound psoralen–oligonucleotide conjugate by UV light to form covalent bonds between the intercalated psoralen moiety and adjacent pyrimidines (Takasugi et al., 1991; Giovannangéli et al., 1992b; Grigoriev et al., 1993a,b). In addition, a number of photoactive intercalators were tested as potential agents for site-specific strand breaks in DNA (artificial photoendonucleases) (Le Doan et al., 1987, 1990, 1991; Perrouault et al., 1990). The applications of cross-link methodology to permanent inhibition of gene expression and photocutting of the DNA target by intercalated agents coupled to triplex-forming oligonucleotides will be described in more detail in Chapter 7.

Thus, the data accumulated to date show that groove-binding agents may destabilize DNA triplex structures. However, some of them induce mixed DNA–RNA triplexes. Some of the intercalating agents have lower affinity to the triplex structure and also destabilize triplexes. However, the structures of intercalators can be designed so as to provide maximum stacking interactions with the bases at the intercalation site. In this way triplex-stabilizing intercalators can be selected. Their conjugation with the triplex-forming oligonucleotides can result in further enhancement of triplex stability. Photoactive intercalative agents can also be used for additional stabilization of the duplex–oligonucleotide complex via cross-linking, or for site-directed DNA cleavage.

Conclusion

In addition to the initially discovered TA*T, CG*C^{+}, CG*G, and TA*A triads, a number of less usual triads, including those with base analogs, have been found to form stable triplexes. The search for new triplex recognition schemes has two general motives. On the one hand, there is the need to extend plausible sequence motifs for intramolecular triplexes which are suggested to participate in the processes of transcription, replication, chromosome packing, etc. On the other hand, the versatility of triplex-mediated strategies in genome regulation, selective DNA purification, development of artificial endonucleases, etc. depends on the ability of triplex-forming oligonucleotides to recognize various types of sequences. A number of nonionic and nuclease-resistant oligonucleotide analogs have been designed which may have increased metabolic stability and improved triplex-forming properties. The interaction of triplexes with proteins has not yet received sufficient attention. A number of well-studied drugs were tested for their interaction with triplex structures. In some cases, triplex-inducing or triplex-stabilizing effects were demonstrated that have potential importance in the development of antigene and diagnostic strategies.

5 The Forces Participating in Triplex Stabilization

Triplex Stabilizing Factors

It is widely suggested that triplexes have a biological significance. Therefore, there should be natural mechanisms to control their stability. Interest in triplex stabilization is also a goal of the numerous applications of triplex technology that require high-affinity complexes and metabolic stability. There is also a cognitive aspect. By somehow making triplexes more stable, it would be possible to gain insight into the underlying interactions that govern the formation and maintenance of triple-stranded nucleic acid structures.

In this chapter we will consider the factors stabilizing both intra- and intermolecular triplexes, describe their nature in terms of physical interactions, and discuss the cellular analogs of model stabilizing conditions.

Reduction of Interstrand Repulsion

Screening the negative charges of phosphate groups on polynucleotide strands facilitates approaching of these strands to each other. As a result, triplex formation is facilitated. Both inorganic and organic cations can be used. The data described in Chapter 3 (phase diagrams) show that at low concentrations of monovalent cations, only the double-stranded complexes can exist. To overcome the repulsion of the phosphate groups of three strands, very high concentrations (up to 1 *M*) of the monovalent cations Na^+, K^+, or Li^+ are required (Felsenfeld and Rich, 1957; Felsenfeld et al., 1957; Rich, 1960; Blake et al., 1967; Felsenfeld and Miles, 1967; Michelson et al., 1967; Krakauer and Sturtevant, 1968). This trend is valid for triple-stranded complexes with neutral bases. However, in the case of cytosine-containing third strands where the cytosines acquire protons and become positively charged, increasing concentrations of monovalent cations stabilize triplexes to a lesser extent compared to triplexes containing only neutral bases (e.g., TA*T triplex) (Wilson et al., 1994). If a solution also contains cations of other valences, increasing concentrations of monovalents may actually destabilize the triplexes which may be formed using other stabilizers (Lipsett, 1964; Lee et al., 1984; Latimer et al., 1989; Maher et al., 1990; Lyamichev et al., 1991; Singleton and Dervan, 1993). The ability of Na^+ and K^+ to promote the formation of tetrastranded complexes in guanine-rich sequences is another triplex-destabilizing effect of monovalent cations. For this reason, triplex formation in such sequences can be difficult (Johnson et al., 1992; Cheng and Van Dyke, 1993; Olivas and Maher, 1995a). To avoid tetraplex formation, Li^+ ions in the buffer (Radhakrishnan et al., 1991a; Johnson et al., 1992) or modified bases in oligonucleotides (Gee et al., 1995; Olivas and Maher, 1995b) can be used.

In the first experiments on triple-stranded polynucleotides, it was found that significantly lesser concentrations (1–10 m*M*) of divalent cations (Mg^{2+}, Ca^{2+}, Mn^{2+}, Zn^{2+}) were equally or more effective in triplex formation compared to monovalent cations (Felsenfeld and Rich, 1957; Felsenfeld et al., 1957). The addition of Mg^{2+} was also shown to abrogate the destabilizing effect of high Na^+ concentrations on cytosine-containing triplexes (Maher et al., 1990; Lyamichev et al., 1991; Soyfer et al., 1992).

Double and triple helices of nucleic acids have long been known to be stabilized by naturally occurring polyamines, the effect being dependent on the polyamine structures (Tabor, 1962; Glaser and Gabbay, 1968; Thomas and Thomas, 1993). Polyamine stabilization of triplexes has been used in many studies. The efficacy of polyamines generally increases with their net charge: spermine > spermidine > putrescine. Submillimolar concentrations of spermine, which carries four positive charges promote the formation of PyPuPy (Raae and Kleppe, 1978; Moser and Dervan, 1987; Hanvey et al., 1991; Lyamichev et al., 1991; Soyfer et al. 1992) and PyPuPu triplexes (Beal and Dervan, 1991; Soyfer et al., 1992). At acidic pH, PyPuPy triplexes containing thymines and protonated cytosines in the third strand form faster in the presence of spermine (Hampel et al., 1991). The preformed triplex can then be transferred into neutral pH media. Spermidine and putrescine that have three and two positive charges, respectively, act similarly to spermine, but only at higher concentration (Hampel et al., 1991; Thomas and Thomas, 1993). Spermine and spermidine derivatives bearing small hydrophobic modifications promote the formation of triple-stranded poly(U) · poly(A) · poly(U), whereas increasing the size of hydrophobic substituents lowers the degree of stabilization of both double- and triple-helices (Glaser and Gabbay, 1968). The triplex-stabilizing effect also depends on the distance between the consecutive charges in the polyamine chains (Thomas and Thomas, 1993). In the case of PyPuPy triplexes containing thymines and methylcytosines in the third strand (which makes triplex existence possible at neutral pH), polyamines may be important only for the association of the oligomer with the duplex, since decreasing the spermine concentration after the triplex formed did not reduce the amount of triplex detected (Hanvey et al., 1991).

An important question concerns triple-stranded complex formation in the presence of a mixture of various cations, which is a characteristic feature of physiological media (Darnell et al., 1986). Data relevant to this question are available for the PyPuPy triplexes. Increasing the Na^+ concentration (in the 50 mM–1 M range) destabilizes PyPuPy triplexes containing protonated cytosines, as shown by UV-melting, restriction nuclease protection, and photofootprinting assays (Lipsett, 1964; Lee et al., 1984; Latimer et al., 1989; Maher et al., 1990; Lyamichev et al., 1991; Soyfer et al., 1992). More detailed studies showed that increasing the concentration of a monovalent cation at a constant concentration of a divalent cation results in a noticeable destabilization of triplexes. The association rate slightly decreases, but the dissociation rate increases to a much higher extent (Maher et al., 1990; Singleton and Dervan, 1993). A similar trend was found for triplexes in the presence of monovalent K^+ and tetravalent spermine, although the magnitude of destabilization was not as great (Singleton and Dervan, 1993). Similarly, increasing the concentration of divalent Mg^{2+} diminishes the triplex-stabilizing effect of spermine (Singleton and Dervan, 1993). By contrast, the increasing concentration of a higher-valence cation enhances triplex stability in the presence of a constant concentration of a lower-valence cation (Maher et al., 1990; Lyamichev et al., 1991; Soyfer et al., 1992; Singleton and Dervan, 1993). This general trend depends, however, on the structure of the multivalent cation. For example, divalent polyamines, in which two positive charges are spatially separated, are more efficient triplex-stabilizers than Mg^{2+}, which could be considered as a point charge (Thomas and Thomas, 1993). Explanations of differential cation effects in terms of cytosine–phosphate interactions and a counterion condensation model have been proposed (Latimer et al., 1989; Maher et al., 1990; Singleton and Dervan, 1993), and will be discussed later in this chapter.

pH Stabilization

As described in Chapter 3, PyPuPy triplexes containing CG*C^+ triads are formed under acidic pH conditions because of the need to protonate the N3 of the third-strand cytosines. This trend is valid for intermolecular triplexes (Lipsett, 1963, 1964; Pochon and Michelson, 1965; Felsenfeld and Miles, 1967; Michelson et al., 1967; Morgan and Wells, 1968; Thiele and Guschlbauer, 1971; Lee et al., 1979; Lyamichev et al., 1988) and H DNA (Lyamichev et al., 1985, 1987; Vojtiskova and Palecek, 1987; Hanvey et al., 1988b; Htun and Dahlberg, 1988; Johnston, 1988; Kohwi and Kohwi-Shigematsu, 1988). Recently, some of the PyPuPu triplexes were also shown to depend on pH when the third-strand sequence is such that it is possible to form the CG*A^+ triads (Malkov et al., 1993a). The protonation of third-strand cytosines and adenines makes it possible to form two Hoogsteen hydrogen bonds between these bases and purine bases in of the duplex target.

The general theoretical model of a pH-dependent structural transition (Lyamichev et al., 1985; Frank-Kamenetskii, 1992) predicts that the lifetime of the protonated triplex (τ_t) is strongly dependent on pH:

$$\tau_t = \tau_o \cdot 10^{-N\mathrm{pH}/r}$$

where r is the number of base pairs in the tract per protonated site, N is the length of the tract in base pairs, and τ_o refers to the geometrical mean value of lifetime for the unprotonated base triad. This expression is valid for the range between the pK_a values for the protonation site in single- and triple-stranded structures. This pH dependence is very sharp: for the third strand containing 10 cytosines, the lifetime changes 10-fold when the pH is changed by 0.1 unit.

The pK_a values for isolated cytidine (pK_a = 4.3) and adenosine (pK_a = 3.8) are relatively low. However, in the ordered polynucleotide structures these values are significantly higher. Two individual poly(dC) strands produce poly(dC) · poly(dC) duplex (so-called self-duplex). The transition (double- to single-stranded form) proceeds with a pK_a of 7.3 to 7.5 (Inman, 1964; Gray et al., 1988). A similar transition for the self-associated structure of poly[d(TC)] has a pK_a near 6.2 (Gray et al., 1988). An apparent pK_a for cytidine in the triple-stranded form was found to be near 5.5 (Xodo et al., 1991; Singleton and Dervan, 1992b). There are no similar data for adenosine-containing triplexes; however, self-associated structures of adenosine-containing polymers are known to form with an apparent pK_a near 6 (Steiner and Beers, 1959; Antao et al., 1988).

In spite of the general increase in the pK_a values of base-paired nucleosides compared to those for their unpaired forms, a kinetic barrier to triplex formation still remains. Protonated triplexes are formed more easily under mildly acidic conditions (Lipsett, 1963, 1964; Morgan and Wells, 1968; Thiele and Guschlbauer, 1971; Lee et al., 1979; Lyamichev et al., 1985; Pochon and Michelson, 1985; Mirkin et al., 1987; Moser and Dervan, 1987; Soyfer et al., 1992; Malkov et al., 1993a). Other factors, such as polyamines, can help in overcoming this barrier in vivo and maintaining the triplex under physiological conditions.

Length-Dependence

Generally, increasing the length of the PyPu tracts facilitates the formation of triplexes. The length of the PuPy tract appropriate for H or H* DNA formation is somewhere around 20 base pairs. Figures 3.7 and 3.8 show that in this case triplex extrusion proceeds at a moderate negative superhelical density ($-\sigma \sim 0.05$) and pH approximately 5.0 (Lyamichev et al., 1985; Htun and Dahlberg, 1989), or in the presence of millimolar concentrations of zinc ions at neutral pH (Soyfer et al., 1992; Martinez-Balbas and Azorin, 1993). A theoretical lower limit of 15 base pairs was estimated for H DNA formation (Lyamichev et al., 1989b). The greater the length of the PuPy tract, the less negative superhelix density was required to induce the triplex (Htun and Dahlberg, 1989; Lyamichev et al., 1989b, 1991; Collier and Wells, 1990; Soyfer et al., 1992). The longer PyPu tracts such as $d(GA)_{37}$ form an intramolecular triplex at neutral pH and a moderate level of negative supercoiling (Collier and Wells, 1990). Formation of an H* form in relatively short stretches of $d(A)_n \cdot d(T)_n$ is prohibited but becomes possible when $n = 69$ (Fox, 1990). However, a comparative analysis of the triplex-forming abilities of the PyPu tracts which are several dozen nucleotides long is complicated by the multiple variables, e.g., formation of either H or H* form DNA as a function of pH and ion concentrations (Kohwi and Kohwi-Shigematsu, 1988), multiple conformers in which a half of either a Py or a Pu strand can be complexed, or a double triplex structure where the looped strand of H DNA is donated to form H* DNA with another PuPy tract (Kohwi-Shigematsu and Kohwi, 1991; Panyutin and Wells, 1992; Beltran et al., 1993; Kohwi and Kohwi-Shigematsu, 1993; Martinez-Balbas and Azorin, 1993).

The studies in which the triplexes were formed from polynucleotide and oligonucleotide components of various lengths showed an increase in triplex stability with increasing oligomer length (Lipsett et al., 1960, 1961; Naylor and Gilham, 1966; Cassani and Bollum, 1969; Raae and Kleppe, 1978). An experimentally determined linear dependence of $1/T_m$ on $1/n$ (where T_m is the triplex melting temperature, and n is the oligonucleotide length) was reported (Cassani and Bollum, 1969). This dependence describes triplex formation from the duplex target and oligonucleotide by analogy with ordinary double-stranded oligomer complexes (Cantor and Schimmel, 1980c). Studies of triplex formation in the experimental system $d(A)_8 \cdot d(T)_8 + d(T)_n$ have shown that the minimum length of the third-strand oligomer that can bind to the duplex is five nucleotides (Sugimoto et al., 1991). In addition to the requirement of forming sufficient numbers of Hoogsteen hydrogen bonds and stacking contacts, triplex stability is influenced by the necessity of adjusting conformations of the duplex and the incoming third strand. Otherwise, structural inconsistencies accumulated at several helical turns can significantly destabilize CG*G and TA*A triplexes (Pochon and Michelson, 1965; Broitman et al., 1987). For example, poly(G), as opposed to oligo(G), did not yield the PyPuPu triplex with poly(C) in $0.15M$ NaCl (Pochon and Michelson, 1965). The triplex was formed after a short alkaline treatment of poly(G) to reduce the molecular weight down to an average degree of polymerization of 15. Formation of poly(U) · poly(A) · poly(A) in the presence of $5\,mM$ Mg^{2+} is limited by the size of poly(A) strands ($n = 28$–150 bases) (Broitman et al., 1987). In both cases, a possible explanation for the intermediate size of the purine strand is that the geometry of the CG*G and UA*A triads and/or the third strand backbone suffers increasing deformation as each succeeding residue of the third strand winds around the target duplex. Eventually, the deformation is accumulated to the point at which third-strand binding becomes unfavorable (Pochon and Michelson, 1965; Michelson et al., 1967; Broitman et al., 1987). However, no attempt was made in these studies to use a stronger triplex-stabilizer, namely, any divalent cation for the CG*G triplex and a transition metal cation for the TA*A triplex. In a later work, 5 m*M* Mg^{2+} included in the incubation mixture allowed the formation of polynucleotide CG*G triplexes between poly(dG) · poly(dC) of an average size of 300 base pairs and third-strand polymers several hundred nucleotides long (Letai et al., 1988). Thus, it seems plausible that complexing multivalent cations to the third purine strand can at least partly alleviate the unfavorable interactions that make the triplex structure unstable at long sequences.

Usually the length of the third-strand oligonucleotide is not less than 9 or 10 nucleotides (Le Doan et al., 1987; Moser and Dervan, 1987; Lyamichev et al., 1988, 1990, 1991; Pilch et al., 1990a; Soyfer et al., 1922; Malkov et al., 1993a; Rubin et al., 1993). The stability of shorter triplexes could be too low to be used at physiological ionic conditions and temperatures. It was the insufficient stability of shorter triplexes that required the development of the strategy of alternate strand recognition at two consecutive PyPu tracts (Horne and Dervan, 1990; McCurdy et al., 1991; Ono et al., 1991a; Sun et al., 1991a; Beal and Dervan, 1992b; Froehler et al., 1992a; Jayasena and Johnston, 1992b, 1993). For example, a quite stable triplex is formed by the 18-mer binding as a third strand under conditions in which the triplex with 9-mer is not formed (Horne and Dervan, 1990).

Differential Effect of Divalent Cations

Although the formation of intramolecular and intermolecular PyPuPu triplexes can proceed in the presence of various multivalent cations, including polyamines and Co^{3+}, most studies have dealt with the effects of divalent metal cations (Cooney et al., 1988; Kohwi and Kohwi-Shigematsu, 1988; Bernues et al., 1989, 1990; Kohwi, 1989b; Collier and Wells, 1990; Lyamichev et al., 1990, 1991; Frank-Kamenetskii et al., 1991; Dayn et al., 1992b; Kang et al., 1992; Panyutin and Wells, 1992; Soyfer et al., 1992). Their effect cannot be explained by simple electrostatic interaction. The general trend has some specific features for intra- and intermolecular triplexes. A stable intermolecular $d(C)_n \cdot d(G)_n \cdot d(G)_n$ triplex is formed in the presence of both alkali earth (Mg^{2+}, Ca^{2+}) and transition cations (Cd^{2+}, Co^{2+}, Mn^{2+}, Zn^{2+}, and Ni^{2+}), but not Ba^{2+} and Hg^{2+} (Frank-Kamenetskii et al., 1991; Soyfer et al., 1992; Malkov et al., 1993b). By contrast, the triplex between mixed A,G purine sequences and complementary pyrimidine sequences is formed only in the presence of one of the transition metal cations (Mn^{2+}, Zn^{2+}, Cd^{2+}, Co^{2+}, and Ni^{2+}) as a universal condition for the formation and/or stabilization of the intermolecular PyPuPu triplex (Soyfer et al., 1992; Malkov et al., 1993b). Mg^{2+}, Ca^{2+}, Ba^{2+}, and Hg^{2+} are ineffective in this case (Soyfer et al., 1992).

Several research groups described the PyPuPu triplex (H* DNA) in $(dC)_n \cdot (dG)_n$ inserts in supercoiled plasmids in the presence of alkali-earth cation Mg^{2+} (Kohwi and Kohwi-Shigematsu, 1988; Lyamichev et al., 1991; Kang and Wells, 1992; Soyfer et al., 1992). Bernues et al. (1990) were unable to detect H* DNA in $d(C)_n \cdot d(G)_n$ inserts in the presence of transition metal cation Zn^{2+} at 25°C. However, this became possible at an elevated temperature (37°C) and prolonged 1.5-h incubation (Kang and Wells, 1992). Kang and Wells also observed H* DNA in this sequence in the presence of another transition metal cation, Mn^{2+}. Bernues et al. (1989, 1990) observed H* DNA in $d(GA)_{22} \cdot d(CT)_{22}$ only in the presence of Zn^{2+}, Cd^{2+}, and Mn^{2+}. At the same 5 mM concentration, the efficiency of cations was in the order: $Zn^{2+} > Cd^{2+} > Mn^{2+}$. Bernues et al. (1990) failed to form a triplex in the presence of Ca^{2+}, Mg^{2+}, Co^{2+}, Ni^{2+}, Cu^{2+}, or Hg^{2+}. Again, plasmid incubation at an elevated temperature resulted in triplex extrusion in a similar sequence in the presence of Mg^{2+}, Ca^{2+}, and Co^{3+} (Panyutin and Wells, 1992). Thus, two preliminary conclusions inevitably arise: (1) divalent cations have specific binding sites on DNA bases, and (2) bound cations can stabilize either the double-stranded or the triple-stranded forms of DNA. The nature of the differential effects of divalent cations on triplex formation is not understood. Plausible explanations will be described in the next section.

The Hydration State of Nucleic Acids

The effect of decreased water activity on double-helical structures of nucleic acids is well known from X-ray studies of polynucleotide fibers, IR and CD studies of polynucleotide films at reduced humidity, and optical studies of nucleic acid solutions containing various proportions of organic solvents (Langridge et al., 1961; Brahms and Mommaerts, 1964; Fuller et al., 1965; Tunis-Schneider and Maestre, 1970; Pohl and Jovin, 1972; Ivanov et al., 1973, 1974; Herbeck et al., 1976). In particular, upon dehydration, the A DNA conformation forms in samples containing Na^+ as a counterion (Brahms and Mommaerts, 1964; Fuller et al., 1965; Ivanov et al., 1974; Malenkov et al., 1975). On the basis of the then-existing model of the triplex structure which implied the B to A conformational transition (Arnott and Selsing, 1974), the role of organic solvents was tested (Moser and Dervan, 1987). It was found that the stability of the $(dT)_{15} \cdot (dA)_{15} \cdot (dT)_{15}$ triplex increased by a factor of 10 upon addition of 40 volume percent of ethylene glycol, methanol, ethanol, dioxane, or dimethylformamide. However, the stability of triplexes with a mixed thymine and cytosine third strand in the absence of additives was higher than in the case of a homothymine third strand, and it even slightly decreased with added ethylene glycol.

During X-ray studies of polynucleotide fibers, they are kept under strict conditions of controlled relative humidity, usually lower than 100%. The only study in which fiber dehydration explicitly resulted in triplex formation was that of Campos and Subirana (1987), who observed a polynucleotide triplex in the dehydrated mixture of poly(dG) and poly(dC) and N-α-acetyl-L-arginine. In other studies (Rich, 1958a; Sasisekharan and Sigler, 1965; Arnott and Bond, 1973; Arnott and Selsing, 1974; Campos and Subirana, 1991), no evident correlation between sample dehydration and efficiency of triplex formation was reported.

Experimental NMR studies of hydration sites in PyPuPy- and PyPuPu-type triplexes (reviewed in Radhakrishnan and Patel, 1994b) show that long-lived water molecules (with lifetimes >1 ns) are detected in all three grooves of the triple helix: the minor groove of the double helix and two grooves between the third strand and either the duplex Pu or the duplex Py strand. The presence of nonstandard triads resulting in localized structural perturbations does not significantly change the patterns of hydration (Radhakrishnan and Patel, 1994a). The data of molecular dynamics simulations show a spine of water molecules that bridge the amino groups of cytosine and guanine in the groove between pyrimidine and the third strand of the PyPuPu triplex (Mohan et al., 1993b; Weerasinghe et al., 1995).

Hydrophobic Substituents in the Third Strand

A number of hydrophobic substituents in the third strand bases have been tested. As described in Chapter 1, the triple-stranded complexes of poly(A) with poly-5-methyluridylate (polyribothymidylate) and 5-methoxyuridylate were more thermostable than polyuridylate (Massoulie et al., 1966; Hillen and Gassen, 1979). In a more recent study, two hydrophobic uridine analogs, ribothymidine and 5-(1-propynyl)-2′-uridine, stabilized the triplex, the effect being more pronounced for the more hydrophobic substituent (Froehler and Ricca, 1992). The use of 5-methylated cytosine in the third strand allows the formation of triple-stranded complexes with duplex targets at neutral pH (which is highly unfavorable for similar complexes including cytosine) (Lee et al., 1984; Povsic and Dervan, 1989; Xodo et al., 1991).

The purine analogs bearing substitutions at positions other than those involved in hydrogen bonds could potentially be triplex stabilizers. Of such analogs, 8-methyl-adenine (Limn et al., 1983) and 2-amino-8-methyl-adenine (Howard et al., 1985; Kanaya et al., 1987) were tested. However, they are not capable of forming triple-stranded complexes, because the bulky substituent at C8 favors *syn* conformation of the base, which is incompatible with the duplex and triplex structures (Howard et al., 1974, 1975; Limn et al., 1983; Kanaya et al., 1987).

Possible Interactions Which Favor and Stabilize Triplexes

The formation and stabilization of triplexes are achieved through several types of interactions. Usually, changes in triplex stability are discussed in terms of electrostatic interactions between phosphate backbones and the number of hydrogen bonds between the duplex and the third strand. More rarely, other interactions are considered. Additionally, many studies have dealt with the resulting triplex structures formed after definite incubation periods. However, the various kinds of interactions could have varying degrees of importance during formation of the structure and maintenance of a preformed triplex. For example, the PyPuPy triplex containing 5-methylcytosine was more easily formed in the presence of polyamines; subsequent removal of polyamines did not lead to decomposition of an already stabilized triplex (Hanvey et al., 1991).

Electrostatic Forces

Electrostatic effects in triplex formation and stabilization could be involved in two variants:

1. Debye–Hückel screening of negative phosphate charges by positive counterions (ionic strength-dependent);
2. Site-specific neutralization of phosphate charges by bound cations.

The cations that neutralize negative phosphate charges are effective in significantly different concentrations. It is possible to form a triple-stranded complex for neutral polymers [poly(A), poly(U), poly(T), poly(I), etc]. using monovalent cation concentrations between 0.1 and $1\,M$ (Felsenfeld and Rich, 1957; Rich, 1960; Fresco, 1963; Stevens and Felsenfeld, 1964; Michelson et al., 1967; Krakauer and Sturtevant, 1968). The concentrations necessary for divalent cations (e.g., Mg^{2+}, putrescine^{2+}) are in the range of 1 to $10\,mM$ (Felsenfeld and Rich, 1957; Tabor, 1962; Glaser and Gabbay, 1968; Maher et al., 1990; Hampel et al., 1991; Singleton and Dervan, 1993; Thomas and Thomas, 1993). Polyamines carrying three and four positive charges, (spermidine and spermine, respectively) are effective at concentrations of $0.1\,\mu M$ to $5\,mM$ (further, they cause significant polynucleotide aggregation) (Tabor, 1962; Glaser and Gabbay, 1968; Maher et al., 1990; Hampel et al., 1991; Lyamichev et al., 1991; Soyfer et al., 1992; Singleton and Dervan, 1993; Thomas and Thomas, 1993). As discussed by Felsenfeld and Rich (1957), the disparity between the triplex-stabilizing ionic strengths created by monovalent, as opposed to divalent, cations implies that Debye–Hückel phosphate screening cannot account for all of the cation-induced stabilization. Moreover, the fact that triplex formation reaction was second-order in the presence of Mg^{2+} or Mn^{2+} suggests that these cations have specific binding sites on the polymers (Felsenfeld and Rich, 1957). The competing effects of cations with different charges show that they can have the same binding sites. Regardless of cation charges, all of them are capable of binding to the phosphate groups to reduce their repulsion.

The details of favorable electrostatic interactions are not completely understood. However, strikingly similar effects are produced by the mixed-valence salt solutions on several processes: thermal denaturation of duplex DNA (Dove and Davidson, 1962; DeMarky and Manning, 1975; Manning, 1975; Record, 1975; Thomas and Bloomfield, 1984), DNA condensation (Wilson and Bloomfield, 1979; Bloomfield, 1991), and triplex formation (Hampel et al., 1991; Lyamichev et al., 1991; Soyfer et al., 1992; Singleton and Dervan, 1993; Thomas and Thomas, 1993). Sodium ions alone increase the melting temperature of duplex DNA (see Figure 1.11). However, in the presence of multivalent cations, increasing Na^+ concentration decreases the melting temperature down to a minimum value at a critical Na^+ concentration (10 and 100 mM Na^+ in the presence of Mg^{2+} and spermine^{4+}, correspondingly). Above the critical Na^+ concentration, the DNA melting proceeds as in oligovalent cation-free solutions. Even very high Na^+ concentrations cannot result in DNA condensation, but they can interfere with the condensing effect of multivalent ions, either divalent metal cations or polyamines (Wilson and Bloomfield, 1979). These effects were reasonably explained on the basis of the counterion condensation theory (Manning, 1978; Record et al., 1978). This implies that mono- and multivalent cations result in different limiting neutralization of phosphates (Na^+ producing lower neutralization than, e.g., Mg^{2+}). When two different cations compete, their occupancies of the binding sites are proportional to their concentrations. Thus, higher Na^+ concentration (as compared to that of a multivalent cation) results in a lower phosphate neutralization and diminishes the melting temperature or impairs the condensation effect of the other cation. At very high Na^+ concentrations, the DNA solution is almost insensitive to the presence of multivalent cations and behaves as if it contained only Na^+ ions. Therefore, the melting temperature of DNA rises again. Yet, Na^+ (in any concentration) cannot result in the achievement of the critical value of phosphate neutralization necessary for DNA condensation. Thus, in this case, it cannot occur.

Intracellular concentrations of various cations have been estimated as 140 mM K^+, 1 mM Mg^{2+}, and 1 mM spermine (up to 5 mM in the nucleus) (Tabor and Tabor, 1976; Darnell et al., 1986; Sarhan and Seiler, 1989). Singleton and Dervan (1993) developed the counterion condensation model for triplex (T) formation from the duplex (D) and oligonucleotide

(O), which takes into account the multiple competing effects of these same mono- (K^+), di- (Mg^{2+}), and tetravalent (spermidine, Spm^{4+}) cations. They considered the reaction

$$O + D + iK^+ + jMg^{2+} + kSpm^{4+} \overset{K_{T^o}}{\leftrightarrow} T \tag{1}$$

where i, j, and k are the numbers of K^+, Mg^{2+}, and Spm^{4+} ions thermodynamically bound per phosphate during the association reaction, and K_{T^o} is the thermodynamic equilibrium constant. In the case of purely electrostatic interactions between the nucleic acid species and counterions, for the apparent equilibrium constant K_T:

$$\ln K_T = \ln K_{T^o} + n\{i \ln[K^+] + j \ln[Mg^{2+}] + k \ln[Spm^{4+}] - \eta' \ln \kappa\} \tag{2}$$

where n is the oligonucleotide length, κ is the Debye–Hückel screening parameter (proportional to the square root of the ionic strength), and the factor η' refers to the difference in screening of the phosphates between the free and bound nucleic acid states – the larger the η', the better the relative screening of repulsive interactions in the triple helix. It is clear that K_T is a function of all three cation concentrations. The effect of a particular cation on K_T results from a combination of the negative enthropy of cation condensation during triplex formation (the first three terms in braces) and the change in electrostatic free engery as a consequense of phosphate neutralization during the transition (the last term in braces).

The association of Mg^{2+} and Spm^{4+} with single-, double-, and triple-stranded nucleic acids was evaluated according to equations for K_{ion} (related to i, j, and k):

$$\log K_{Mg} = 2\psi(1 - 4\theta_{Spm}) \cdot \log[K^+] + \log K_{Mg^o} \tag{3}$$

$$\log K_{Spm} = 4\psi(1 - 2\theta_{Mg}) \cdot \log[K^+] + \log K_{Spm^o} \tag{4}$$

where ψ is the thermodynamic parameter for monovalent cation binding to a particular nucleic acid conformation, values of θ are oligovalent cation-binding densities, and values of K^o refer to oligovalent cation-binding constants at $1\,M$ K^+. These equations show the competing character of multivalent cation binding: binding of one cation decreases the binding constant of another cation. The cation-binding densities were calculated according to the model of McGhee and von Hippel (1974) for ligands binding to overlapping sites on a uniform linear lattice of phosphates, using K_{Mg} and K_{Spm} as adjustable parameters. The single-strand binding constants for spermine were found to be 10-fold lower than for duplex binding, and spermine–triplex binding constants were 1.5-fold those for spermine–duplex binding. The competition with Spm^{4+} reduces by 10-fold the estimated binding constants for Mg^{2+}. Using the calculated affinities of multivalent cations to various forms of nucleic acids and differentiated forms of equation (2), the influences of different cations on the association constant were considered.

Monovalent cations are triplex stabilizers at high concentrations. However, at lower concentrations, they can destabilize triplex by competing for the binding sites with other more effective stabilizers and thereby shifting the triplex formation equilibrium. The stabilizing efficiency of cations increases in the series $K^+ < Mg^{2+} < Spm^{4+}$, the effect of Spm^{4+} being so great that its micromolar concentrations promote complete dismutations of DNA and RNA polymers into the corresponding triplexes plus single strands (Glaser and Gabbay, 1968; Hampel et al., 1991; Thomas and Thomas, 1993).

At low ionic strengths, spermine has the highest affinity to all three, single–, double–, and triple–stranded, DNA forms and dominates counterion condensation. All three DNA forms are almost saturated by Spm^{4+} at a concentration of 1 mM and at K^+ concentration of 15 mM. Under these conditions, the triplex is stable and $K_T \sim 10^8 M^{-1}$. As $[K^+]$ increases, the equilibrium binding constants for Spm^{4+} binding to all DNA forms decrease. Yet spermine has the highest cation–triplex binding constant and, therefore, remains bound to phosphates. The enthropy of Spm^{4+} condensation becomes more negative. In addition, the net difference in counterion binding to the triplex, as opposed to the duplex or oligonucleotide, results in a decrease in the electrostatic free energy of association. Although this energy is still triplex-stabilizing, its effect decreases as ionic strength increases. Thus, increasing $[K^+]$ in the presence of Mg^{2+} and Spm^{4+} results in a reduction of the apparent equilibrium constant, K_T. At high $[K^+]$ ($>0.5 M$), Spm^{4+} and Mg^{2+} have low affinities to all three DNA forms. K^+, significantly contributing to net counterion condensation, should increase K_T. Note, however, a special case: no protonated cytosine-containing PyPuPy triplex was experimentally found at $[K^+] = 0.5 M$ (Singleton and Dervan, 1993). Thus, the presence of positively charged bases modulates the external cation neutralization of phosphates.

For experimentally employed concentrations up to 4 mM, spermine dominates cation condensation on phosphates, and the highest value of K_T was at the highest concentration of Spm^{4+}. The effect of $[Mg^{2+}]$ on the triplex is intermediate between those of $[K^+]$ and $[Spm^{4+}]$. When $[Mg^{2+}] < [Spm^{4+}]$, the condensed cations are predominantly spermines, making the effect of Mg^{2+} negligible. At higher Mg^{2+} concentrations, its competition with Spm^{4+} reduces triplex stability, but to a lesser extent than K^+. At $[Mg^{2+}] > 10$ mM, the triplex should be stabilized by the increasing concentration of Mg^{2+}, as was experimentally observed (Maher et al., 1990).

Thus, the major trends of polyamine triplex stabilization, especially as it concerns their effect in the presence of mono- and divalent cations, can be reasonably explained by the modified counterion condensation model, which treats various cation types as point charges. However, polyamines are not point-charged molecules, and their effects (in particular, for the equivalently charged analogs) really depend on their structure. Thus, a question arises about the interactions of positive charges distributed over polyamine molecules with negatively charged phosphate groups distributed over nucleic acid strands.

Among a series of polyamine analogs, diaminopropane and spermidine were the most efficient compounds in the induction and thermal stabilization of the poly(dT) · poly(dA) · poly(dT) triplex (Thomas and Thomas, 1993). These compounds share a common structural determinant: the trimethylene bridging region separating two positive charges. However, for spermidine homologs, the distance between the other pair of positive charges is even more important, since it results in a maximum stabilizing effect at four methylene groups between them. Bearing in mind also the three to four methylenes between the four charges in spermine (the strongest triplex stabilizer whose homologs were not studied), it can be concluded that such intercharge spacings are optimal for interactions with some specific groups in triple-helical nucleic acids. However, there is no clear understanding of the details of polyamine binding not only to the triple-stranded but also to the double-stranded nucleic acids.

Double-stranded B DNA shows no significant specificity to stabilization against thermal denaturation by putrescine or spermidine homologs, although the diaminoethane homolog of putrescine was slightly less effective than other homologs (Tabor, 1962). Charge separations by two and three methylenes in spermidine homologs were also less stabilizing (Thomas and Bloomfield, 1984). This could be interpreted as an insufficient fit between polyamine and phosphate charges on DNA. Molecular modeling suggested a binding mode involving interactions of spermine with phosphate groups and hydrogen bonding with N7 of purines of the opposite DNA strands (Feuerstein et al., 1986). This agrees well with spermine binding across the major groove of the B DNA-forming dodecamer, which is accompanied by interactions with phosphates and guanine bases (Drew and Dickerson, 1981). On the other hand, calculations of Tsuboi (1964) and Liquori et al. (1967) resulted in models in which spermine is bound in the minor groove. The low-resolution X-ray fiber data of Suwalsky et al. (1969) were also interpreted in a framework of the model of spermine lying across the minor groove of DNA binding two strands together. Experimental data on photoaffinity cleavage with polyaminobenzenediazonium salts showed a strong preference of polyamines for the minor groove of B DNA and for the major groove of A DNA (Schmid and Behr, 1991). The latter data fit with the crystal structure of the complex of spermine with the A DNA-forming self-complementary octamer d(GTGTACAC) in which spermine molecules are bound to the floor of the deep major groove of A DNA and interact with the bases but not the phosphates (Jain et al., 1989). This contradicts the earlier claim of Minyat et al. (1978), who interpreted the data on spermine and spermidine-induced B-A transition as due to interactions of polyamine charges with phosphate charges in the minor groove.

Polyamines are also known to stabilize Z DNA by neutralizing the repulsion of the phosphate groups attached to guanosines which are closely spaced across the minor groove (Thomas et al., 1985; Basu and Marton, 1987; Vertino et al., 1987); the order of efficacy is the same as for the poly(dT) · poly(dA) · poly(dT) triplex (Thomas and Thomas, 1993). In this case, spermine in crystal complexes with Z-forming hexanucleotide demonstrates complex behavior (Egli et al., 1991; Bancroft et al., 1994). Two types of spermine contributions include stabilization of lateral contacts between neighboring duplexes and stabilization of the stacked helices by simultaneous binding in the minor groove to different oligomer molecules.

In the triplex structure, the major groove is occupied by the third strand; therefore, it is more probable that polyamines may be localized in the minor groove. In such a case, their role would be to neutralize the phosphates of the duplex strands to facilitate third-strand binding. However, such a binding mode could hardly explain the homolog specificity experimentally found for triplex stabilization. Some contacts of polyamines with the third strand (e.g., in the new grooves between the duplex purine and the third strand or the duplex pyrimidine and third strand) could be suggested, but no experimental or theoretical efforts have been made in this direction. Thus, the problem of polyamine localization on double-stranded and triple-stranded DNA conformation has not been definitely solved. Yet, it is clear that polyamines can electrostatically interact with the phosphates and make hydrogen bonds with the accessible groups of nucleobases. In addition to stabilization caused by phosphate neutralization, polyamines can tie the polynucleotide structure when the opposite ends of the polyamine molecule interact with different strands.

Another feature of triplex stabilization by mixed valence cations which is not readily explained by the modified counterion condensation theory is the instability of cytosine-containing protonated triplexes at high concentrations of monovalent cations. A similar effect was observed for protonated polycytidylic acid (Akinrimisi et al., 1963; Inman, 1964a), which was explained by the screening of favorable interactions between protonated cytosine and the phosphate backbone by high sodium concentrations. The same explanation was also invoked for triplex structures (Latimer et al., 1989).

Dehydrating agents, which reduce effective charges on the polymer DNA, promote greater phosphate neutralization, thereby driving it to the necessary critical value (Wilson and Bloomfield, 1979; Votavova et al., 1986; Bloomfield, 1991). For instance, Mg^{2+}, which is ineffective in lateral condensation of DNA duplexes in aqueous solutions, causes the condensation of DNA in 50% methanol. Association of the duplex and the third strand and subsequent triplex stabilization could be due, at least in part, to such kinds of interactions. This could follow from the data of Moser and Dervan (1987), which indicate that the $(dT)_{15} \cdot (dA)_{15} \cdot (dT)_{15}$ triplex is stabilized by the addition of 40% (volume) ethylene glycol, methanol, ethanol, dioxane, or dimethylformamide. However, their data on the cytosine-containing PyPuPy triplexes, where these same solvents resulted in lower triplex stability, show that other (hydrophobic) interactions are of great importance (see below).

Stacking Interactions

Experimental studies of mismatches and bulges in triplex structures show the general importance of stacking interactions (Belotserkovskii et al., 1990; Roberts and Crothers, 1991; Rougée et al., 1992; Xodo et al., 1993). These authors found that the free energy penalty for introducing a single mismatch ranges from 2.5 to 6 kcal/mol, which is close to the corresponding values for duplex DNA and RNA (Gralla and Crothers, 1973; Tibanyenda et al., 1984). Theoretical studies of the influence of mismatches on triplex stability (van Vlijmen et al., 1990; Mergny et al., 1991b) are consistent with experimental results. The mismatched bases are partially or completely expelled from their helix positions, resulting in a loss of Hoogsteen hydrogen bonds and a distortion of two vertical contacts with neighboring bases. The dependence of the triplex structure on stacking interactions is also confirmed by the finding that a single abasic site, which eliminates two stacking interactions, results in a decrease in affinity similar to that observed for imperfectly matched natural triads (Horne and Dervan, 1991). Enhanced base stacking interactions of 5-methylcytosine with its neighbors are a likely source of the triplex-stabilizing influence of C5-methyl substituted cytosine (Hausheer et al., 1992; Singleton and Dervan, 1992b). Stacking interactions of the terminal bases of two oligonucleotides occupying immediately adjacent target sites on DNA result in a severalfold increase (3.5 to 20) in the equilibrium association constant (Strobel and Dervan, 1989; Colocci et al., 1993). The estimated increase in the binding free energy is about 1.8 kcal/mol for terminal thymine–thymine stacking of oligonucleotides (Colocci et al., 1993). However, it is probably base-dependent, as in the case of duplex DNA and RNA (Belintsev et al., 1975; Freier et al., 1986; Delcourt and Blake, 1991). The enthalpy of triplex formation in sequences containing one or two bulges is the same as in a perfect triplex (Roberts and Crothers, 1991). The lack of effect of bulges on enthalpy suggests that the bulges do not significantly disturb stacking in the helix and are probably extruded, as shown by Wang and Patel (1995).

The experimental data show that the stacking interactions in the duplex are at least as important as interbase hydrogen bonding (Turner et al., 1988). However, a quantitative aspect of the contribution of stacking interactions to triplex stabilization is far from being understood. In spite of frequent use of an RNA secondary structure predictive algorithm based on the melting data for various combinations of nearest neighbors in RNA strands (Freier et al., 1986), similar attempts for duplex DNA have not yet resulted in a consensus (see Cheng and Pettitt, 1992a, for discussion). The combinations of triad nearest neighbors are more numerous than the 10 combinations of base pairs in a duplex, and relevant interactions are complex enough to be determined experimentally. Therefore, an evaluation of stacking contributions to the triplex stability could be done using computer simulations, but theoretical approaches still need some improvements, as is the case for duplex DNA (Cheng and Pettitt, 1992a).

Hoogsteen Hydrogen Bonds

The sequence specificity of triplex formation is the major argument in favor of the importance of Hoogsteen hydrogen bonds. Although other contributions (e.g., base stacking and hydrophobicity) are also triplex-stabilizing, the absence of Hoogsteen hydrogen bonds in a triad exerts a strong triplex-destabilizing effect. Stable triads include TA*T (UA*U) and CG^*C^+ for PyPuPy triplexes, and CG*G, TA*A, TA*T, and CG^*A^+ for PyPuPu triplexes. All of them have two Hoogsteen hydrogen bonds. Third-strand bases providing one Hoogsteen hydrogen bond and no significant interference with nonbonded groups on the purine target can sometimes be accommodated in the triplexes. The common-type triad GC*T and AT*G, with inversion of duplex purine and pyrimidine bases, demonstrated sufficient stability in the PyPuPy triplex (Figure 4.2) (Griffin and Dervan, 1989; Radhakrishnan et al., 1991b, 1992a,b, 1994a,b; E. Wang et al., 1992; Yoon et al., 1992). In the PyPuPu triplex, the base triads with one Hoogsteen bond each (CG*A, TA*C, and GC*T) were also sufficiently stable to be detected (Figure 4.3) (Beal and Dervan, 1992a). In those studies in which the relative triplex stabilities were evaluated, the introduction of a triad with one Hoogsteen bond consistently resulted in lower stability (Griffin and Dervan, 1989; Beal and Dervan, 1992a; Yoon et al., 1992). In a recent study, Pei et al. (1991) showed that TA*C, CG*U, and TA*A triads do not drastically destabilize hybrid triplexes between third-strand polyribonucleotide and double-stranded DNA. It should be noted that all of these triads were studied as single-point perturbations in stable triplexes rather than as tracts.

Although some base triads (e.g., TA*G and CG^*A^+ triads in the PyPuPy triplex; Johnson and Morgan, 1978) can be simply drawn as containing two Hoogsteen hydrogen bonds, the resulting orientations of glycoside bonds do not suit the appropriate conformation of the sugar–phosphate backbone. This makes the observation of such triads in the triplexes impossible or highly unlikely. For example, the melting temperature drops from 68°C for the perfect oligomer PyPuPy triplex to 44°C for a triplex with a single CG^*A^+ triad which sterically constrains the structure (Macaya et al., 1991). Significant triplex destabilization by hypothetical triads with two Hoogsteen hydrogen bonds, which cannot be smoothly accommodated in the PyPuPy triplex geometry, also follows from the results of Belotserkovskii et al. (1990), who tested all the possible triads in H DNA.

The relative strength of Hoogsteen hydrogen bonds shows its importance in two cases of protonated triads. As was discussed in Chapter 1, the length of the hydrogen bond can vary, and it is noticeably shorter for the $\geqslant N^+—H \cdots O=C$ case than for $>N—H \cdots O=C$ (Table 1.1). Experimentally, this is confirmed by the strong enhancement of the CG^*C^+ triad as the degree of protonation increases. This results in the formation of H DNA in the absence of the stimulating effect of supercoiling (Lyamichev et al., 1985), or in a significant stabilizing contribution of the CG^*A^+ triad in the PyPuPu triplex, which in this case does not require any multivalent cation (Malkov et al., 1993a).

The importance of strong Hoogsteen hydrogen bonds also manifests itself in the need of divalent metal cations for PyPuPu triplex formation.

Hoogsteen Hydrogen Bond Enhancement

The divalent alkaline earth metal cations Mg^{2+}, Ca^{2+}, and Ba^{2+} bind primarily to the phosphate groups (Martin and Mariam, 1979; Sigel, 1989) but can also be involved in base coordination. However, the proportion of alkaline earth metal cations simultaneously participating in intramolecular chelate with the base and phosphate in adenosine-5′-monophosphate does not exceed 10% of bound cations (Sigel, 1989). The binding sites on nucleobases include the N7 and O6 positions of purines (Wang et al., 1979; Gessner et al., 1985; Bhattacharyya et al., 1988; Jean et al., 1993).

Divalent transition metal cations (Mn^{2+}, Co^{2+}, Ni^{2+}, Cu^{2+}, Zn^{2+}, Cd^{2+}, Pd^{2+}, Pt^{2+}), in addition to phosphate binding, have significant affinities for certain sites in nucleobases (Eichhorn, 1981; Saenger, 1984; Martin, 1985; Sigel, 1989; Gao et al., 1993). The following overall affinity order for the metal ion-binding sites has been proposed for complexes of transition metal cations with nucleic acid monomers at neutral pH: N7(G) > N3(C) > N7(A) > N1(A) > N3(A,G) (Sigel, 1989). For instance, the stability constants for Cu^{2+} and Zn^{2+} binding by N7 of guanine are more than an order of magnitude higher than those for N7 of adenine (Martin, 1985). Although N3 atoms may coordinate metal ions in crystal complexes of Pt(II) with guanine derivatives (Raudaschl-Sieber et al., 1985) and Rh(I) with 8-azaadenine derivatives (Sheldrick and Gunther, 1988), the affinities of metal ions for N3 of purines are much lower than those for N7 (Saenger, 1984).

Various modes of cation binding to the bases have been suggested. Zn^{2+} binds primarily via N in GC regions and via PO_4 in AT regions (Jia and Marzilli, 1991). It was also proposed that a Zn atom can simultaneously bind to both N7 of guanine and the phosphate group (Richard et al., 1973; Zimmer et al., 1974). Direct coordination of the Zn^{2+} ion to N7 and PO_4 (chelate complex) would require a significant distortion of the deoxyribose ring; however, indirect coordination of Zn^{2+} to PO_4 mediated by the water molecule is still possible (Granot et al., 1982; Reily and Marzilli, 1986). Other models suggested that the cation can be sandwiched between two adjacent G residues (Richard et al., 1973) or form the N7(G)-Me^{2+}-N7(G) cross-link (Miller et al., 1985). However, the strikingly similar effects of Zn^{2+}, Co^{2+}, and (although less pronounced) Mg^{2+} on circular dichroism spectra of calf thymus DNA containing adjacent GG bases and poly[d(AC)] · poly[d(GT)], where such a sandwich is impossible, imply the absence of any significant amount of sandwich or cross-linked G-Me^{2+}-G complexes (Jia and Marzilli, 1991). Recent crystallographic results imply that Co^{2+} and Cu^{2+} ions tightly bind mainly to N7 of guanine (Gao et al., 1993). Raman studies of DNA–metal complexes also confirm that the N7 sites of purines are the primary binding sites for many transition metal cations (Duguid et al., 1993). Thus, there appears to be a consensus that the primary binding sites of transition metal cations are the phosphate groups and/or the N7 of purines (mainly, guanine) regardless of how the exact structure is formed.

Divalent metal ions have been classified in decreasing order $Mg^{2+} > Co^{2+} > Ni^{2+} > Mn^{2+} > Zn^{2+} > Cd^{2+} > Cu^{2+}$ according to their relative ability to bind to the phosphate groups rather than to the bases (Jia and Marzilli, 1991). Many studies (see Duguid et al., 1993, for references) indicate that divalent cations bind preferentially to the bases in the following order: $Hg^{2+} > Cu^{2+} > Pb^{2+} > Cd^{2+} > Zn^{2+} > Mn^{2+} > Ni^{2+}$, $Co^{2+} > Fe^{2+} > Ca^{2+} > Mg^{2+}$, Ba^{2+}. They have a dual effect on DNA stability. At a high metal/DNA(P) ratio ($R > 4$), such cations facilitate thermal denaturation of DNA, presumably because of coordination to N3(C) and N1(A), which are exposed to the medium within the melting temperature interval (Langlais et al., 1990; Jia and Marzilli, 1991; Duguid et al., 1993). This destabilizing effect should be accumulated over a sufficient length of the duplex. At least, destabilization of a 160-base-pair DNA fragment was less pronounced then for a much longer 23-kilobase-pair DNA (Duguid et al., 1993). On the other hand, divalent cations stabilize the double-helical structure of DNA against thermal denaturation by neutralizing the negative charges on the polynucleotide backbone and reducing their repulsion. Coordination of transition metal cations to the sites in bases (N7, N3) that are not involved in hydrogen bonding at low R ($R < 2$–3) increases the melting temperature of double-stranded DNA (Minchenkova and Ivanov, 1967; Shin and Eichhorn, 1968; Jia and Marzilli, 1991).

Stabilization of PyPuPu triplexes by divalent metal cations could originate at least partly from Hoogsteen hydrogen bond enhancement (Figure 5.1). This assumption is an extrapolation to the triplex structure from a similar explanation for double-stranded DNA. In addition to the phosphate group neutralization, the stabilizing effect of transition metal cations on the double-helical structure of DNA was tentatively suggested in hydrogen bond stabilization when bound to the bases (Langlais et al., 1990; Jia and Marzilli, 1991). Quantum mechanical calculations indicate that the polarization effects of the metal ion on the ligand caused by guanine and adenine N7 coordination increase the stability of hydrogen bonds between complementary bases, and that the value of the additional net stabilization energy is comparable to that of hydrogen bond energy without a bound cation (Anwander et al., 1990). This qualitatively agrees with the increase in the melting temperature of DNA that accompanies the initial binding of Zn^{2+} to the bases (Jia and Marzilli, 1991) and the shift of Raman spectroscopic bands relevant to interbase hydrogen bonds during Zn^{2+}–DNA complexing (Langlais et al., 1990).

According to a recently developed hypothesis (Potaman and Soyfer, 1994), in the case of triplexes composed of CG*G triads, the alkaline earth metal cations Mg^{2+} and Ca^{2+} are bound largely to phosphates and, to a small extent, to N7 of guanines, both in the duplex and in the single strand. Since the binding of cations to guanine N7 is weak, they are readily expelled from this position in the duplex, and the formation of Hoogsteen-type hydrogen bonds between the duplex and the third strand occurs. Although weakly bound to N7 of the third strand, Mg^{2+} and Ca^{2+}, due to the above-mentioned polarization effect, can stabilize Hoogsteen-type hydrogen bonds to some extent. Those ions bound to phosphates reduce the overall duplex–third strand repulsion. The process is similar when transition metal cations participate in triplex formation. These cations are expelled from the duplex N7 more slowly because of their tighter binding, and triplex formation should be also slower compared to that mediated by Mg^{2+} or Ca^{2+}. On the other hand, because of a greater affinity of Zn^{2+}, Cd^{2+}, etc. to N7 of guanine, their stabilization of hydrogen bonds should be more prominent and such a triplex should be more stable.

In the case of TA*A triads, either alkaline earth or transition metal cations have lower affinities for adenine N7. Neutralization of phosphate charges by alkaline earth metal cations with relatively small (compared to guanine) contributions to hydrogen bond enhancement is not sufficient to promote triplex formation. In turn, transition metal cations partially neutralize phosphate charges, and, due to a still noticeable affinity for N7, bind to adenine bases. This affinity is probably large enough to exert considerable stabilization on Hoogsteen-type hydrogen bonds necessary for triplex formation. Since cation affinities for adenine N7 are generally lower than those for guanine N7, the minimum concentration of the

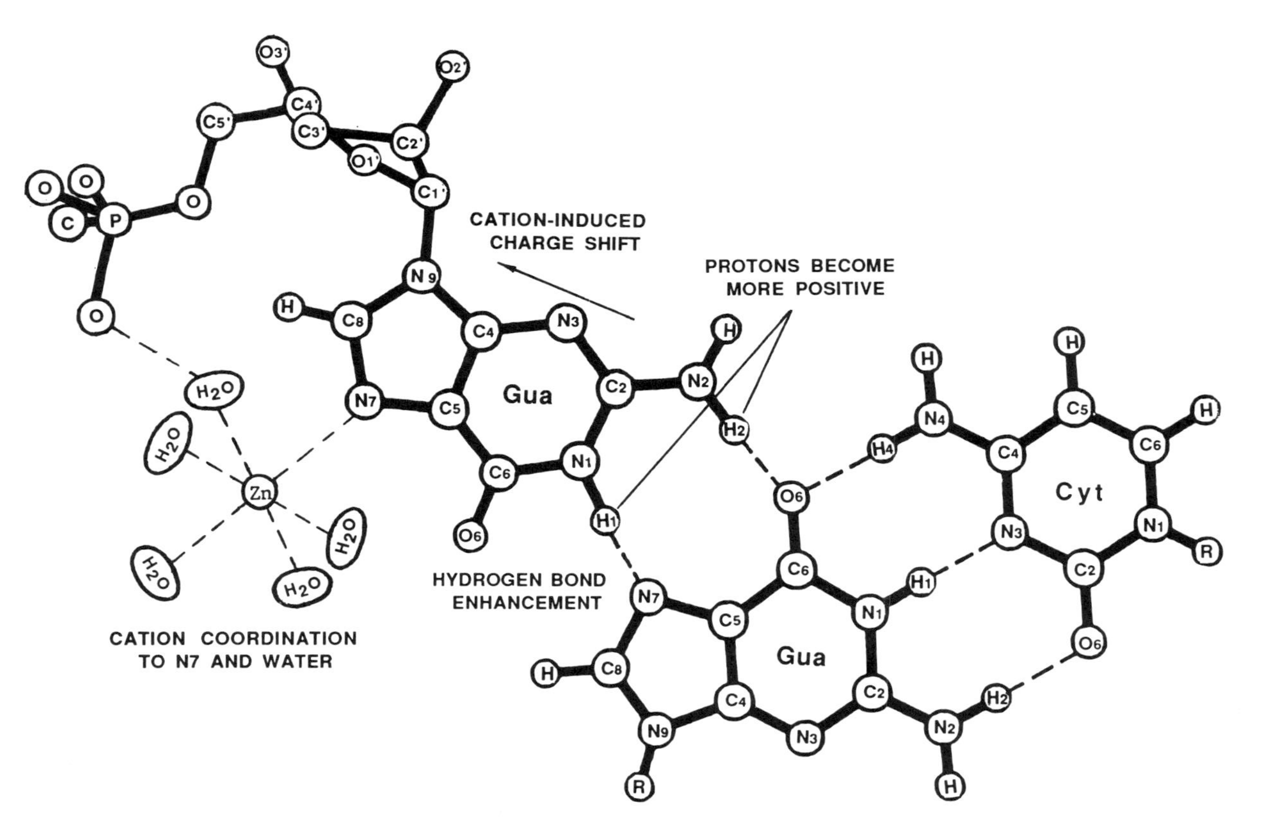

FIGURE 5.1. Schematic presentation of the suggested Hoogsteen-type hydrogen bond enhancement when divalent metal cations bind to the N7 position of the third strand purines. From Potaman and Soyfer (1994). Reprinted with permission.

specific cation stabilizing the TA*A triad should be significantly larger than that stabilizing the CG*G triad.

The Hoogsteen bond enhancement hypothesis is confirmed by some available experimental data:

1. The particular effect of cation coordination to N7 of third-strand purines is manifested by the introduction in the third strands of 7-deaza-purines that are unable to coordinate cations. Such oligonucleotides could bind to the duplex target, but the reverse-Hoogsteen hydrogen bonds between duplex purines and 7-deaza-purines of the third strand are expected to be weak because of the absence of cation-induced stabilization. This suggestion is partly confirmed by the data of Milligan et al. (1993), who demonstrated that the third purine-rich strands containing 7-deazaguanine and 7-deazaxanthine formed unstable PyPuPu type triplexes.
2. Cations with greater affinities to the purine N7 should be better triplex stabilizers, at equal concentrations. Zn^{2+}, which binds to N7 of purines stronger than Mn^{2+} (Duguid et al., 1993), is more effective than Mn^{2+} in promoting triplex formation in plasmid DNA containing $(GA)_n \cdot (CT)_n$ insert (Bernues et al., 1990).
3. The differences in affinities of metal cations for the N7 sites suggest that the kinetics for the formation of the G*GC triplex (when cations are slowly displaced from their positions in duplex) should be slower compared with that for the A*AT triplex. In support of this prediction, Panyutin et al. (1989) showed slow kinetics of intramolecular triplex formation in plasmid DNA containing $(G)_n \cdot (C)_n$ inserts.

However, the effect of divalent metal cations in the case of H* DNA may be more complex. Whereas in the case of intermolecular triplexes, annealing of triplex-forming strands from higher (denaturation) to lower temperature allows observation of mainly triplex-stabilizing cation effects, for H* DNA the structure of the duplex is also important. It should be kept in mind that in this system, divalent cations stabilize not only the triple-stranded structure plus the single strand, but also the double-stranded structure. Therefore, the double helix-stabilizing effect of divalent cations must be overcome to form the necessary denaturation bubble in order to initiate a formation of a new complex. In line with this suggestion are the data of Panyutin and Wells (1992) who showed the triplex-promoting effect of various cations, including spermidine and Co^{3+}, on the $d(GA)_{37} \cdot d(CT)_{37}$ sequence when the opening of the intraplasmid PyPu tract was facilitated by the incubation at 50°C. Yet, the scarcity of reliable data on cation influence on H* DNA stability, and the significant bias toward different H* DNA isomers dependent on the loop sequence, do not allow a prediction of specific details at the present time.

Cation stabilization can be only a particular case of charge redistribution in the third-strand purines that "gives rise to a dipole with [increased] positive charge at the hydrogen end of the X–H bond" (Jeffrey and Saenger, 1991). Another example is N1 protonation of the third-strand adenine (Malkov et al., 1993a). When it binds to the duplex guanine, the resulting hydrogen bond of the type $\geqslant N^{+}—H \cdots O=C$ is rather strong: its length in nucleic acids is approximately 0.1 Å less than that of the $>N—H \cdots O=C$ bond (Watson et al., 1989; Jeffrey and Saenger, 1991). As a result, the enhancement of Hoogsteen hydrogen bonds is so pronounced that the triplex forms even in the absence of divalent cations. The latter, however, can additionally stabilize the triplex at higher pH when adenine becomes less protonated (Malkov et al., 1993a).

Some substitutions in the purine bases which shift the electron density out of the N1 atoms could potentially strengthen the hydrogen bond in which the N1 takes part. For instance, an electron-withdrawing bromine atom in the eighth position of adenine stabilizes the adenine–uracil base pair (Kyogoku et al., 1967). One could expect that a PyPuPu triplex between an ordinary PyPu tract in DNA and a third strand containing 8Br–Ade would be more stable than an unmodified PyPuPu triplex. However, this suggestion cannot be experimentally verified: the mentioned substitution favors a *syn* conformation of the adenine base relative to a sugar–phosphate backbone, which is inappropriate for triplex formation (Howard et al., 1974, 1975).

Hydration Forces

Some insight into triplex stabilization principles can be gained by considering the hydration states of DNA or oligonucleotide strands. Experimentally, the role of dehydration is illustrated by the data of Campos and Subirana (1987), who showed that dehydration of poly(dG) · poly(dC) in the presence of *N*-α-acetyl-L-arginine ethylamide leads to triplex formation.

It is known that the primary hydration shell around DNA consists of several dozen water molecules per nucleotide pair (Tao et al., 1989). The distribution of water molecules along the sequence is not even. For instance, the homopurine sequence is hydrated to a higher extent relative to the alternating purine/pyrimidine sequences (Marky and Macgregor, 1990; Rentzeperis et al., 1992), poly(dA) · poly(dT) being hydrated more than other polynucleotides (Buckin et al., 1989a,b). The hydration of the AT pair in the B form of DNA is larger than that of the GC pair, and the substitution of monovalent Cs^{+} cation for a divalent Mg^{2+} cation results in a decrease of the hydration of the polynucleotide with bound cation (Buckin et al., 1989a). Experimental measurements of the interactions between parallel B form DNA double helices led to the conclusion that polyvalent ligands bound to DNA can act by reorganizing the water structure and polarizing the water molecules surrounding the macromolecular DNA surfaces, creating attractive long-range hydration forces

(Rau and Parsegian, 1992). For a wide variety of "condensing agents" from divalent Mn^{2+} to polymeric protamines, the resulting intermolecular force varies exponentially with a decay rate of 1.4 to 1.5 Å, one-half that obtained previously for hydration repulsion. One can suggest that the multivalent cations displace some molecules of water and reorganize the others around the duplex and single strand, thus inducing triplex formation.

The NMR data show that long-lived water molecules with lifetimes >1 ns are immobilized in all three grooves of PyPuPy and PyPuPu triple helices (Radhakrishnan and Patel, 1994a,b). As discussed by Radhakrishnan and Patel (1994b), these water molecules may stabilize the triplex structure by solvating the complex. They may also screen repulsive electrostatic interactions between phosphate groups across the narrow groove between the third strand and the duplex purine strand of the PyPuPy triplex. They may also contribute by bridging polar groups belonging to different strands (Radhakrishnan and Patel, 1984b). The latter suggestion is in agreement with the data of molecular dynamics simulations for the $(dC)_7 \cdot (dG)_7 \cdot (dG)_7$ triplex showing the presence of a series of water molecules bound between the NH_2 of the third-strand G and the NH_2 of C at each base plane (Mohan et al., 1993b; Weerasinghe et al., 1995). These water molecules were not placed in these positions initially. Formation of this spine of water was complete after 10 ps of molecular dynamics simulation. The spine of water is unique to the CG*G triad-containing triplex because of the presence of appropriate amino groups in the first-strand C and the third-strand G. The spines of water were found previously in double-stranded DNA and were implicated in stabilization of specific conformations and transitions between them (Drew and Dickerson, 1981; Saenger, 1984). In the case of triple-stranded nucleic acids, organized water molecules may play a significant role in differentiating the stabilities of CG*G triad-containing triplexes and those containing TA*A and TA*T triads (Mohan et al., 1993b). For example, the hydration states of various sequences in DNA and its affinity for multivalent cations may be relevant to the overall triplex-forming ability of the PyPu tracts. Based on the measurements of ultrasonic velocity during a course of DNA titration with Mg^{2+}, Buckin et al. (1994) came to the conclusion that magnesium ions form two types of complexes with the AT and GC base pairs. In outer-sphere complexes, Mg^{2+} does not penetrate into the hydration shell of a highly hydrated oligomer stretch (AT base pairs), and the resulting dehydration effect of Mg^{2+} is small. By contrast, Mg^{2+} penetrates deeper into the hydration shell of less hydrated sequences (GC base pairs), forming inner-sphere complexes, and the dehydration effect is large. Whether this observation is relevant to triplex formation mediated by divalent metal cations remains to be elucidated. In this view, similar data on hydration states of DNA complexes with transition metal cations would be of high value. Thus, the hydration effects are influenced by the nucleotide sequence; however, the limited available data do not give enough basis for a specific hypothesis.

Contribution of Hydrophobicity

It is clear that hydrophobic interactions should stabilize the triplex structure. For example, an increase in hydrophobicity of 5-(1-propynyl)-2′-uridine compared to thymidine results in relatively greater stabilization when the triplexes carrying these nucleotides are compared (Froehler and Ricca, 1992). Triplex stabilization by a methyl group in position 5 of cytidine can be due to the hydrophobic effect, since the methyl group is less solvated (Xodo et al., 1991; Cheng and Pettitt, 1992a; Roberts and Crothers, 1992; Singleton and Dervan, 1992a; Wang and Kool, 1995b) and does not significantly change the pK_a of nucleoside (Xodo et al., 1991; Singleton and Dervan, 1992). However, a quantitative evaluation of this effect is complicated by the multiple influences a single substitution can produce. The stabilization can also be due to an increase in stacking interactions between a methylated residue and its neighbors, as suggested by Hausheer et al. (1992) and Singleton and Dervan (1992b). It is difficult to quantify the proportion of triplex-stabilizing increase in hydrophobicity of 5-(1-propynyl)-2′-deoxycytidine, compared to cytidine, and the destabilizing decrease in the basicity of the analog (Froehler and Ricca, 1992). In addition, the introduction of hydrophobic substituents can destabilize not only the triplex, but also the duplex, due to perturbation of the *syn–anti* equilibrium in the targeted strand, as illustrated by poly(m^8A) (Limn et al., 1983).

The hydrophobic contribution seems to manifest itself in the case of triplex formation in the presence of a number of organic cosolvents (Moser and Dervan, 1987). It has long been known that water-soluble organic liquids weaken hydrophobic interactions in the systems protein–protein, protein–DNA, and nucleic acid strands (Millar, 1974; Shifrin and Parrott, 1975; Schwartz and Fasman, 1979; Freifelder, 1987). The decrease in stability of cytosine-containing triplexes under the influence of ethylene glycol, methanol, ethanol, dioxane, and dimethylformamide (Moser and Dervan, 1987) is probably caused by a reduction of hydrophobic contacts between the duplex and the third strand. There is no apparent contradiction with the opposite trend for the $(dT)_{15} \cdot (dA)_{15} \cdot (dT)_{15}$ triplex. In the latter case, increasing concentrations of organic solvents may serve to distort the rigid B′ conformation of $(dA)_{15} \cdot (dT)_{15}$ (Lyamichev, 1991) so that it may accommodate the incoming third strand.

Interrelation of Different Triplex-Stabilizing Contributions

As can be seen in this chapter, several types of interactions are responsible for the formation and stability of triplexes. Some of them (e.g., reduction of phosphate repulsion) are of paramount importance during the association of the duplex and the third strand, but, once removed, can leave the triplex in a metastable state (as illustrated by the effect of polyamines; Hampel et al., 1991; Hanvey et al., 1991). Other contributions are very important in the formation and stabilization of triplexes (e.g., supercoiling promotes triplex formation in circular DNA, but a single nick relieving superhelical stress results in a rapid distortion of the triplex structure; Hanvey et al., 1989). Different kinds of contributions can be cumulative or interchangeable (e.g., supercoiling and low pH for protonated triplexes, and effects of purine-coordinated metal cations and phosphate-bound polyamines) (Soyfer et al., 1992).

We should mention at this point that sometimes it is difficult to distinguish the contributions of hydrogen bond enhancement, electrostatic effect, and hydration effects. For instance, divalent metal cations can be involved in all of them. This problem is further complicated by the fact that the factors which promote triplex formation are also favorable for DNA aggregation. In some respects, this is similar to other conformational transitions in DNA solutions, namely, the B–A and B–Z transitions. Both require either dehydration by organic solvents or the presence of multivalent cations (Pohl and Jovin, 1972; Ivanov et al., 1974; Malenkov et al., 1975; Xu et al., 1993). However, these same agents also promote DNA condensation (reviewed by Bloomfield, 1991). It was even suggested that the lateral interactions of aggregated DNA molecules promote transition from the B to the A form of DNA (Skuratovskii and Bartenev, 1978). Parallel measurement of circular dichroism spectra (as indicators of the B to A transition) and sedimentation coefficients (as indicators of DNA aggregation) provided evidence that the two processes are independent (Potaman et al., 1980). In the case of triplex formation, the condensing factors (i.e., facilitating approach of the duplex and the third strand) would be of great importance (Soyfer et al., 1972). Yet, they influence the structures of complexes far beyond the conditions that promote gross DNA condensation. Thus, triplex formation can be considered as a strictly local process that has important implications for consideration of the biological roles of H DNA (Chapter 6) and numerous biotechnological and therapeutic applications of intermolecular triplex technology (Chapter 7).

Conclusion

The formation and maintenance of triplexes is governed by several types of interactions. Triple-helical order is due to a combination of stacking interactions, Hoogsteen hydrogen bond formation, hydration forces, and hydrophobic interactions, whose relative contributions are difficult to evaluate at the moment. Sometimes, they can be cumulative and the lack of one contribution can be compensated by another. However, because of the greater complexity of the subject, triplex stability has been studied much less than duplex stability. Triplex stability depends on a number of factors whose influence can be explained in terms of different types of interaction, so that the design and interpretation of experiments requires a certain care. Yet, clear understanding of triplex formation and stabilizing conditions is of paramount importance. On the one hand, it can help in identifying stabilizing factors in vivo, thus providing a solid basis for the description of intracellular events where the triplexes are presumed to play a role. On the other hand, the most attractive triplex-based therapies are targeted against primary causes of cell malfunction, and should also be applied under triplex-stabilizing conditions.

6
In Vivo Significance of Triple-Stranded Nucleic Acid Structures

The PyPu tracts necessary for triplex formation occur in the genomes of various organisms. In eukaryotes, PyPu tracts a few dozen base pairs long constitute up to 1% of the entire genome (Birnboim et al., 1979; Hoffman-Liebermann et al., 1986; Manor et al., 1988; Wong et al., 1990; Bucher and Yagil, 1991; Tripathi and Brahmachari, 1991). They are less frequent in prokaryotic DNA but nevertheless are statistically overrepresented (Bucher and Yagil, 1991). Many of them have the potential for intramolecular triplex formation (Table 6.1). The nonrandom distribution of the PyPu tracts in eukaryotic genomes has suggested their involvement in various biological processes. The previous discussion dealt with several types of triple-helical nucleic acids: intramolecular triplexes in single (or shortly interrupted) PyPu tract (H DNA), intramolecular triplexes between the PyPu tract and the third strand donated by another distant PyPu tract, intermolecular triplexes between the PyPu tract and external poly(oligo)nucleotides, etc. It is possible that because they differ in structure, the various types of triplexes also have different biological roles. It is also interesting that a number of single-strand-binding proteins specific to homopurine (A,G)- or homopyrimidine (T,C)-containing regions have been found and are suggested to participate in the stabilization of triplex structure. A number of PyPu sequence-binding proteins specific for the double-stranded B form have also been found. Competition between protein binding and H form extrusion at the same site would result in interference of the effects of protein and H form DNA with each other. Before discussing the biological roles of triplexes, it is worth examining the possibility of their existence in vivo.

In Vivo Existence of Triplexes

Search for Triplexes in the Cell

Several attempts have been made to determine whether the PyPu sequences really do form triple-stranded structures in the cell. Information from these studies may be separated into results from direct determinations and indirect indications of the triplexes in vivo.

Immunological Assays

To directly probe the triple-stranded structures in chromosomes, a monoclonal antibody (Jel 318) was produced by immunizing mice with poly[d(Tm^{5}C)] · poly[d(GA)] · poly[d(m^{5}CT)], which forms a stable triplex at neutral pH (Lee et al., 1987). This antibody was demonstrated by numerous criteria to be specific for the TAT-rich PyPuPy triplex DNA. It did not bind to calf thymus DNA or other non-PyPu DNAs, such as poly[d(TG)] · poly[d(CA)]. In addition, this antibody did not recognize PyPu DNAs containing m^{6}A (e.g., poly[d(TC)] · poly[d(Gm^{6}A)], which cannot form a triplex, since the methyl group at position 6 of adenine prevents Hoogsteen base-pairing. The interaction of Jel 318 with chromosomes was tested by immunofluorescent microscopy of mouse myeloma cells fixed in methanol/acetic acid. A duplex DNA-specific antibody (Jel 239) was used as a control. The Jel 318-induced fluorescence was weaker than that of Jel 239, but binding to metaphase chromosomes and interphase nuclei was observed. Chromosome staining by Jel 318 was the same in the presence of *E. coli* DNA, but it was eliminated by addition of polymer triplex. To avoid a possible ambiguity in interpretation arising from the fact that acid fixation itself can change the structure under study (Hill and Stollar, 1983), nuclei were also prepared from mouse melanoma cells by fixation in cold acetone. Again, Jel 318 showed weak but consistent staining of the nuclei (Lee et al., 1987). Unfixed, isolated mouse chromosomes also reacted positively with the antibody, particularly when they were gently decondensed by exposure to low ionic conditions at neutral pH, indicating that fixation is not mandatory for antibody staining (Burkholder et al., 1988). In addition to the Jel 318 antibody of the γ2b isotype, another monoclonal antibody (Jel 466) of the γ2a isotype was prepared using mice immunization with poly[d(Tm^{5}C)] · poly[d(GA)] · poly-[d(m^{5}CT)] (Agazie et al., 1994). Contrary to Jel 318, which is specific for the TAT-rich triplexes, Jel 416 was specific for the triplex form of poly[d(TC)] · poly[d(GA)]. A comparison of the immunofluorescent staining of mouse and human chromosomes with two antibodies showed that Jel 466 was positive for the GC-rich R band and negative for the AT-rich C band and G band, whereas the opposite was characteristic of Jel 318 and Hoechst 33258 binding. Thus, the staining patterns agree well with the sequence preferences of two triplex-binding antibodies. Additional evidence that triplexes exist in vivo is provided by immunoblotting of triplexes in crude cell extracts (Lee et al., 1989). Thus, there is a growing body of immunological evidence that triplexes are present in eukaryotic chromosomes.

TABLE 6.1. Naturally occuring PyPu tracts appropriate to form triple-helical DNA structures.

Source	Sequence	Reference
Ad2 major late promoter	GGGGGCTATAAAAGGGGG	Goding and Russell, 1983
Ad7 major late promoter	GGGGGTATAAAAGGGGG	Kilpatrick et al., 1986
Ad12 major late promoter	GGGCTATAAAAAGGG	Kilpatrick et al., 1986
Bermuda land crab satellite	$(CCT)_n$	Fowler and Skinner, 1986
Blood coagulation factor IX	$(GA)_{16}$	Wu et al., 1990
Chicken repetitive DNA	$(AGAGG)_n$	Dybvig et al., 1983
Chicken β-globin gene	GGGGAAGAGGAGGGG	Schon et al., 1983
Chicken β-globin 5′-end	$(G)_{16}$CGGG	Nickol and Felsenfeld, 1983
Chicken α2(1) collagen promoter	TCCCTCCCCTTCCTCCCTCCCT	McKeon et al., 1984
		Finer et al., 1987; Koller et al., 1991
Chicken embryo myosin heavy chain	147 bp PyPu	McCarthy and Heywood, 1987
Chicken histone H5 gene family	CTCCYTGTCY	Ruiz-Carillo, 1984
Drosophila hsp26	AGAGAGAGAAGAGAAGAGAGAGA	Siegfried et al., 1986
Drosophila hsp22	TCTCTCCCACTCTCT	Mace et al., 1983
Drosophila hsp70, pHTΔ5	CTCTCTGTACTATTGCTCTCTC	Mace et al., 1983
Drosophila hsp70, pHT1	CTCTCTTTTCTTTTTGGGTCTCTC	Mace et al., 1983
EGF-R		Johnson et al., 1988
Erythropoietin	$(CACCC)_n$	Imagawa et al., 1994
ets-2		Mavrothalassitis et al., 1990
HIV-1	31 bp and 38 bp PyPu	McShan et al., 1992
HSV-1 DR2 repeats	$(GCGAGGAGGGGGG)_n$	Wohlrab et al., 1987
Human α2–α1 globin intergenic	CCTCCTCCACCTCCTCC	Shen, 1983
Human β-globin gene (upstream)		Choi and Engel, 1988
Human δ-globin gene (upstream)		O'Neill et al., 1991
Human γ-globin gene (upstream)		Ulrich et al., 1992
Human c-*myc* gene	CCCTCCCCATAAGCGCCCCTCC	Boles and Hogan, 1987
Human decorin gene	150 bp PyPu	Santra et al., 1994
Human dihydrofolate reductase	GAGGGAGAGAGGCAGAGAGGG	Caddle et al., 1990
	$(AG)_{27}$, $(GAGA)_4$, etc.	Caddle et al., 1990

Human herpesvirus 6	AGGGAGGTGTGGCCTGGGCGGGA	J. Wang et al., 1994
Human hypervariable locus D8S210		Brereton et al., 1993
Human IgA switch region	$(AGAGG)_n$	Collier et al., 1988
Human *mdr*1	GAAAAAGATAAGAAGGAAAAGAAA	Chen et al., 1990
Numan N-*myc* gene		Krystal et al., 1990
Human Na, K-ATPase	$GGAGGA(G)_5AAGGC(G)_5AGA(G)_5AGAAGGA$	Shull et al., 1989
Human papilloma virus-11	$(A)_6GAGGAGGGAGCG(A)_4$	Hartman et al., 1992
Human platelet-derived growth factor A-chain		Z.Y. Wang et al., 1992
Human retroposon cluster		Liu and Chan, 1990
Human t(10,14) breakpoint cluster	CCTTCCTCTCCCCCTCCCCTCCCCTCCCTC	Lu et al., 1992
Human thyroglobulin gene (upstream)	209 bp PyPu	Christophe et al., 1985
Human U1 RNA gene 3′ flank	$(CT)_n$	Htun et al., 1984
6-16 interferon-responsive element	GGGAGAGAGGGAAAATGAAA	Porter et al., 1988
JCV 96-bp repeats	$(A)_8$	Amirhaeri et al., 1988
Maize *Adh*1	36 bp PyPu	Lu and Ferl, 1992
Mouse Cγ2a and Cγ2b	$(TC)_n$	Weinreb et al., 1990
Mouse c-Ki-*ras*	$T(CCCT)_3CCT(TCCC)_3$	Hoffman et al., 1990
Mouse c-*pim*-1	CCCCTCCCCCTCC	Svinarchuk et al., 1994
	GGAGGGGGAGGG	Svinarchuk et al., 1994
Mouse metallothionein-I	128 bp PyPu	Bacolla and Wu, 1991
Neisserial *opa*	$(CTTCT)_n$	Belland et al., 1991
Quail T64	$(TTCCC)_{48}$	Michel et al., 1992
Rabbit β1 globin gene	$(TC)_{24}$	Margot and Hardison, 1985
Rat NCAM	178 bp PyPu	Chen et al., 1993
Rat preproinsulin II gene	GGGGTCAGGGG	Evans et al., 1984
Rat rRNA genes	$(CT)_n$ and $(CCCT)_n$	Financsek et al., 1986
Rat thyroglobulin gene	115 bp PyPu	Musti et al., 1987
Rat L1 repetitive element		Usdin and Furano, 1989
Sea urchin histone gene repeat h22	$(CA)_{10}(CT)_{22}$	Hentschel, 1982
	$AGAGGAAGG(GA)_{16}(G)_{11}AGGGAGAA$	Hentschel, 1982
SV40 72-bp repeat	GGGGAGCCTGGGG	Evans et al., 1984
Tetrahymena telomeres	$(TTGGGG)_n$	Henderson et al., 1987

Chemical Probing

Kohwi et al. (1992) used choloroacetaldehyde (CAA) to probe H* DNA in *E. coli* cells at neutral pH. The PyPu tract-containing plasmid was grown in *E. coli* cells in the presence of Mg^{2+}, necessary for triplex formation, and chloramphenicol, which increased the level of supercoiling, thereby also promoting triplex formation. For a $(dG)_{30}$ triplex-forming sequence, the in situ CAA modification pattern was similar to that for the CG*G triplex in vitro. In a recent paper, Kohwi and Panchenko (1993) described the formation of the CG*G triplex in *E. coli* cells under the influence of the transcriptional activation of a downstream gene. As is well established, transcription increases the local level of negative DNA supercoiling upstream of the transcribed gene (Liu and Wang, 1987; Wu et al., 1988; Frank-Kamenetskii, 1989; Tsao et al., 1989; Wang et al., 1990; Dayn et al., 1992a; Droge, 1993; Bowater et al., 1994).

The method of direct chemical probing was also applied to detect triplexes in the *E. coli* plasmid pEJ4, which contains PuPy tracts (Karlovsky et al., 1990). *E. coli* cells were preincubated in the media with pH 4.5. To reveal triplex formation, the authors used osmium tetroxide and bipyridine as a probe and showed a modification pattern characteristic for H DNA at pH 4.5 and 5.0. Increasing the intracellular superhelical density by cultivating the cells in higher-ionic-strength media (0.35 *M* NaCl) resulted in a stronger site-specific modification. A more detailed analysis of another triplex-forming plasmid pL153 revealed significantly different osmium, bipyridine modification patterns in vitro and in situ (Figure 6.1). This suggested the prevalence in situ of the H-y5 isomer which is formed at lower superhelical densities (Htun and Dahlberg, 1989), whereas the H-y3 isomer was dominant in vitro. In addition to low pH, the described experiments required a 30-min cell exposure to osmium, bipyridine at 37°C. These conditions could seriously reduce the viability of *E. coli* cells. Thus, the data of Karlovsky et al. (1990) show the triplex formation in the cells under very specific conditions.

Glaser et al. (1990) used diethylpyrocarbonate to probe for H DNA in the *Drosophila hsp26* promoter in nuclei isolated in neutral pH buffer. They failed to detect any modification pattern consistent with the presence of the H form, whereas it was detected in the experiments in vitro.

Although it is possible to suggest a high transient level of DNA supercoiling (see discussion later in this chapter, section "Factors That Could Be Responsible for Triplex Formation In Vivo"), neither 2 m*M* Mg^{2+} nor pH 5 is a common condition in intracellular media. *E. coli* cells are known to maintain homeostasis well in the presence of relatively small deviations of extracellular pH from the value of 7.6 characteristic for the intracellular medium and even during short-term incubation at pH 5.5 or 9 (Slonczewski et al., 1981). However, *E. coli* cells are no longer capable of maintaining intracellular pH during prolonged acid or alkali

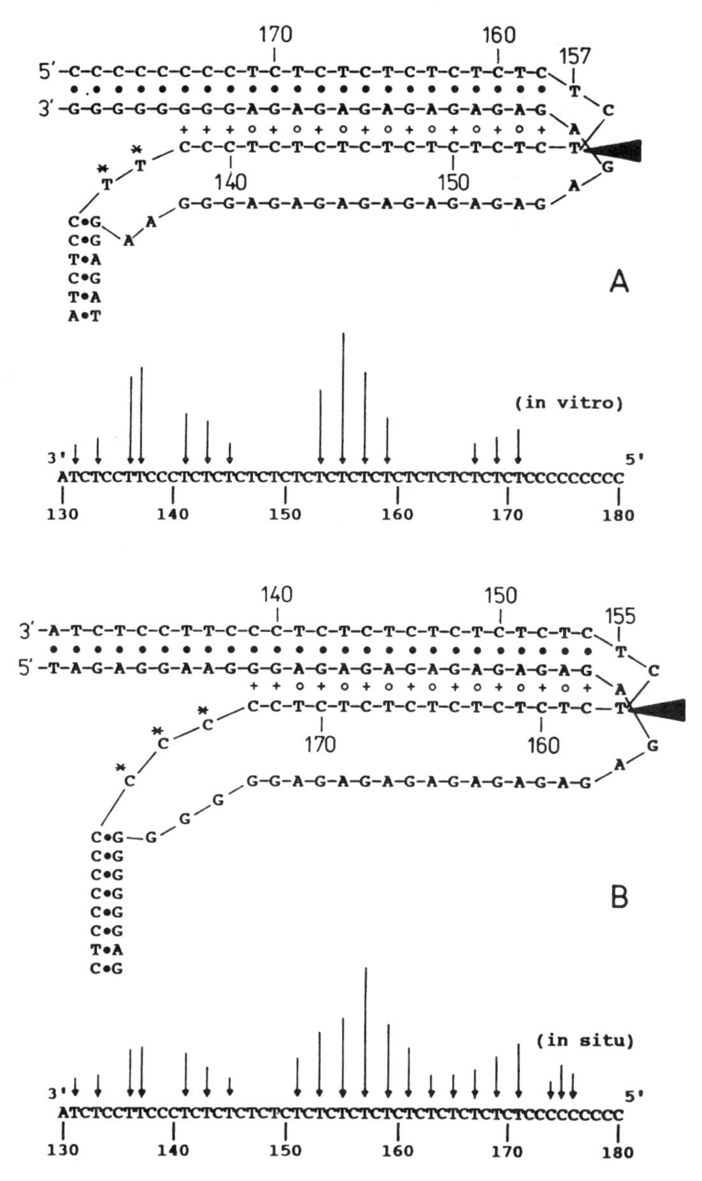

FIGURE 6.1. According to osmium, bipyridine (Os, bipy) probing of thymines in the PyPu tract of the pL153 plasmid, H DNA formation results in different conformers in vitro (panel A) and in situ (panel B). The arrow length is proportional to the extent of chemical cleavage (according to the densitometry of the sequencing gel) in vitro and after 30-min cell incubation at an external pH of 5.0. The triangle shows the strongest modified base in the triplex, and asterisks correspond to modification at the B–H junctions. The modification patterns in panels A and B are consistent with the H-y3 isomer and H-y5 isomer, respectively. Reprinted from Palecek (1991). © CRC Press, Boca Raton, Florida.

incubation (Slonczewski et al., 1981). Therefore, in the above-described studies, intracellular pH values might be close to those outside the cells. For this reason, Karlovsky et al., (1990) argued that at least some bacteria may grow at pH 5. Thus, the above data from a direct search for the triplexes in the cells show that triplexes may in principle exist in such a multicomponent medium as the bacterial cell. A high level of DNA supercoiling and proper environmental conditions are the major limiting factors in the formation of these structures.

Photochemical Probing

Ussery and Sinden (1993) described experiments on trimethylpsoralen (Me_3psoralen) photobinding to the PyPu tracts in plasmid DNA in *E. coli* cells. Their results showed that the formation of the H-y3 isomer of H DNA in cells was dependent on DNA superhelicity and extracellular pH. These authors found a prominent difference for Me_3 psoralen photobinding patterns when the cells were grown in a Luria broth (LB) of pH 8 and in K medium, which acidifies over time down to pH 5. No triplex was found in LB, whereas the Me_3 psoralen photobinding pattern for incubation in K medium was consistent with the presence of H DNA. Me_3 psoralen photobinding to a central TA separating two PyPu tracts was reduced in situ compared to Me_3 psoralen photobinding to a TA outside the triplex region. This indicated that TA in a triplex-forming sequence was in a single-stranded loop that connects the two halves of the folded pyrimidine strand. In addition, Me_3 psoralen reactivity at triplex–duplex junctions was different from that for B DNA and matched the reactivity for H DNA in vitro. The use of topoisomerase I mutant cells with a higher level of supercoiling in vivo was the triplex-promoting condition in these experiments.

Indirect Assays

Several indirect studies have shown the possibility of triplex existence in vivo. Parniewski et al. (1990) used the well-known fact that the sequence GATC is methylated by an enzyme, *Dam* methylase, when it is in the double-stranded B form DNA, but not in an alternative non-B conformation. A GATC site at the center of or adjacent to a PyPu mirror repeat was undermethylated in the plasmid grown in JM 101 *E. coli* strain. This result could be explained by the participation of the PyPu tract in H DNA in vivo. Two problems should be solved to make this suggestion a definite conclusion. First, *Dam* methylase in vitro methylated the GATC sites under conditions appropriate for the existence of H DNA (or a mixture of H and B DNA). The authors suggested, therefore, that some proteins might be involved in triplex stabilization in vivo, because the addition of chloramphenicol (an inhibitor of protein synthesis) relieved the undermethylation. However, chloramphenicol, besides inhibiting total protein synthesis, also results in an increase of superhelical density, which should promote triplex formation (see, e.g., Karlovsky et al., 1990). Second, GATC undermethylation was observed when the plasmid was grown in JM-type cells but not when it was grown in other types of *E. coli* cells (A. Jaworski, unpublished data). This selectivity was not related to the $RecA^+$ or $RecA^-$ type of these cells. Therefore, other unknown factors involved should be identified to reliably interpret the results of the *Dam* undermethylation pattern in triplex-forming plasmids.

Deletion analysis of the PyPu tracts inserted in the tetracycline resistance gene (Jaworski et al., 1989) showed a significant instability of those (longer) inserts which can form intramolecular triplexes (H or H* DNA) in vitro. These data may reflect the existence and mutational role of triplexes in vivo, provided that there is a satisfactory model for the underlying process (see discussion later in the chapter "Possible Role in Mutational Processes").

Sarkar and Brahmachari (1992) showed that the PyPu tract cloned into the transcribed region of a bacterial gene significantly decreased gene expression. A kind of unusual structure in the PyPu tract might be responsible for the premature termination of transcription. Rao (1994) reported that the cloned $d(GA)_n \cdot d(TC)_n$ sequences that can potentially adopt triplex structures could slow down the movement of the DNA replication fork. In both of these cases, H DNA was suggested to be responsible for preventing polymerases from progressing along the DNA template in the cellular system.

Reaban and Griffin (1990) found that transcription of a supercoiled plasmid containing the $(AGGAG)_{28}$ tract from the murine IgA switch region resulted in loss of superhelical turns. The RNase H capable of degrading double-stranded RNA or RNA–DNA hybrids eliminated this effect. These authors suggested that supercoil relaxation involves the interaction of newly synthesized mRNA with the single strand of H DNA extruded by a local wave of supercoiling upstream of an RNA polymerase.

Although none of these experiments is conclusive, further work will hopefully provide more convincing evidence of the existence of triplexes in vivo.

Factors That Could Be Responsible for Triplex Formation In Vivo

DNA in the cell is thought to be under torsional stress from supercoiling. The simplest example is the closed circular DNA of plasmids. The bacterial or prokaryotic chromosomal DNA may be organized into separate topological domains by proteins and/or RNA molecules that bind simultaneously to distant parts of the same DNA molecule (Stonington and Pettijohn, 1971; Worcel and Burgi, 1972; Pettijohn and Hecht, 1973; Sinden and Pettijohn, 1981) or by DNA interaction with membrane-associated proteins (Sindern, 1994). The DNA topoisomerases can create definite levels of (unrestrained) DNA supercoiling within these domains (see Gellert, 1981; Wang, 1985; Sinden, 1994, for reviews). Packaging of DNA in bacteria and eukaryotic chromatin results in the formation of toroidal coils around histones or histone-like proteins (Drlica and Rouvbiere-Yaniv, 1987; van Holde, 1989; Wolffe, 1992). This corresponds to a restrained part of DNA supercoiling. Nucleosomal dissociation during transcription may convert a restrained supercoiling into an unrestrained supercoiling (Weintraub, 1983). The increased level of unrestrained supercoiling may require relaxation via some unwound DNA conformation, e.g., Z DNA, H DNA, etc. (Weintraub, 1983; van Holde and Zlatanova, 1994; Sinden, 1994).

Liu and Wang (1987) worked out the theory of changes in unrestrained supercoiling during transcription within two divergent topologically closed domains. The large size of an RNA polymerase, as well as the interaction of the DNA and RNA transcript-bound proteins with the cell membrane, may result in the absence of significant rotation of the RNA polymerase relative to the transcribed DNA molecule. This creates two subdomains in the DNA molecule, and the linking difference cannot be redistributed between them. During the movement of the RNA polymerase, a negative supercoiling ($Lk < Lk_o$) arises in DNA behind the RNA polymerase and a positive supercoiling ($Lk > Lk_o$) in front of the RNA polymerase (see Liu and Wang, 1987; Sinden, 1994; for more detail). A number of studies experimentally confirmed the difference in the level of supercoiling upstream and downstream from the transcribed site (Liu and Wang, 1987; Wu et al., 1988; Frank-Kamenetskii, 1989; Rahmouni and Wells, 1989; Tsao et al., 1989; Wang et al., 1990; Zheng et al., 1991; Dayn et al., 1992a; Droge, 1993; Bowater et al., 1994). An assortment of other mechanisms may create localized or transient torsional stress in eukaryotic DNA: binding of transcription factors or other proteins, activity of helix-tracking proteins, looping of DNA by protein binding at two distant locations, histone acetylation, and gyrase activity of topoisomerases (van Holde and Zlatanova, 1994). A torsionally strained DNA molecule may be partially relaxed when it extrudes an unwound structure such as an H form.

An important aspect of triplex stabilization arises from the necessity of reducing the very high negative charge density originating from the phosphate groups in three strands. The repulsion of these phosphates should be screened by some counterions. Phosphate repulsion in vitro can be alleviated by high concentrations (up to 1 *M*) of monovalent cations (e.g., Na^+, K^+, Li^+) (Felsenfeld and Rich, 1957; Rich, 1960; Krakauer and Sturtevant, 1968). However, real concentrations of monovalents in the cell (0.1–0.2 *M*; Darnell et al., 1986) are far from stabilizing and therefore may only be partially responsible for triplex formation and maintenance in vivo.

Much lower (millimolar) concentrations of divalent metal cations (Mg^{2+}, Zn^{2+}, etc.) stabilize the triplexes in vitro (Felsenfeld and Rich, 1957). Although concentrations of magnesium up to 1 m*M* can be found in the cell (Darnell et al., 1986), such concentrations may be sufficient for only partial stabilization of the PyPuPy triplexes and the PyPuPu-type CG*G triplexes (Felsenfeld and Rich, 1957; Kohwi and Kohwi-Shigematsu, 1988; Lyamichev et al., 1991; Soyfer et al., 1992; Singleton and Dervan, 1993), and they are insufficient for stabilization of the PyPuPu-type triplexes with mixed CG*G and TA*A triads (Malkov et al., 1993a). The TA*A triad-stabilizing zinc ions are abundant in the cell, but they are in a protein-bound form (e.g., in the active centers of numerous metalloenzymes) and thus are not readily available to provide a triplex-stabilizing effect.

Multivalent cations (polyamines) are known to play important roles in cell proliferation and differentiation (Tabor and Tabor, 1976, 1984). Polyamines (double-charged putrescine, triple-charged spermidine, four-charged spermine, and their analogs) can be considered a general class of stabilizers of nucleic acid helices. Polyamine concentrations in the nucleus may be as high as 5 m*M* (Sarhan and Seiler, 1989), but they are largely in a macromolecule-bound form, whereas free polyamine concentrations are in the micromolar range (Davis et al., 1993). The most efficient triplex stabilizer (spermine) at concentrations of 0.1 μ*M* to 5 m*M* promotes the formation of PyPuPy (Glaser and Gabbay, 1968; Raae and Kleppe, 1978; Moser and Dervan, 1987; Hampel et al., 1991; Hanvey et al., 1991; Lyamichev et al., 1991; Maher et al., 1992; Soyfer et al., 1992; Singleton and Dervan, 1993; Thomas and Thomas, 1993) and PyPuPu triplexes (Beal and Dervan, 1991; Soyfer et al., 1992). Hopefully, the concentration of free spermine in the cell is sufficient to produce a significant triplex-stabilizing effect.

Investigations of triplex formation (Singleton and Dervan, 1993) under conditions close to those in the living cell (140 m*M* potassium, 1 m*M* magnesium, 1 m*M* spermine) (Tabor and Tabor, 1976; Darnell et al., 1986; Sarhan and Seiler, 1989) show that these conditions are not ideal, at least, for the intermolecular PyPuPy triplexes studied (see Chapter 5). Indeed, under other constant ionic parameters, the decrease in potassium concentration to 5 m*M* results in a 100-fold increase of an equilibrium association constant for triplex formation. Therefore, in addition to triplex stabilization by the intracellularly available small cations, whose phosphate-screening effects may interfere with each other, it is desirable to indicate some other macromolecular sources of triplex stabilization.

Due to the relatively rare occurrence of perfect PyPu mirror repeats, some nonstandard types of triads could be used to form H (H*) DNA structures. Section "Natural Bases in Unusual Triads" (Chapter 4) described stable PyPuPy triplexes accomodating AT*G and CG*T triads with single Hoogsteen hydrogen bonds, and PyPuPu triplexes accommodating GC*T and CG*A triads. Therefore, in the mouse c-Ki-*ras* promoter sequence, either a protonated PyPuPy triplex with eight CG*C$^+$ triads could be formed in the presence of Na^+ at pH 4.5 (Pestov et al., 1991), or an alternative PyPuPu triplex containing two protonated CG*A$^+$ triads could be formed (Raghu et al., 1994). For some other genes, an examination of the 5′ flanking sequences shows that the triplex structures can be formed in promoter sequences themselves or between promoter sequences and some distant sequence (e.g., enhancer), provided some nonstandard base triads, including protonated triads, are allowed. In addition to triplex-stabilizing factors discussed earlier, other possibilites of triplex stabilization in the cell consist of the participation of RNA (Miller and Sobell, 1966; Minton, 1985; Reaban and Griffin, 1990; Reaban et al., 1994) or the PyPu tract-binding proteins (Biggin and Tjian, 1988; Davis et al., 1989; Gilmour et al., 1989; Kiyama and Camerini-Otero, 1991; O'Neill et al., 1991; Yee et al., 1991; Hoffman et al., 1992; Kolluri et al., 1992; Muraiso et al., 1992; Goller et al., 1994; Hollingsworth et al., 1994; Horwitz et al., 1994) whose exact roles have not been elucidated so far (see discussion later in this chapter: section "Possible Regulation of Transcription").

A regulatory oligonucleotide could participate in the repression of transcriptional initiation by binding alone (Minton, 1985) or as a part of a ribonucleotide complex (Miller and Sobell, 1966) upstream of the gene. In that case, the RNA strand would have a dual importance: it would serve as a third strand and as a stabilizing factor for the newly forming triplex. The latter suggestion may be justified by experimental data showing that the RNA third strand targeted to the DNA duplex results in a triplex at least as stable as a triplex completely composed of DNA (Escudé et al., 1992; Roberts and Crothers, 1992; Han and Dervan, 1993). Except for a single report on a PyPu sequence-binding factor with the properties of a ribonucleoprotein (Davis et al., 1989), no experimental data showing the existence of mixed DNA–ribonucleoprotin complexes are available.

Minton (1985) hypothesized that the newly transcribed mRNA can be involved in a triple helix, thereby leading to a premature termination of transcription at the PyPu stretches in DNA. It was shown that the RNA product may lead to transcriptional inhibition (Champoux and McConaughy, 1975; Mizuno et al., 1984, Adeniyi-Jones and Zasloff, 1985; Green et al., 1986; Krystal et al., 1990; Wu et al., 1990; Celano et al., 1992). The newly synthesized homopyrimidine or homopurine mRNA could fold on the available PyPu tract in front of the RNA polymerase, form the hybrid RNA–DNA triplex, and thereby prevent the RNA polymerase from further movement along the DNA template. However, the RNA-induced transcriptional inhibition may well be explained, and is experimentally supported, by antisense binding of one RNA transcript to homologous mRNA transcribed from another gene (Green et al., 1986; Krystal et al., 1990; Celano et al., 1992).

Formation of hybrid complexes of newly synthesized RNA with the duplex or single-stranded DNA near a moving RNA polymerase (Reaban and Griffin, 1990; Reaban et al., 1994) may result in stabilization of an unwound local DNA structure that may play a role in supercoil regulation (Wang and Lynch, 1993). In one of the models proposed by Reaban and Griffin (Reaban and Griffin, 1990; Reaban et al., 1994), mRNA associates with the single strand of preformed H DNA and may therefore prevent the overall structure (triplex + RNA-bound single strand) from flipping back into the duplex conformation. The in vitro data of Belotserkovskii et al. (1992), who were able to stabilize H DNA at neutral pH with a single-strand-binding oligonucleotide, provide support for this model.

Another hypothesis proposed that an H DNA structure formed through interaction of two distant PyPu tracts of the same DNA duplex was stabilized by single-strand-binding proteins that could fix an unpaired single strand (Lee et al., 1984). Proteins that preferentially bind to the homopyrimidine (T,C)-containing single strands (Yee et al., 1991; Kolluri et al., 1992; Muraiso et al., 1992; Goller et al., 1994) or to the homopurine (A,G)-containing single strands (Aharoni et al., 1993; Hollingsworth et al., 1994) have been identified. In the absence of conclusive evidence on the role of such proteins, it is widely speculated that they can also bind to DNA structures with a single-strand character that may be formed as the PyPu tract breathes or is denatured under a torsional stress. Thus, single-strand fixation commences, and the other unpaired strand may fold on its target PyPu duplex spontaneously, or other factors may bind and further stabilize or induce triplex formation (Kolluri et al., 1992; Aharoni et al., 1993).

In spite of the attractiveness of this hypothesis, it should be verified by a careful examination of the available experimental data. The length of single-stranded DNA required for efficient binding of the above-mentioned proteins is not known. The binding properties of these proteins were studied using oligonucleotides most of which were more than 30 nucleotides long. In one of the studies, the reduction of the oligonucleotide length from 24 to 6 nucleotides resulted in a fourfold reduction of the binding affinity (Aharoni et al., 1993). No attempt has been made to study protein binding to single strands flanked by duplex DNA fragments. However, efficient binding of such proteins to DNA might require a binding site that is quite long. For example, in the very well-studied case of the *E. coli* single-strand-binding protein, four 20-kD subunits forming a functional protein cover approximately 70 nucleotides of a single-stranded DNA (Krauss et al., 1981). Binding of this protein to supercoiled plasmids where an H DNA region has been formed results in further DNA unpairing and triplex elimination (Klysik and Shimizu, 1993). Thus, a protein creates a single-stranded site of appropriate length by depleting a triple helix. In a similar manner, a duplex depletion with concomitant supercoil relaxation was observed earlier for interactions of single-strand-binding proteins with locally denatured regions that appeared due to other structural rearrangements (Glikin et al., 1983; Langowski et al., 1985). Yet, as seen in Table 6.1, many naturally occurring PyPu tracts that are suggested to form part of H DNA can hardly provide single strands more than 10 to 20 nucleotides long. Thus, if the proteins specific for single strands of PyPu tracts also require rather long binding sites, this will result in a virtual disappearance of the triplex instead of the suggested stabilization. Clearly, the issue of triplex stabilization by single-strand-binding proteins requires additional investigations dealing with the PyPu tract-specific proteins.

There has been a single report on the triple-helix-binding protein (Kiyama and Camerini-Otero, 1991). A protein with a molecular mass of 55 kDa has a two- to fivefold higher affinity for the poly(dT) · poly(dA) · poly(dT) triplex than the poly(dA) · poly(dT) duplex. The authors suggested that this protein might promote triplex formation. However, no experimental data on this subject have been published yet.

To summarize, there is an assortment of various factors that may induce and/or stabilize the triple-helical structures in the cell. When considered separately, these factors contribute to triplex stability in vitro. However, many of them are available in cells simultaneously, and their combined effects should be carefully examined. Although done in vitro, an excellent example of such work is the study of the interplay of triplex stabilization by various cations (Maher et al., 1990; Singleton and Dervan, 1993). The stabilizing effects of cations of various valences interfere with each other, since these cations compete for the same binding sites. Clearly, more data are necessary to understand how triplexes could be formed and maintained in living cells.

Possible Biological Roles of Triplexes

Several possible roles of triplexes in DNA structure and function have been suggested. The data published to date suggest an involvement of triple helices in transcription, replication, recombination, chromosome organization, and mutational processes. Single base triads in RNA also play structural and functional roles.

Possible Regulation of Transcription

Studies show that numerous genes contain *cis*-acting promoter elements that are nuclease-hypersensitive. These regions are nucleosome-free and are therefore hypersensitive to DNase I. Many of these elements are dually sensitive, that is, they are also sensitive to single-strand-specific nucleases. The PyPu tracts often occur in the 5′ flanking regions of eukaryotic genes, and a number of them have been shown to be sensitive to the single-strand-specific nuclease S1 (Larsen and Weintraub, 1982;

Schon et al., 1983; Christophe et al., 1985; Siegfried et al., 1986; Boles and Hogan, 1987; Davis et al., 1989; Gilmour et al., 1989; Hoffman et al., 1990; Bucher and Yagil, 1991; Tripathi and Brahmachari, 1991). Detailed structure–function analyses for a number of eukaryotic genes have shown that the PyPu sequences are important for the promoter function. Partial loss of transcription efficiency was found when the PyPu tracts were deleted from promoter regions of human epidermal growth factor receptor (EGFR) (Johnson et al., 1988), c-*myc* (Davis et al., 1989; Postel et al., 1989), *ets*-2 (Mavrothalassitis et al., 1990) and decorin (Santra et al., 1994) genes, mouse c-Ki-*ras* (Hoffman et al., 1990) and TGF-β3 (Lafaytis et al., 1991) genes, *Drosophila hsp26* (Gilmour et al., 1989; Glaser et al., 1990) and actin (Chung and Keller, 1990) genes, and other genes. Many naturally occurring PyPu tracts being cloned in plasmids, adopt an intramolecular triplex/single strand structure (H form) in experiments in vitro (Lyamichev et al., 1985, 1990, 1991; Kinniburgh, 1989; Usdin and Furano, 1989a,b; Glaser et al., 1990; Pestov et al., 1991; Raghu et al., 1994).

The hypothetical models for the involvement of the H form in transcriptional regulation have been summarized by Sinden (1994). Since higher superhelicity generally enhances transcription efficiency (Wang and Lynch, 1993), this might mean that the free energy of template supercoiling would assist RNA polymerase in overcoming an energetic barrier for DNA unwinding (Parvin and Sharp, 1993). The unwound structure of the H form was suggested to be appropriate for the binding of RNA polymerase, which otherwise should itself locally unpair the DNA duplex to initiate transcription (Lee et al., 1984). Thus, the H form could serve as the RNA polymerase entry point (Figure 6.2). Single-strand-binding proteins might transiently occupy a single strand, repressing RNA polymerase binding when transcription is not required. The H DNA structure appropriate for RNA polymerase entry could be formed

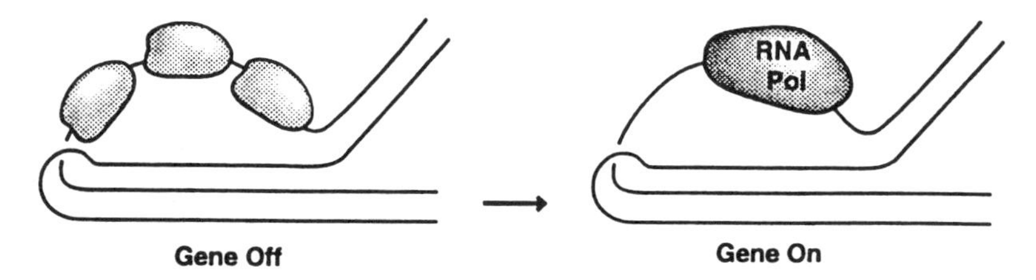

FIGURE 6.2. The H-form structure may serve as an entry point for RNA polymerase. If the H DNA is stable, its single strand may be used for binding either single-strand-specific proteins (SSB) or RNA polymerase. SSBs could hinder RNA polymerase binding whenever gene expression is unnecessary (gene off, left). Some regulatory stimulus may act to clear the site of SSB binding for RNA polymerase (gene on, right). From Sinden (1994). Reprinted with permission.

in the course of chromatin reorganization during (or just before) transcription. The negative superhelicity created by the transcription of the adjacent gene (Wang and Lynch, 1993) may be another factor stimulating DNA unwinding at the promoter site in question. If the H DNA has been formed in a topological domain in a steady state (at a cellular stage when transcription is absent), then the negative superhelicity, single-strand-binding proteins, etc. might be responsible for its maintenance before initiation of transcription.

There are no direct experimental data showing how H DNA may facilitate RNA polymerase binding, formation of the functional DNA–RNA polymerase complex, and efficient RNA synthesis. Yet the results obtained using two other models support the above suggestion.

Daube and von Hippel (1992) used a synthetic RNA–DNA bubble duplex to mimic the nucleic acid framework of a functional transcription complex. Their construct had a DNA duplex with a noncomplementary sequence 12 nucleotides in length to which an RNA oligonucleotide primer was hybridized. Both *E. coli* and T7 RNA polymerases bound to this construct, synthesized RNA with good efficiency, and possessed other properties of a normal promoter-initiated transcription elongation complex. The authors were convinced that in their experiments the core RNA polymerase (without a promoter-specific σ subunit) was able to form a functional transcription complex, suggesting that binding to a locally unwound DNA is easier.

In another model (Mollegaard et al., 1994), a locally unwound DNA structure was created by homopyrimidine peptide nucleic acid (PNA) binding that resulted in the displacement of one DNA strand (D loop, see Chapter 3). The *E. coli* RNA polymerase holoenzyme bound to this looped DNA structure and initiated synthesis of RNA. The transcript length was that of the DNA template. The PNA does not possess a 3′-hydroxyl group to be used for elongation, therefore, RNA polymerase must have started RNA synthesis without any primer. It is important that transcription initiated from a PNA-created loop, where the σ subunit recognition has no role, had an efficiency comparable to that for the strong *E. coli lacUV5* promoter, where the σ subunit is responsible for RNA polymerase binding. Again, the fact that the σ subunit is not necessary for the creation of a functional transcription complex, suggests facilitated RNA polymerase binding to an unpaired DNA site (RNA polymerase entry point).

Generally, the higher the superhelical density of the templates, the more efficient transcription is (see Wang and Lynch, 1993 for review). However, excessive superhelical tension may result in some reduction of transcription level compared to the maximum level at native levels of supercoiling. This effect was observed for bacterial (Borowiec and Gralla, 1985; Brahms et al., 1985) and eukaryotic (Parvin and Sharp, 1993) systems. For some promoters such as the DNA gyrase A, B promoters, lower superhelical density is favorable for gene expression (Menzel and Gellert, 1983; Dorman et al., 1988). Thus, H DNA formation could serve to maintain the optimum template topology by partially relieving excessive superhelical density, as suggested by Kato and Shimizu (1992).

A positive role of H DNA itself in transcriptional regulation is conceivable. The presence of the H-forming sequence in the promoter region may be favorable for transcription. Indeed, the insertion in a model plasmid of the H-forming PyPu tract upstream of the β-lactamase promoter resulted in greater transcriptional stimulation than in the plasmid with no PyPu tract (Kato and Shimizu, 1992). Any triplex-distorting variations in the sequence should reduce transcriptional efficiency. Deletions of the PyPu tracts mentioned above have led to transcriptional inhibition. The fact can be considered as evidence in favor of the regulatory role of H DNA. Point or multiple-point mutations resulting in the destabilization of H DNA may also lead to a reduction in transcriptional efficiency. Firulli et al. (1994) studied the consequences of several mutations in the S1-nuclease-sensitive promoter of the *c-myc* gene. Those mutations in the repeating sequence motif $(ACCCTCCCC)_4$ that would result in the appearance of more mismatches within suggested that the triplex led to reduced transcriptional activity of the gene.

However, an active role of the PyPu tracts and H DNA in transcription is not always found. The PyPu domains essential for transcription may be too short to form the H DNA structure (Chung and Keller, 1990; Lafyatis et al., 1991), or they may form the H DNA with a number of mismatches that should significantly weaken the entire structure (Kinniburgh, 1989; Lu and Ferl, 1992; Ulrich et al., 1992; Firulli et al., 1994). Even if the length of the PyPu tract is sufficient for extrusion of the H form, in some cases there is a lack of correlation between the H-forming potential of the DNA sequence and the transcriptional efficiency of the promoter, Glaser et al. (1990) introduced a point mutation into the PyPu tract in the promoter of the *Drosophila hsp26* gene. Although this mutation led to the destabilization of the H DNA, it had no effect on the level of gene expression. In another experiment, these authors substituted a triplex-forming tract $(GA)_n \cdot (TC)_n$ with another one, $(G)_n \cdot (C)_n$. The transcription efficiency in the latter case was indistinguishable from that for the promoter sequence without the essential $(GA)_n \cdot (TC)_n$ region. In this case, the H DNA itself played no role in maintaining high transcriptional efficiency. In a detailed structure-function study of the c-Ki-*ras* promoter, Raghu et al. has shown that various point mutations in the PyPu tract that did not affect its capability to form stable H DNA in vitro and those destabilizing the triplex lead to a comparable drop in transcriptional activity relative to the original promoter (Raghu et al., 1994). Both Glaser et al. (1990) and Raghu et al. (1994) suggested that the PyPu sequences contain binding sites for specific proteins, and that the mutations they introduced might not affect H-forming potential, but reduce the protein affinity of the sequence. In the case of the *Drosophila hsp26* gene, an immediate candidate protein is the GAGA transcription factor (Biggin and Tjian, 1988), since there are three sequence elements closely matching the GAGA binding site consensus (Glaser et al., 1990). Using the electrophoretic band shift and Southwestern hybridization assays, Raghu et al. (1994) detected two major proteins binding to the PyPu tract under investigation: a 97-kDa protein, which is likely the transcription factor Sp1, and another protein of 60 kDa in human cells and of 64 kDa in mouse cells.

A number of other proteins have been identified that bind to the PyPu tracts in promoter regions: the $(G)_n \cdot (C)_n$ tract-specific BPG1 protein in the case of the chicken β-globin gene (Clark et al., 1990), the nuclear transcriptional factor Sp1 in the case of the dihydrofolate reductase promoter (Gee et al., 1992), NFκB in the case of the interleukin-2 receptor α gene (Grigoriev et al., 1992), NSEP in the case of the c-*myc* gene (Davis et al., 1989), unidentified DNA-binding factors for the human δ-globin gene (O'Neill et al., 1991) and for the interferon-responsive promoter element of the human gene 6-16 (Roy and Lebleu, 1991), etc. Transcription may be regulated by these proteins only and therefore may not require H DNA formation.

As an example, the abnormal persistence of human fetal hemoglobin synthesis after birth may be considered. The molecular basis of this disease stems from the absence of a transcriptional switch from the fetal γ-globin gene to the β-globin gene. Ulrich et al. (1992) found that mutations in the PyPu tract upstream of the γ-globin gene that should result in destabilization of a putative imperfect intramolecular triplex led to the expression of the gene when it should be turned off. Thus, if H DNA is involved, its instability and concomitant failure to repress transcription may lead to the inappropriate expression of a gene and the development of the disease. This H DNA might be stabilized by a protein that preferentially binds to a single-stranded DNA probe whose sequence corresponds to the PyPu tract (Horwitz et al., 1994). As an alternative explanation, the PyPu tract, where H DNA presumably forms, contains sequences which resemble recognition sites for B DNA-binding proteins such as transcription factors Sp1, octamer-binding protein OTF-1 and erythroid regulatory factor NFE-1 that may play a role in regulation of gene expression (Fischer and Nowock, 1990; O'Neill et al., 1990; Sykes and Kaufman, 1990; Jane et al., 1993). Mutations in the PyPu tract make imperfect sites more appropriate for protein binding, thereby activating transcription and resulting in abnormal γ-globin synthesis. The experimental data accumulated to date gives no basis for preferring to either of these explanations.

Another line of hypotheses considers the PyPu tract as a possible site of H DNA formation, or alternatively as a binding site for the B DNA-specific protein, which may activate or repress transcription. For example, such a protein may be a transcription factor that promotes RNA polymerase binding to a promoter (Figure 6.3A). The gel comigration data of Kohwi and Kohwi-Shigematsu (1991) show that the $(dG)_n \cdot (dC)_n$ tract might be a binding site for an activator protein present in human cells. The formation of H* DNA makes the structure of the PyPu tract inappropriate for protein and subsequent RNA polymerase binding, thereby inhibiting gene expression.

If a protein itself is a repressor that hinders favorable interactions of RNA polymerase with its binding site, the formation of H DNA at the repressor binding site might be similar to the action of the inducer protein that serves to displace the repressor and facilitate RNA polymerase binding at the adjacent site (Figure 6.3B; Sinden, 1994). No experimental data are available to illustrate this attractive hypothetical option in the utilization of H DNA by living cells.

Transcription in eukaryotes is assumed to be regulated mainly at the initiation and termination stages. Yet, as a complex, multistep process, transcription may be regulated at any of the numerous steps: binding of RNA polymerase and accessory proteins, initiation, elongation, and

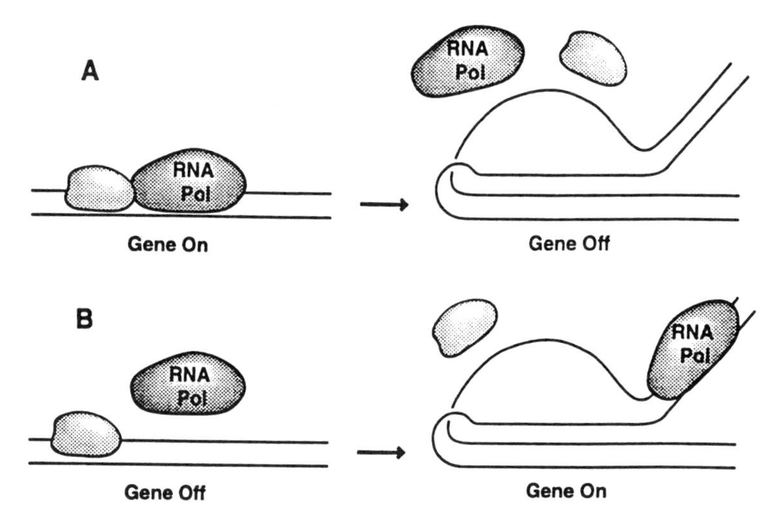

FIGURE 6.3. A possible interplay between B DNA-binding proteins and the H form in the stimulation or repression of a gene. (A) A B DNA-binding protein as a positive transcription factor could bind to a promoter region and facilitate RNA polymerase binding (gene on, left). An H form at the same site would preclude binding of the transcription factor and, as a result, of the RNA polymerase (gene off, right). (B) In another option, a DNA-binding protein could serve as a repressor (gene off, left). In this case, extrusion of the H form would be similar to the binding of an inducer protein which displaces repressor, thereby making RNA polymerase binding possible (gene on, right). From Sinden (1994). Reprinted with permission.

termination (Spencer and Groudine, 1990). A blockage or pausing during elongation has been identified during transcription of HIV-1, human and mouse *c-myc* and ADA, and *Drosophila hsp70* genes (see Spencer and Groudine, 1990, for review). The possible role of triplex at the elongation stage may be illustrated by the experiments of Sarkar and Brahmachari (1992), who used codon degeneracy to engineer a 38-base-pair-long H-forming sequence into the β-galactosidase gene of the pBluescriptIISK+ plasmid. *E. coli* JM109 cells transformed with the H DNA-forming construct had an 80% lower expression of the β-galactosidase gene than another plasmid in which other codons that did not constitute the PyPu tract coded for the same amino acid sequence. This down regulation of gene expression can be explained by the formation of a triplex in the PyPu tract located inside the gene. The structure of H DNA is assumed to block the RNA polymerase progressing along the DNA template.

The DNA–RNA hybrids could also take part in the regulation of gene expression (Figure 6.4). Similar to the B DNA-specific protein, a ribonucleoprotein complex could act either as a transcription factor facilitating RNA polymerase binding or as a repressor (Miller and Sobell, 1966). In this case, the specificity of binding would be determined by the RNA strand–PyPu duplex, rather than protein–DNA interactions. The RNA–DNA hybrid duplex between the RNA strand of the ribonucleoprotein complex and a single strand of the H DNA could provide enough specificity for the ribonucleoprotein complex-mediated

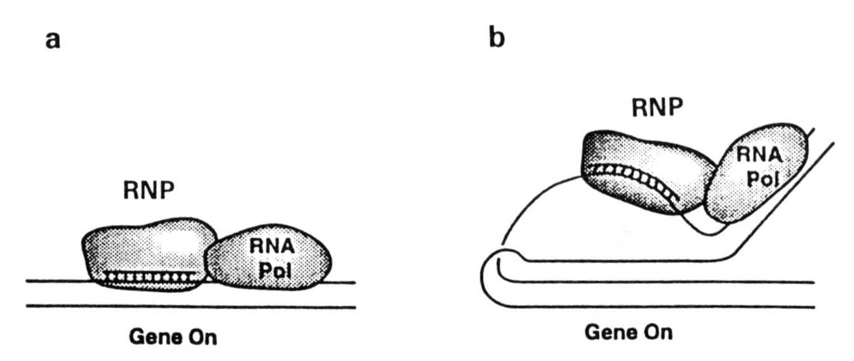

FIGURE 6.4. Potential interactions of DNA with an RNA strand. A ribonucleoprotein (RNP) complex may bind DNA using its RNA molecule for sequence-specific recognition. (a) If the RNP complex plays a role of a transcription factor, bound via a triplex-recognition principle, it may interact with the RNA polymerase facilitating transcriptional initiation. Such an RNP complex may also act as a classical transcriptional repressor, preventing the RNA polymerase from binding to the promoter (not shown). (b) The RNP complex may form a duplex with the single strand of a preformed H DNA and then recruit the RNA polymerase to initiate transcription at the loop–duplex junction. From Sinden (1994). Reprinted with permission.

binding of RNA polymerase at the loop–duplex junction (Sinden, 1994). The RNA strand itself may form a duplex with a looped DNA strand to stabilize the entire H DNA structure. In this case, H DNA may serve to relax a few superturns in order to maintain optimal superhelical density of a topological domain during transcription.

Although there is a definite basis to suggest a regulatory role of the PyPu tracts in 5′ flanking regions of various genes, there has been no direct evidence that it is the structure of H DNA which plays a role of transcriptional regulator in vivo. The suggestion that B DNA-specific proteins binding to the PyPu tract play an active role in gene expression also does not contradict experimental data. There is also a difficulty originating from the rare correspondence in the position of the tracts in the same gene in different species. For example, in analyzing histone H4 genes from *X. laevis*, chicken, mouse and two from sea urchin, Bucher and Yagil (1991) found the PyPu tracts in all of the five genes. However, the PyPu tract in the promoter region was found only in the mouse H4 gene. Thus, the regulatory role of the PyPu tracts (if any) must be different for H4 genes from different species.

Possible Regulation of Replication

A phenomenon similar to transcription elongation control was observed for another vital cellular process, DNA replication. Murray and Morgan (1973) found that replication of a model polynucleotide duplex was inhibited by third strand binding. Manor and coworkers (Manor et al., 1988; Rao et al., 1988; Lapidot et al., 1989; Baran et al., 1991) studied the influence of PyPu tracts on the replication of single- and double-stranded DNA. They found that the $(GA)_n \cdot (TC)_n$ tracts pause several DNA replication enzymes in vitro under conditions appropriate for triplex formation. Termination of DNA replication by the triplex-forming PyPu tracts has also been found in other studies (Dayn et al., 1992b; Samadashwily et al., 1993; Rao, 1994; see Mirkin and Frank-Kamenetskii, 1994, for review). There are several different triplex structures that might be involved in replication blockage.

In the experiments of Dayn et al. (1992), the H* DNA consisting of CG*G and TA*T triads could form either H-r3 or H-r5 isomers, depending on the specific sequences designed. The termination sites were located differently for specific H*-forming regions but were mapped precisely by chemical probing of the triplex-forming sequence. In this case, the replication-terminating structure was formed prior to DNA synthesis. Figure 6.5 shows sites where the movement of DNA polymerase may be hindered by the preexisting triple-stranded DNA structure. Depending on the H* DNA isomer, DNA polymerase moving in the 3′ to 5′ direction along the template strand might stall either at the end or in the middle of the PyPu tract.

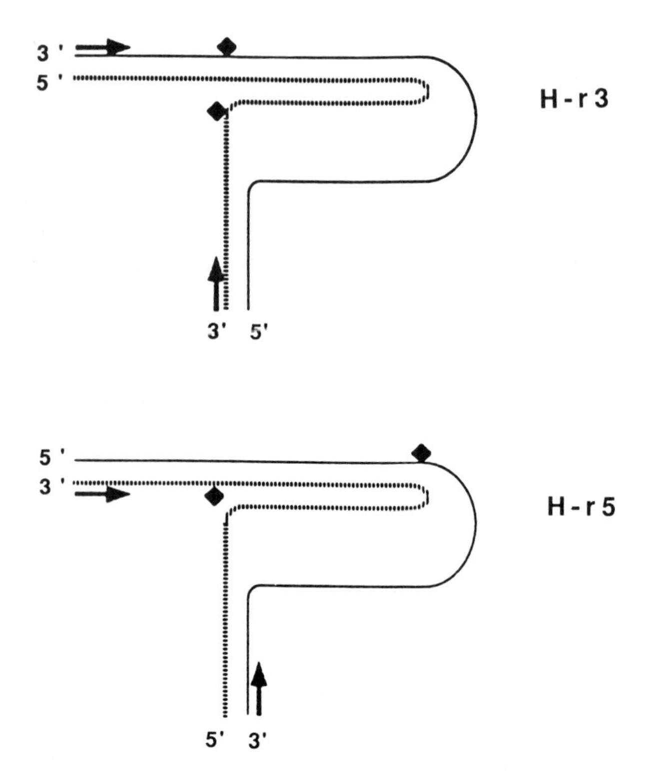

FIGURE 6.5. The progression of a DNA polymerase along one strand of the double-stranded template may be hindered by the preexisting triple-stranded DNA structure. Depending on the H* DNA isomer, DNA polymerase advancing in the 3′ to 5′ direction faces the triplex either at the end or in the middle of the PyPu tract. From Mirkin and Frank-Kamenetskii (1994). Reproduced, with permission, from Annual Review of Biophysics and Biomolecular Structure, Vol. 23. © 1994 by Annual Reviews Inc.

An active DNA polymerase itself may create a DNA structure that blocks polymerization. During the replication of single-stranded DNA, a DNA polymerase stalls in the center of the homopyrimidine or homopurine tract (Lapidot et al., 1989; Baran et al., 1991; Samadashwily et al., 1993). Baran et al. (1991) suggested that when the new DNA chain has been synthesized through the middle of the homopolymer template, another portion of the homopolymer sequence folds back and forms a triplex (Figure 6.6). The DNA polymerase is trapped in this structure, preventing further DNA synthesis. If the DNA is double-stranded, the DNA polymerase uses only one strand as a template, and the other, nontemplate strand is displaced during synthesis. The latter strand containing the purine sequence folds on itself downstream of the replication fork. The resulting triplex presents one more type of replication

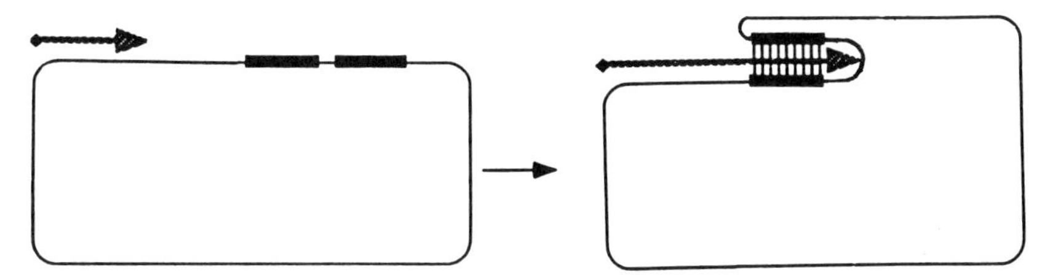

FIGURE 6.6. When the new DNA chain is synthesized through the middle of the homopolymer template, another portion of the homopolymer sequence folds back and forms a triplex (Baran et al., 1991). This triplex serves as a trap for the DNA polymerase, which cannot continue DNA synthesis. From Mirkin and Frank-Kamenetskii (1994). Reproduced, with permission, from Annual Review of Biophysics and Biomolecular Structure, Vol. 23. © 1994 by Annual Reviews Inc.

block (Figure 6.7). In accordance with this model, in experiments on open circular DNA, which cannot form a triplex before replication, T7 DNA polymerase stalled exactly in the middle of the PyPu tract when the purine-rich strand was displaced (Samadashwily et al., 1993). Further evidence of replication termination by the induced triplex came from the mutational analysis. The H* DNA-destabilizing mutations relieved the DNA polymerase from the polymerization block, whereas the compensatory mutations restoring the H* DNA-forming potential restored the replication block.

The above-described models of DNA replication blockage by triplex structures were elaborated in experiments in vitro. The actual replication fork contains, in addition to DNA polymerase, a number of accessory

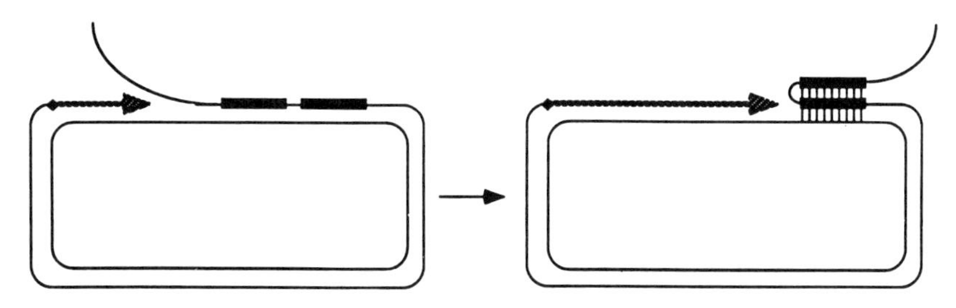

FIGURE 6.7. Some DNA polymerases may act without accessory proteins and synthesize a new DNA chain using one strand as a template and displacing another, nontemplate strand. The latter strand containing a PyPu sequence may fold on itself downstream of the replication fork. During the replication of open circular DNA under conditions favorable for both replication and H* DNA formation, the T7 DNA polymerase stalled exactly in the middle of the PyPu tract when the purine-rich strand was displaced. Reproduced from Samadashwily et al. (1993) *EMBO J* by permission of Oxford University Press.

proteins (helicases, topoisomerases, single-strand-binding (SSB) proteins, etc). These proteins may destroy the preformed triplex, prevent DNA single strands from folding onto their target duplex, etc. Little is known about the influence of accessory proteins on triplex-caused termination. Model studies show that SSB easily restores replication by unwinding the intramolecular triplex, but it is less effective in disrupting intermolecular triplex (Samadashwily and Mirkin, 1994). In in vitro experiments the DNA helicase from *E. coli* was able to unwind a model intermolecular triplex (Maine and Kodadek, 1994); however, the helicase activity of the SV40 T-antigen was strongly inhibited by triplex formation (Peleg et al., 1995). The reason for this discrepancy is not clear. The in vivo data on a role of the PyPu tracts in replication is limited. The $(GA)_{27} \cdot (TC)_{27}$ tract-containing region of DNA located 2 kb from the integration site of the polyoma virus was shown to be a strong terminator of DNA replication in rat cells (Baran et al., 1983). The involvement of this PyPu tract in replication termination was confirmed by the DNA polymerase pausing when a corresponding fragment was cloned into SV40 DNA (Rao et al., 1988). The replication of SV40 DNA in CV-1 cells was inhibited by the triplex-forming oligonucleotide-acridine conjugate (Birg et al., 1990). Thus, at least in some cases triplex formation can block DNA replication in vivo.

Brinton et al. (1991) found an unusual cluster of simple repeats, including a Z DNA-forming region $(GC)_5(AC)_{21}$ and a long PyPu tract with a potential for H DNA formation, to have a significant effect on replication of a plasmid shuttle vector. One copy of this cluster, when cloned on either side of the origin of replication of SV40, reduced the amount of DNA replicated in COS cells up to twofold. Two copies on both sides of the origin reduced replicated DNA down to 5% of that in a vector without the cluster.

Possible Triplex-Mediated Chromosome Folding

Long chromosomal DNA can be segregated in separate domains which have the form of loops somehow tied at the bases (Stonington and Pettijohn, 1971; Worcel and Burgi, 1972; Pettijohn and Hecht, 1973; Sinden and Pettijohn, 1981). Distribution of PyPu tracts over the whole length of genomic DNA (Birnboim et al., 1979; Hoffman-Liebermann et al., 1986; Manor et al., 1988; Wong et al., 1990; Bucher and Yagil, 1991; Tripathi and Brahmachari, 1991) have led to the suggestion that triplexes may promote chromosome compact packaging (Lee and Morgan, 1982; Hampel et al., 1993).

Lee and coworkers designed and conducted a series of experiments to show the DNA condensation due to triplex formation between different DNA molecules or distant PyPu tracts of the same molecule (Lee et al., 1989; Hampel et al., 1993; Hampel and Lee, 1993; Hampel et al., 1994).

Using a plasmid containing a long $(GA)_{45} \cdot (TC)_{45}$ insert and a long copolymer of d(Tm^{5}C), they were able to observe "rosettes" in an electron microscope that were produced when the same molecule of poly[d(Tm^{5}C)] formed three-stranded structures with the PyPu tracts of several plasmids (Lee et al., 1989). This was the simplest model of plasmid DNA packaging. In another experiment, Hampel et al. (1993) used the linear plasmids containing single PyPu tracts at the ends to observe the linear dimerization of DNA molecules via triplex formation. In the case of plasmids with PyPu tracts at both ends, structures with electrophoretic mobilites of relaxed circular DNA were formed. When the PyPu tracts in the same molecule were relatively far from its ends, Hampel et al. (1994) observed Ω-shaped loops. Indirect evidence that may be interpreted in favor of triplex-mediated DNA condensation was obtained in two-dimensional pulsed-field electrophoresis of yeast chromosomes (Hampel and Lee, 1993). Whereas the chromosomal DNA had normal mobility in the first dimension run at pH 8, its mobility was drastically reduced in the second dimension run at acidic pH in the presence of polyamines as triplex stabilizers. Many agents, including polyamines, may be used to nonspecifically condense DNA (Bloomfield, 1991), but only a mechanism in which triplex formation occurs is expected to be pH-dependent. Once formed at pH 4 to 6, DNA aggregates then stably exist at neutral pH. Some other factors (e.g., proteins) may substitute pH as a factor lowering the barrier between the B and H forms in vivo. In the experiments of Hampel et al. (1994), a triplex structure is formed between the PyPu duplex and one strand of another unwound PyPu tract. Another strand of the second PyPu tract remains unpaired (Figure 6.8). In one of the possible structures, the acceptor duplex should pass through the bubble of the unpaired PyPu tract once for each turn of the triplex, forming a so-called braided knot (which is not a true knot but rather a hydrogen-bonded knot). With the short linear ends, it is conceivable to untie this knot without breaking covalent bonds. In vivo formation of such knots would require the presence of topoisomerase or other nicking-closing activity. Another structure consists of two interwound tracts. Here one strand of the looped tract is a third strand of a triple helix, whereas another (fourth) strand is extruded but must wrap around the triplex without interacting with it.

The biological significance of the interaction of two DNA duplexes (or remote parts of the same duplex) via the triplex mechanism may be suggested in chromosome condensation and gene expression. The quasi-cyclization of chromosomal DNA presents at least one level of DNA condensation. Although an efficient multimerization (condensation) of DNA in vitro can be produced by the triplex-promoting factors (e.g., polyamines), the interaction between PyPu tracts may not be sufficient for DNA packaging in vivo and may require the participation of some proteins. Mammalian viral DNA, which is as rich in PyPu tracts as

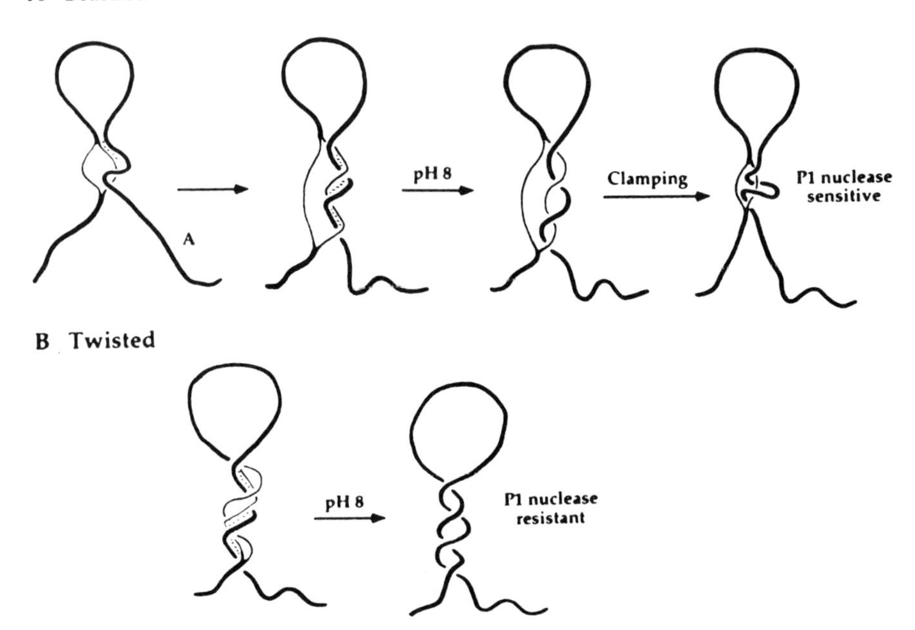

FIGURE 6.8. Models for the junctions between two distant PyPu tracts of the same plasmid mediated by triplex formation. Similar junctions are probably also formed between long chromosomal DNAs. Thick and thin lines denote duplex and single-stranded regions, respectively. Dashed lines stand for Hoogsteen bonds between the duplexes and looped single strands. In a braided junction (panel A), one DNA end (designated A) passes through the PyPu loop once for each turn of the forming triplex. Once formed at lower pH, this structure is stable at pH 8; but, due to the presence of the single strand, it is sensitive to a single-strand-specific nuclease P1. In a twisted model (panel B), two PyPu tracts are interwound, one of them donating a single strand for triplex formation. The remaining unpaired strand appears to be wrapped around the triplex sufficiently tightly and is resistant to nuclease P1. Reprinted with permission from Hampel et al. (1994). © 1994 American Chemical Society.

nuclear genes, is generally packed loosely, as contrasted with the tight packaging of nuclear genes in a chromosome. Therefore, at least for viruses, a compacting role of PyPu tracts themselves is not clear, and some proteins may be important.

Interaction of distant PyPu tracts required for DNA condensation may also be relevant to the transcriptional process. In some cases, two PuPy tracts are located upstream of the start of transcription (Boles and Hogan, 1987; Gilmour et al., 1989; Kohwi, 1989; Firulli et al., 1992; Lu et al., 1993). Both stretches determine in concert the transcription efficiency of the *Drosophila hsp26* gene (Lu et al., 1993). The interaction of two PyPu stretches could be to some extent similar to the better-known case in which the initiation of transcription requires some distant sequence (for example, an enhancer or upstream activating sequence) to be brought in close proximity to a promoter (Guarente, 1988; Collado-Vides et al., 1991). Such an interaction of distant sequence elements is thought to be accomplished through a protein-mediated double-stranded DNA looping (reviewed by Schleif, 1992). At least theoretically, specific DNA sequences may also be linked over long distances by a direct interaction between PyPu tracts, resulting in triplex formation, as suggested by Hampel et al. (1994). It should be noted, however, that interactions of distant PyPu tracts may be mediated by B DNA-specific proteins as well.

Structural Role at Chromosome Ends

Telomeres are structures that stabilize the ends of eukaryotic chromosomes. They consist of a very long repetition of a motif consisting of six to eight nucleotides with the general sequence $(T/A)_mG_n$ (Blackburn and Szostak, 1984; Zakian, 1989). The secondary structure of telomeres consists of the DNA duplex and the single-stranded overhang. Model oligonucleotide telomeric structures form inter- and intramolecular quadruplexes in the presence of monovalent sodium and potassium cations (Henderson et al., 1987; Sen and Gilbert, 1989; Sundquist and Klug, 1989; Williamson et al., 1990).

The study of a synthetic model of the *Tetrahymena* chromosome telomeric terminus, consisting of the DNA duplex and the single-stranded overhang $(T_2G_4)_2$, showed that in the presence of divalent Mg^{2+} cations and physiological pH, the overhang folds back to form a triplex (Figure 6.9). These authors suggested that the triplex structure of telomeres provides a plausible explanation for the in vivo resistance of chromosome ends against degradation and recombination.

3'-CAGTTCGAACCAACCCCAACCCCAACCCC
5'-GTCAAGCTTGCTTGGGGTTGGGGTTGGGG—T
3'-GGGGTTGGGG—T

FIGURE 6.9. Single-stranded overhang $(T_2G_4)_2$ and DNA duplex of the same sequence common to telomeric termini may form a triplex in the presence of Mg^{2+} cations and physiological pH. Since the triple-stranded structure is not an appropriate substrate for many proteins, this triplex structure of telomeres may provide a plausible explanation for the in vivo resistance of chromosome ends against degradation and recombination. Reprinted with permission from *Nature*, Veselkov et al. (1993). © 1993 Macmillan Magazines Limited.

Recombination

Specific kinds of triple-stranded structures have been suggested to mediate homologous DNA strand recombination in the presence of RecA protein (see Radding, 1991; Stasiak, 1992; Camerini-Otero and Hsieh, 1993; Rao and Radding, 1994; Zhurkin et al., 1994a,b, for models, reviews, and discussion). However, these triple-stranded structures do not require the PyPu tracts; the pattern of hydrogen bonding in them drastically differs from that in triplexes formed in the PyPu tracts; the axial spacing between base pairs is stretched to 5.1 Å; and strands of like polarity are aligned in parallel fashion.

Recently, a role of more conventional intramolecular triplexes in recombination was suggested. The state of DNA supercoiling has an effect on homologous and site-specific DNA recombination in vivo. For example, recombination is activated by transcription, which results in a negative superhelical strain upstream of the transcribing gene (Yancopoulos and Alt, 1985; Kim and Wang, 1989; Thomas and Rothstein, 1989). The topoisomerases relaxing supercoiled regions of intracellular DNA may suppress supercoil-stimulated recombination (see Wang et al., 1990, for review). Supercoiling may change the local chromatin structure, giving recombination proteins access to their substrate. The supercoil-induced DNA conformation may also be directly involved in recombination processes.

The conformational aspect of recombination has some experimental basis. The DNA rearrangement in the immunoglobulin class from IgM to IgA, IgG, or IgE occurs in the switch regions, which are complex, highly repetitive regions of DNA. For example, the sequence $(AGGAG)_{28}$ capable of forming H DNA in vitro is located in the switch region of murine IgA (Collier et al., 1988). The possibility exists that this unusual structure may provide a single strand that could pair with a homologous region of a second chromosome initiating recombination, which results in the splicing of antibody genes from a number of individual gene segments. In addition to a common H DNA, some other conformation may be involved. Experiments in vitro showed that transcription through the immunoglobulin switch region results in the formation of some local structure with a nascent mRNA chain as its integral part (Reaban and Griffin, 1990).

A possible role of unusual structures (H DNA or Z DNA) in recombination may be implied in an unequal sister chromatid exchange, in which part of the gene is duplicated on one chromosome and deleted from the other chromosome (Weinreb et al., 1990). The recombination region contains simple repeats of $(TC)_n$ capable of forming H DNA, followed by stretches of the Z-forming $(TG)_n$ sequence.

Kohwi and Panchenko (1993) described a transcription-induced recombination between two direct repeats separated by the sequence containing the H*-forming PyPu tract (Figure 6.9). No recombination was detected in the control plasmid. The $(dG)_n \cdot (dC)_n$-containing plasmid constructs allowed the efficient recombination between homologous sequences of *lac* and *tac* promoter sequences separated by either 200- or 1,000-base-pair regions. The recombination was RecA-independent, the recombination rate being dependent on the length and orientation of the $(dG)_n \cdot (dC)_n$ tract with respect to the gene. Under active transcription conditions in *E. coli*, the plasmids formed CG*G type triplexes, as shown by chemical modification. The H* DNA in this study was suggested to bring two remote sequences in close proximity to make recombination favorable.

The role of H DNA in recombination was implied in two other studies in which no mechanism was suggested. Formation of dimer molecules in recombinant plasmids carrying the H-forming PyPu tracts occurs six times more often than in control plasmids (Kato, 1993). In human cells, homologous recombination between the plasmids containing H DNA-forming sequences occurs three times more often than in controls (Rooney and Moore, 1995).

Generally speaking, the above-described hypothetical models of H DNA involvement in genetic recombination include a displacement mechanism (Figure 6.10), interaction of the H and H* forms (Figure 6.11), and an H form-induced duplex bend that brings homologous sequences into close proximity (Figure 6.12). An RNA–DNA hybrid formed during transcription through the immunoglobulin switch regions was also suggested to be relevant; however, it is not as clear how the proposed structure (see also Reaban et al., 1994) may participate in a recombination event.

A few results showing increased rates of plasmid dimerization, transcription-driven recombination between direct repeats, and the presence of H-forming tracts close to recombination points are indirect evidence in favor of the triplex playing a role in recombination. More experimental and theoretical considerations are needed to fully establish this role.

Possible Role in Mutational Processes

Sequences capable of forming an H DNA structure are often prone to deletion. For example, the simple repeating sequence $(GAA)_n \cdot (TTC)_n$ is genetically unstable when cloned as a long region in plasmid in *E. coli* (Jaworski et al., 1989). Several reasons for the instability of triplex-forming sequences have been suggested (Wells and Sinden, 1993). First, the H DNA was shown to pause or terminate DNA replication in vivo and in vitro (Lapidot et al., 1989; Baran et al., 1991; Brinton et al., 1991; Dayn et al., 1992b; Sarkar and Brahmachari, 1992; Samadashwily et al., 1993; Rao, 1994). The pausing of a replication fork has been associated with a high frequency of mutagenic events (Bebenek et al., 1989). A

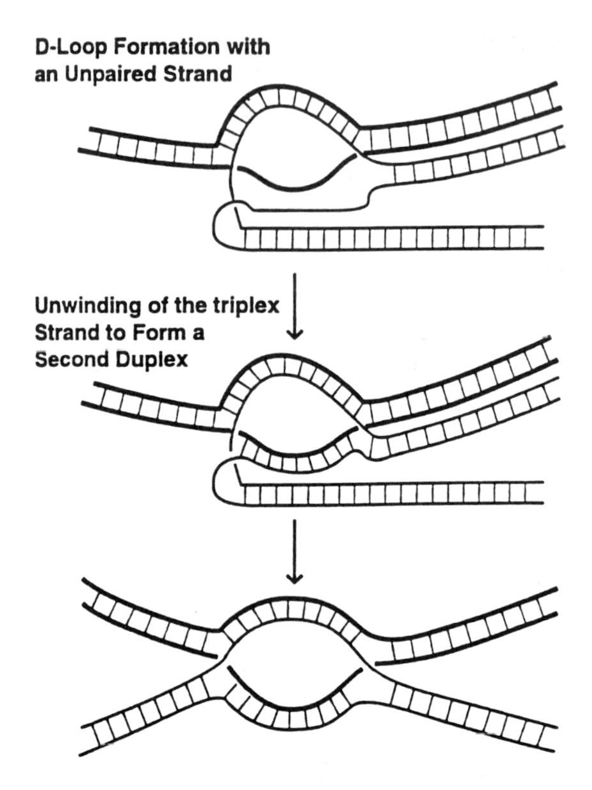

FIGURE 6.10. Possible mechanism of H DNA-mediated genetic recombination. A free single strand of H DNA pairs a complementary strand of a second homologous duplex (top). Following D-loop formation, the third strand of the triplex unpairs. Being complementary to the displaced D-loop strand, it initiates the formation of the Watson–Crick duplex (middle). Rotation of the bottom duplex from right to left shows that a classical recombination intermediate has been formed (bottom). From Sinden (1994). Reprinted with permission.

stalled DNA polymerase could generate long simple repeated sequences in reiterative synthesis, leading to the expansion of the PyPu tract. If this sequence can form another folded structure (e.g., cruciform), it may be deleted (Sinden et al., 1991). Second, simple repeating PyPu sequences themselves i.e., $(G)_n \cdot (C)_n$ or $(GA)_n \cdot [(TC)_n]$ have a high probability of slipped misalignment, which would result in deletions or duplications during DNA replication (reviewed by Sinden and Wells, 1992; Wells and Sinden, 1993).

Do PyPu Tracts Play a Role in RNA Splicing?

The presence of the PyPu tracts in introns could suggest their role in mRNA splicing. Indeed, changes from a pyrimidine to a purine tract at the 3′ end of transcripts of mammalian intervening sequences can interfere with mRNA splicing (Ruskin and Green, 1985). However, most of the PyPu tracts were found more than 100 bases away from the 3′ ends of the introns, and for 163 genes studied, the frequency of the PyPu tracts in introns does not exceed the statistically expected frequency (Bucher and Yagil, 1991). However, the role of base-triad-containing structures in splicing could be to maintain the tertiary interactions that stabilize the structure of the ribozyme, as suggested by the phylogenetic and genetic studies of group I intron sequences (Michel et al., 1990; see next section).

Elements of Triple-Stranded Structure in RNA

In RNA, base triads may form when an unpaired nucleotide forms hydrogen bonds with a purine nucleotide that is already base-paired. The

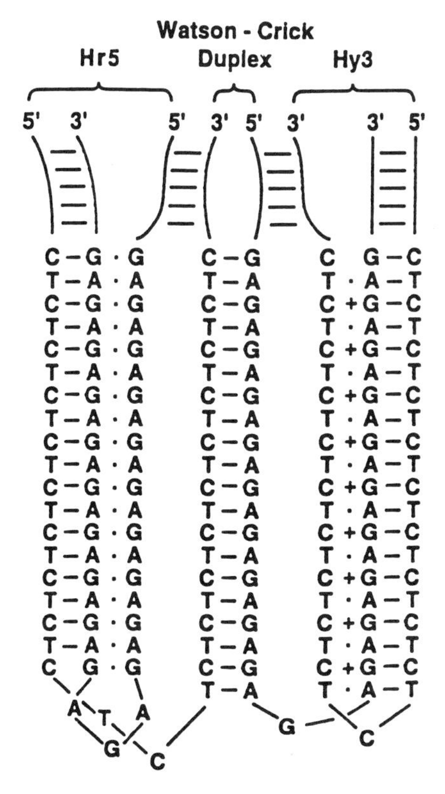

FIGURE 6.11. Recombination may involve two H DNA structures. This model suggests the formation of different (H-r5 and H-y3; or H-r3 and H-y5) isomers in different DNA molecules containing the same PyPu sequence. Single strands of these H and H* forms are complementary and form a Watson–Crick duplex. The unwinding of the strands forming triple helices and their pairing in the Watson–Crick duplexes results in a classical recombination intermediate. From Sinden (1994). Reprinted with permission.

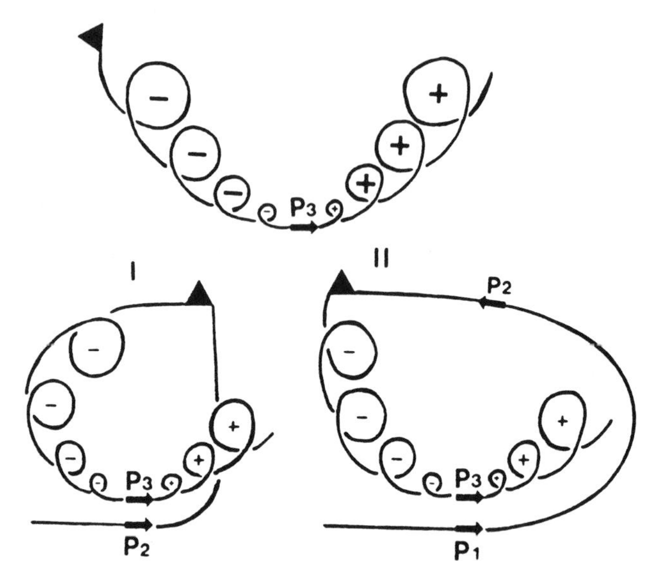

FIGURE 6.12. Models for the triplex-induced approach of different homologous sequences in promoter regions (designated P1, P2 and P3). Once the promoter containing P3 is activated, the transcriptional process results in two supercoiled domains: positive supercoiling accumulates ahead of the transcribing RNA polymerase, whereas negative supercoiling accumulates behind RNA polymerase. Negative supercoiling-induced H* DNA (designated by a filled triangle) significantly bends the DNA helix, bringing P2 and P3, which are separated by 200 base pairs, into close proximity to stimulate recombination. Similarly, P1 and P3, which are 1,000 base pairs apart, may be brought together. From Kohwi and Panchenko (1993).

data accumulated to date do not present any evidence of RNA folding that include long stretches of base triads; however, separate RNA triads may have important roles (reviewed by Chastain and Tinoco, 1991). For the latest extensive analysis of RNA base triads, see Gautheret et al. (1995).

The first class of RNA molecules containing base triads is tRNA. Triads were found at the hinge region of tRNA between the two helical domains, and they presumably stabilize the tRNA in its three-dimensional L shape. Three base triads have been found in $tRNA^{Phe}$, all of which ($U_{12}A_{23}{*}A_9$, $C_{13}G_{22}{*}G_{46}$, and $C_{25}G_{10}{*}G_{45}$) involve nucleotides in the junction loop binding to the Watson–Crick pairs of the D stem (Saenger, 1984). Base triads in the structure of $tRNA^{Asp}$ include the $\Psi_{13}G_{22}{*}A_{46}$ triad and an unusual triad at the beginning of the D loop, where A_{21} binds to the reverse-Hoogsteen U_8A_{14} pair by forming base–base and base–sugar hydrogen bonds (Westhof et al., 1985).

Another prominent example of base triads in RNA is the self-splicing intron from *Tetrahymena*. There is strong evidence that the CG*G triad participates in intron splicing. The cleavage at the 5′ exon involves binding of a free guanosine, whereas the cleavage at the 3′ exon requires binding of an internal guanosine. It was suggested that these guanosines form base triads with the $C_{311}G_{264}$ base pair in the P7 stem (Michel et al., 1989). Replacement of the CG pair with a UA pair, which is not able to form a base triad, suppresses the splicing reaction. The use of a 2-aminopurine-uracil pair, which is able to form the base triad isomorphic to the CG*G pair, restores cleavage at the 5′ exon.

The formation of a base triad was also suggested to explain the structure of *Xenopus laevis* 5S rRNA (Westhof et al., 1989). In this system, chemical modification and model building data were interpreted in terms of formation of a UG*A triad between a nucleotide in a junction loop and an adjacent helix.

The formation of base triads in junction regions where helices stack coaxially may be a recurring RNA structural element (Chastain and Tinoco, 1991). Unlike the triplex structures of nucleic acids considered to date, where only the major groove location of the third strand was found, when two helical regions in a junction stack coaxially, an unusual triplex combination results. The free 3′ end should enter the major groove of the adjacent helix, and the free 5′ strand should enter the minor groove. In the case of RNA, the evidence for minor groove binding of the third strand is that a triple helix did not form when the poly(rG) strand was replaced by poly(rI), which lacks the minor groove amino group capable of hydrogen bonding to poly(rA) (Chastain and Tinoco, 1992).

Other Roles of the PyPu Tracts

Coding of Charged or Hydrophobic Amino Acid Clusters

It is probable that the PyPu tracts can play a role in coding of relatively long, specialized protein structures (Bucher and Yagil, 1991). This can be illustrated by an examination of the SV40 genome. The early proteins of SV40 are coded by that DNA fragment which contains PyPu tracts, the vast majority of which are of the purine type in the coding strand. Bearing in mind that the purine triplets AAA and AAG code for the amino acid lysine, and AGA and AGG code for the amino acid arginine, one can expect that the resulting protein sequences will be rich in charged (mainly positive) residues. This is in accordance with the need of these early proteins to interact with negatively charged DNA.

Can the PyPu Tracts Exclude Nucleosomes from Certain Gene Regions?

Accommodation into the nucleosome core and participation in a non-B conformation seem to represent alternative options for sequences capable of forming unusual structures (see van Holde and Zlatanova, 1994, for review). The presence of long PyPu tracts might play a role in the exclusion of nucleosomes from certain gene regions. For example, long (several dozen base pairs) segments of (dA) · 7(dT), which may form a specific non-B conformation, were not readily incorporated into nucleosomes (Simpson and Kunzler, 1979; Rhodes, 1979; Struhl, 1985; Beasty and Behe, 1987). In addition, theoretical considerations provided evidence that the PyPu sequences in DNA bend less readily and therefore resist packaging in nucleosomes more than mixed sequences (McCall et al., 1985; Zhurkin, 1985).

However, detailed studies with a number of different sequences showed that conditions can be found that allow reconstitution of poly(dG) · poly(dC), poly(dA) · poly(dT), poly[d(AG)] · poly[d(TC)], and even the mixed duplex poly(rGdC) · poly(rGdC) into nucleosomes (Jayasena and Behe, 1989; Hayes et al., 1991; Puhl et al., 1991). Reconstitution of the PyPu tracts into nucleosomes may be less favorable compared to random sequence DNA [e.g., the corresponding loss of free energy for $(dA)_n \cdot (dT)_n$ tracts is approximately 1.1 kcal/mol (Hayes et al., 1991)], but not impossible. It should also be noted that in early studies of micrococcal nuclease-resistant regions of bulk mouse L-cell DNA, Birnboim et al. (1976) came to the conclusion that PyPu tracts occur in nucleosome sequences as frequently as in internucleosome regions.

Thus, at the present state of knowledge, one can affirm that in spite of somewhat unfavorable free energy relative to heterogeneous sequence DNA, PyPu tracts can be accommodated in nucleosomes.

Conclusion

A number of attractive hypotheses for a biological significance of triple-stranded nucleic acids have been put forward. The limited amount of collected experimental data provide a definite basis for suggesting the involvement of triple-helical structures as key participants in some vitally important processes in the living cell. Another, nondirect, approach to elucidating the role of triplexes in the cell is to use DNA-binding proteins and study their influence. This approach may play a role in determining the involvement of triplexes in these processess. The DNA-binding proteins themselves may participate in the same processes. Finally, the application of immunological methodology can be potentially useful for these goals. Unfortunately, due to a shortage of adequate data and an absence of direct evidence for triplexes in living cells, the question of the existence of triplexes in vivo is still open, and more experimental data are needed to understand the biological roles of triple-stranded nucleic acids.

7
Possible Spheres of Application of Intermolecular Triplexes

Applications of Intermolecular Triplex Methodology

Triple helix formation represents the basis for numerous site-specific manipulations with duplex DNA. Appropriately designed third-strand oligonucleotides that hybridize to the targeted duplex domains have been suggested in the development of several practical applications of triplex methodology:

1. Extraction and purification of specific nucleotide sequences
   a. Triplex affinity capture (Ito et al., 1992a–c)
   b. Affinity chromatography (Pei et al., 1991; Kiyama et al., 1994)
   c. Stringency clamps (Roberts and Crothers, 1991)
2. Quantitation of polymerase chain reaction products (Vary, 1992)
3. Nonenzymatic ligation of double-helical DNA (Luebke and Dervan, 1991, 1992)
4. Triplex-mediated inhibition of viral DNA integration (Mouscadet et al., 1994a)
5. Site-directed mutagenesis (Havre et al., 1993; Havre and Glazer, 1993)
6. Detection of mutations in homopurine DNA sequences (Olivas and Maher, 1994a)
7. Mapping of genomic DNA
   a. Triple Helix Vector (Moores, 1990)
   b. Artificial endonucleases (Pei et al., 1990; Perrouault et al., 1990; Dervan, 1992; see Povsic et al., 1992, for a short review)
   c. Electron microscopic mapping (Cherny et al., 1993b)
8. Control of gene expression (Cooney et al., 1988; see Hélène et al., 1992, for review)

Extraction and Purification of the Specific Nucleotide Sequences

Most experiments in molecular biology include the preparation of DNA as an important and necessary step. Isolated DNAs vary widely in their size. Many current methods employ labor-intensive procedures such as extraction and centrifugation or in situ filter hybridization to colonies and plaques (Sambrook et al., 1989). In addition, filter hybridization procedures include prior denaturation steps and other treatments that destroy the integrity of the target DNA molecules, so that one has to reisolate the corresponding clones from the original plates to obtain intact DNA molecules. These drawbacks can be eliminated using different variants of triplex affinity capture (TAC).

Hybridization of Biotinylated Oligonucleotide

In one of the variants of isolation of specific DNA sequences originally termed triplex affinity capture (Ito et al., 1992a), target DNA is bound via intermolecular triplex formation by a biotinylated oligonucleotide, bound to streptavidincoated magnetic beads, and then recovered in double-stranded form under triplex-destabilizing conditions (Figure 7.1) (Ji and Smith, 1993). This approach was tested using both oligopyrimidine and oligopurine triplex-forming oligonucleotides (Ito et al., 1992a,b; Takabatake et al., 1992), and proved to be highly sequence-specific. It was possible to isolate single-copy clones from λ phage derivatives (46 kb; Ji and Smith, 1993) and a yeast genomic library (340 kb; Ito et al., 1992b).

In another variant of TAC, the target DNA was complexed with a biotinylated probe and separated in a gel containing a trap of immobilized streptavidin (Ito et al., 1992c).

Affinity Chromatography

Significant progress can be achieved in TAC methodology using affinity chromatography. A variant of this technique, with commercially available agarose-linked single-stranded homopolynucleotides, was used to isolate duplex fragments containing AT and GC clusters in some viral and eukaryotic DNA (Flavell and Van den Berg, 1975; Zuidema et al., 1978). A similar approach was employed in the studies of sequence specificity of polymer triplex formation (Letai et al., 1988). The characteristic feature of this technique is that on-column hybridization of duplex fragments with the triplex-forming polymer is conducted under triplex-promoting conditions: low pH, presence of high salt concentrations or organic solvents, etc. Bound DNA is recovered under triplex-destabilizing conditions. However, the routine application of this method to the isolation or enrichement of natural DNAs, which do not have simple polymeric sequences, requires relatively simple and reliable methods of oligonucleotide attachment to the column.

A strategy that tests large numbers of diverse sequences for binding to a defined DNA duplex has been applied to a selection of individual RNA sequences from the library of randomized RNA molecules capable of

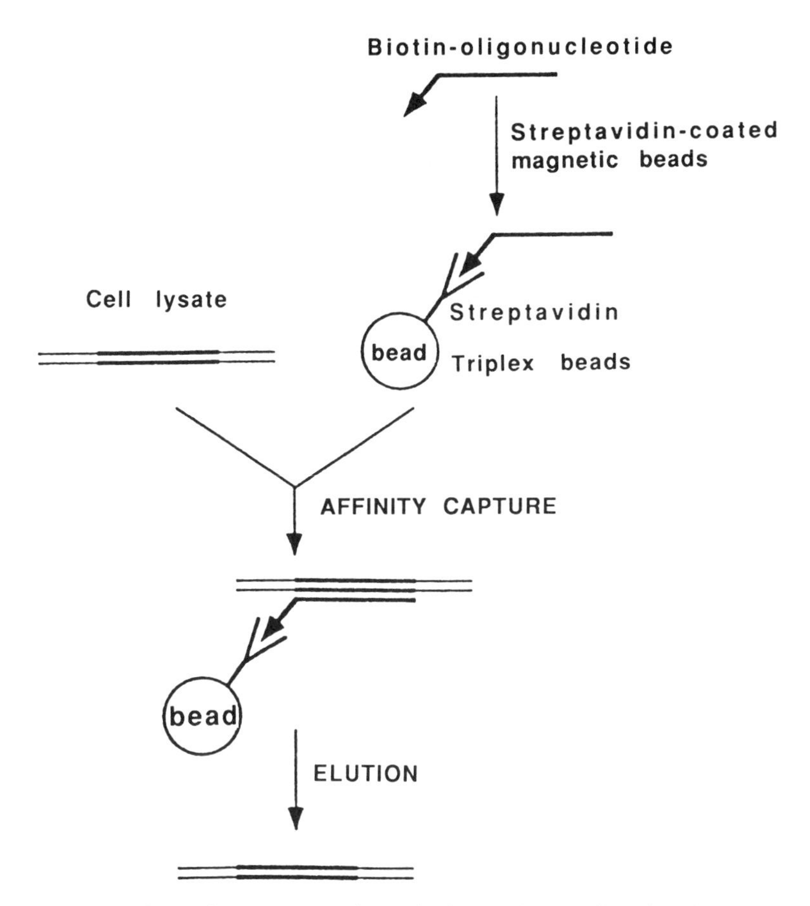

FIGURE 7.1. Schematic representation of the triplex-mediated affinity-capture method for double-stranded DNA purification. A biotinylated oligonucleotide binds to the streptavidin-coated magnetic beads. After DNA from the lysate cell has been allowed to hybridize to this oligonucleotide under triplex-forming conditions, the magnetic beads are removed from this complex mixture, and double-stranded DNA is recovered under triplex-destabilizing conditions. Reprinted with permission from Ji and Smith (1993). © 1993 American Chemical Society.

triplex-binding to a double-stranded 16-base-pair PyPu tract in DNA (Pei et al., 1991). A duplex target was immobilized onto a thiol–Sepharose through a disulfide bond at the 3′ end of the pyrimidine strand.

A number of other examples exist that show the different methods of oligonucleotide immobilization on solid chromatographic supports. Virtually any biotinylated DNA substrate can be linked to a chromatography matrix containing an immobilized avidin (Fishel et al., 1990). Commercially available oligonucleotide derivatives and modified supports can be used to prepare an experimental system appropriate for binding DNA that contains specific sequences.

Oligonucleotides with a specially introduced amino or aminoalkyl group on the 5′-terminal phosphate can be coupled to a silica gel using N-hydroxysuccinimide chemistry (Goss et al., 1990, 1991; Solomon et al., 1992). Using this high-performance variant with micron-size silica particles, single-stranded polynucleotide fragments in the kilobase range were separated rapidly and with good resolution (Goss et al., 1991). However, the utility of such a support for the separation of double-stranded fragments remains to be investigated. The most evident problem that may arise is a relatively small pore size (300–500 Å) that may impose an upper limit on the size of DNA fragments (Massom and Jarrett, 1992).

TAC procedures can be performed in the liquid phase. This allows one to screen a large number of clones and several libraries simultaneously in a parallel manner by using different triplex-forming oligonucleotides. In addition, DNA molecules retain double-strandedness and can be readily used in subsequent biological manipulations. Similar to hybridization, the stringency of the TAC reaction may be controlled by varying the pH, temperature, and composition of the medium. The general use of TAC is somewhat limited by the constraint of PyPu target sequences. Its applicability and utility could be extended by finding new recognition schemes (new triads, artificial base analogs, alternate-strand triple-helix formation). Recognition of other sequence motifs beyond the PyPu tracts could be also extended using triple-stranded DNA complexes promoted by RecA protein (Honigberg et al., 1986; Rigas et al., 1986).

Although the triplex formation between the PyPu duplex and a perfectly matched third strand is 10 to 1,000 fold favorable than with a third strand containing mismatched bases, the formation of imperfect triplexes may be significant if there is an excess of third strands with small differences in sequence. Therefore, the triplex affinity capture is expected to result in a selection of a number of similar sequences. Further, the systematic evolution of ligands by exponential enrichment (SELEX) procedure (Tuerk and Gold, 1990; Irvine et al., 1991) may be employed to isolate the exact desired sequence. The SELEX method is applicable to DNA, RNA, or peptide enrichment (Figure 7.2), which share three general steps: (1) selection of ligand sequences by binding to a target molecule, (2) partitioning of bound and unbound ligand sequences, and (3) polymerase chain reaction (PCR) amplification of ligand sequences in the desired fraction. New rounds of sequence enrichment may employ more stringent conditions for triplex formation, resulting in a homogeneous target product.

The application of this generally outlined strategy to the selection of DNA sequences containing oligo(A) · oligo(T) and oligo (GA) · oligo(TC) tracts via triplex formation with $(dT)_{34}$ was recently described (Kiyama et al., 1994; Nishikawa et al., 1995). For example, chromosomal DNA from HeLa cells was digested with restriction nuclease and amplified by PCR. DNA fragments were mixed with biotinylated $(dT)_{34}$ in the presence of

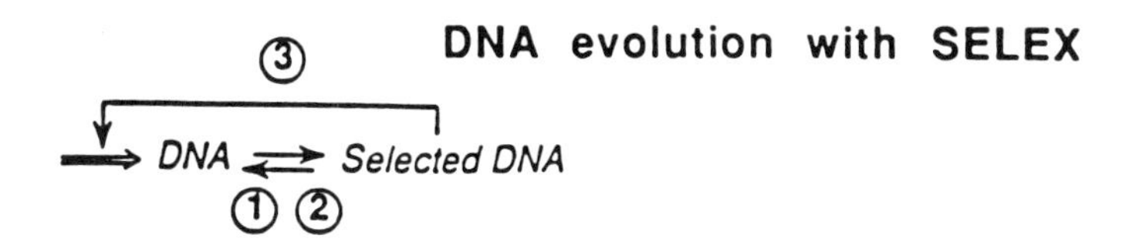

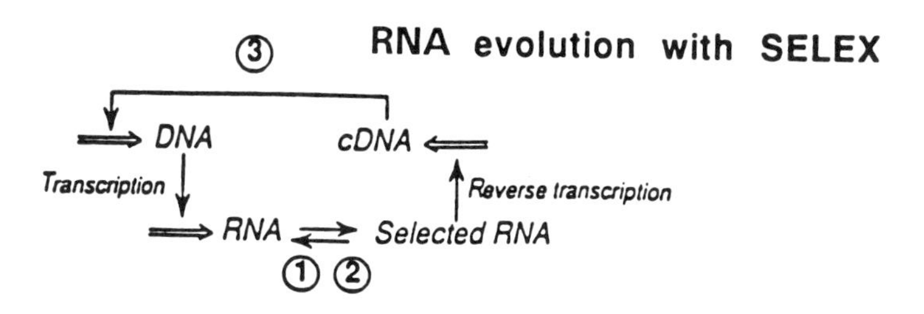

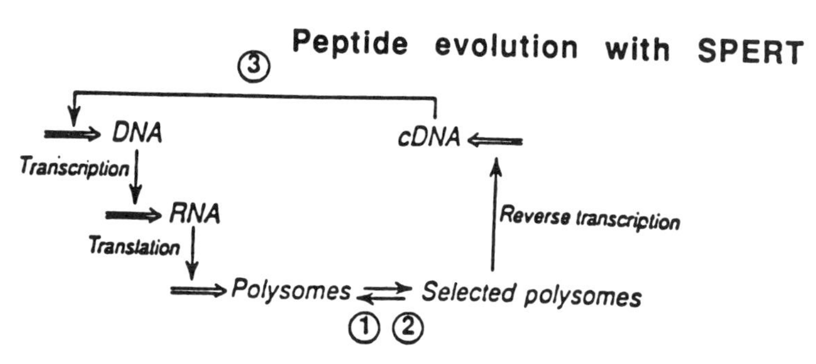

FIGURE 7.2. Enrichment of nucleic acid ligand sequences using SELEX (Systematic Evolution of Ligands by Exponential enrichment; Tuerk & Gold, 1990) or enrichment of peptide sequences using SPERT (Systematic Peptide Evolution by Reverse Translation; Tuerk and Gold, 1990). The three general steps in each process are: (1) selection of ligand sequences by binding to a target molecule; (2) partitioning of bound and unbound ligand sequences; and (3) PCR amplification of ligand sequences in the desired fraction. Reverse transcription of an RNA molecule associated with a selected nascent peptide on a polysome gives the desired effect of reverse translation. From Irvine et al. (1991). Reprinted with permission.

Mg^{2+} and the triplex was adsorbed onto streptavidin-coated magnetic beads. After that, oligo(A)·oligo(T) containing DNA fragments that formed triplexes were eluted from the beads with a buffer containing ethylenediaminetetraacetic acid (EDTA). After PCR amplification, the sample was subjected to another cycle of affinity enrichment. Judging from the amount of DNA recovered after the PCR step, enrichment of triplex-forming DNAs containing PyPu tracts 14 to 37 base pairs long was estimated to be roughly 100-fold after each cycle. However, after the fourth cycle, the enrichment reached a plateau.

The significance of triplex-mediated TAC methodologies is not limited to isolation of naturally existing sequences with sufficiently long PyPu tracts. Long DNA sequences cloned in cosmid vectors and yeast artificial chromosomes at sites internal to triplex-forming tracts may be recovered as somewhat longer fragments created by digestion with restriction endonucleases at sites external to triplex-forming tracts when triplex formation is used to extract these sequences from the bulk of the DNA (see discussion later in section Mapping of Genomic DNA).

Stringency Clamps

An additional segregation of perfect and imperfect triplex-forming duplexes can be based on a technique termed "stringency clamping" (Roberts and Crothers, 1991). It relies on the difference in stability between the complexes formed by a third-strand oligonucleotide with a perfectly matched duplex and with a duplex containing at least one mismatch. It is clear that the latter triplex will be less stable. However, if it is possible to design an alternative secondary structure where a third strand takes part with a stability between that of perfect and defect complexes, this alternative structure would act to extend the stringency over a broad range of conditions by means of competition. Thus, this effect was termed stringency clamp. Several types of clamps were considered. Among them were Watson–Crick and C^+C-containing hairpins, which were more stable than defect triplexes, and the Watson–Crick duplex with an additional complementary strand. Figure 7.3 shows the physical separation of DNA based on triplex formation and stringency clamps. In this experimental design, the pyrimidine third strand PY12 can participate in the more stable structure of the perfect triplex (WC32+PY12), the less stable structure of the duplex (PY12+R12), and the least stable structure of imperfect triplex (WCAT28+PY12). Biotinylated PY12 was annealed in the presence of labeled WC32 and WCAT28 and isolated by elution through a streptavidin-agarose column. The four-nucleotide size difference between WC32 and WCAT28 made it possible to analyze them with a denaturing polyacrylamide gel. In the absence of a stringency clamp, a column may retain WC32 (which formed a perfect triplex) and part of WCAT28 (which, being in excess to WC32, was able to form an imperfect triplex). When the mixture of WC32 and WCAT28 was supplemented with a stringency clamp R12, the available PY12 bound WC32 to form a perfect triplex, and the remaining PY12 bound R12 to form a duplex of intermediate stability. At the proper concentrations of components, there would be no PY12 left to form the least stable triplex structure with WCAT28, which would not be retained on the column. During the pH8 wash, only the perfect triplex-forming hairpin WC32 was

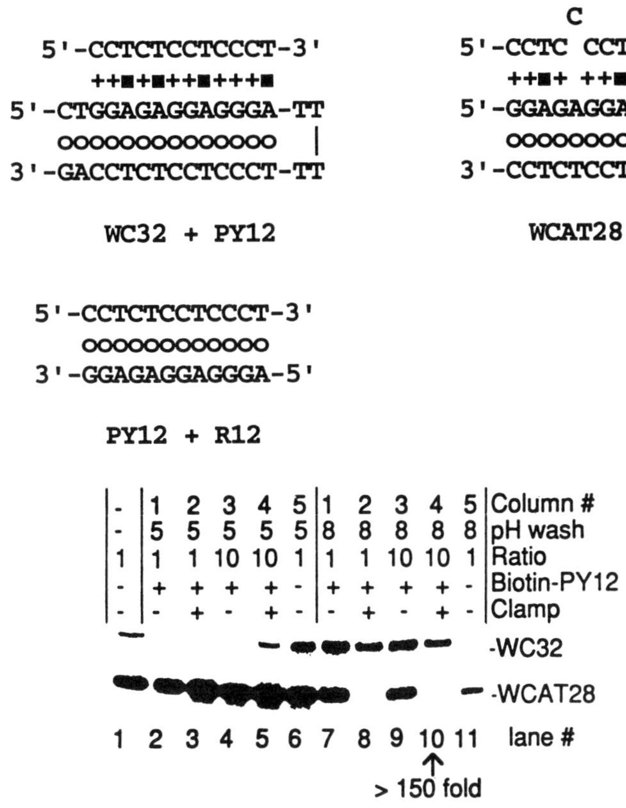

```
5'-CCTCTCCTCCCT-3'
++■+■++■++++■
5'-CTGGAGAGGAGGGA-TT
oooooooooooooo   |
3'-GACCTCTCCTCCCT-TT
```

WC32 + PY12

```
        C
5'-CCTC CCTCCCT-3'
++■+ ++■+++■
5'-GGAGAGGAGGGA-TT
oooooooooooo   |
3'-CCTCTCCTCCCT-TT
```

WCAT28 + PY12

```
5'-CCTCTCCTCCCT-3'
oooooooooooo
3'-GGAGAGGAGGGA-5'
```

PY12 + R12

-	1	2	3	4	5	1	2	3	4	5	Column #
-	5	5	5	5	5	8	8	8	8	8	pH wash
1	1	1	10	10	1	1	1	10	10	1	Ratio
-	+	+	+	+	-	+	+	+	+	-	Biotin-PY12
-	-	+	-	+	-	-	+	-	+	-	Clamp

-WC32

-WCAT28

1 2 3 4 5 6 7 8 9 10 11 lane #

> 150 fold

FIGURE 7.3. Purification of a desired fragment from a mixture by affinity chromatography and stringency clamping. After the hybridization of a biotinylated pyrimidine oligonucleotide PY12, forming a perfect triplex with a Watson–Crick hairpin WC32 and an imperfect triplex with a hairpin WCAT28, the mixtures were chromatographed over streptavidin-agarose columns. The off-column eluates were analysed by denaturing polyacrylamide gel electrophoresis (PAGE). Lane 1 shows the relative radioactivity of the 1:1 mixture of ^{32}P-labeled hairpins. During the column wash with a triplex-stabilizing buffer at pH 5, the WCAT28 forming imperfect triplex was eluted off the column (lanes 2–6), while WC32 was retained on the column. A column wash with the triplex destabilizing buffer at pH 8 released WC32 and part of WCAT28 which was able to form imperfect triplex (lane 7). The purine oligonucleotide R12 is complementary to PY12, and the duplex PY12+R12 has an intermediate stability between the WC32+PY12 triplex and the WCAT28+PY12 triplex. When the mixture of WC32 and WCAT28 was supplemented with R12, available PY12 bound WC32 to form a perfect triplex, and the remaining PY12 bound R12 to form a duplex of intermediate stability. With a proper choice of component concentrations, there would be no PY12 left to form the least stable triplex structure with WCAT28. This situation corresponds to lane 3, which shows that WCAT28 is not retained on the column at pH 5. During the pH 8 wash, only the perfect triplex-forming hairpin WC32 is eluted off the column (lane 8). A similar trend was observed for the hairpin mixture containing a 10-fold molar excess of WCAT28 over WC32 (lanes 4–5 and 9–10, respectively). The resulting ratio of WC32/WCAT28 in lane 10 is more than 150. Reproduced from Roberts and Crothers (1991).

eluted off the column. The resulting enrichment of WC32 over WCAT28 was more than 150 times greater in this experiment.

Quantitation of Polymerase Chain Reaction Products

Primers containing 5′-oligopyrimidine sequences were used to develop an automated analysis of PCR products (Vary, 1992). This approach is based on a two-step PCR format that uses flanking and nested primers. A set of flanking primers is used for a primary amplification of the desired target sequence to a large amount. The second round of amplification with nested primers bound to target sequences inside positions of the flanking primers provides further amplification of the desired sequences and incorporation of the primers carrying various affinity labels to capture and detect the PCR product (Mullis and Faloona, 1989). The triplex-based approach (Figure 7.4; Vary, 1992) uses two different 5′-oligopyrimidine-containing nested PCR primers to capture and label PCR products for quantitation by fluorescence concentration assay (Jolley et al., 1984).

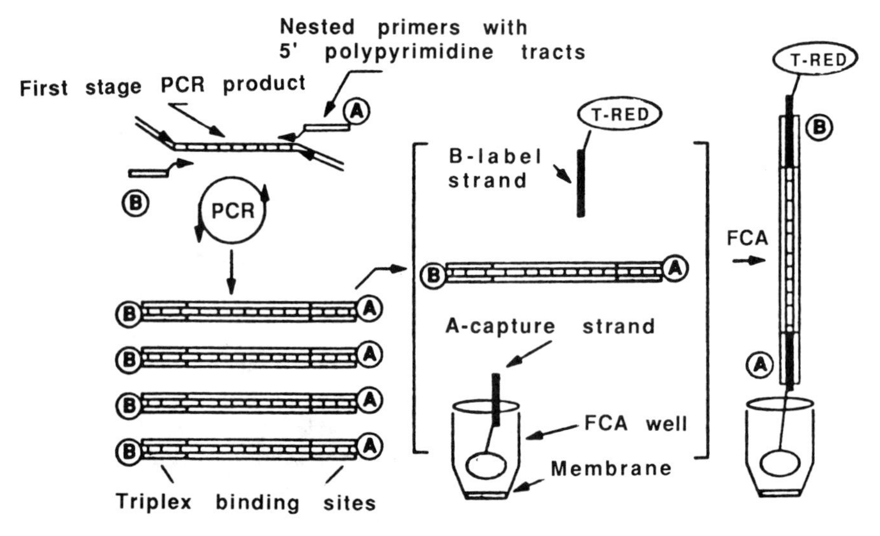

FIGURE 7.4. Capture and detection of PCR products using triplex-forming sequences. After a primary amplification of the desired sequence to a large amount, nested primers that bind inside positions of flanking primers (employed in the first step) are used for further amplification and incorporation of the oligopyrimidine terminal sequences. One of these sequences is used to capture PCR products via triplex formation with oligopyrimidine bound to latex beads. Another sequence, the Texas Red-labeled oligopyrimidine strand homologous to the second pyrimidine end, is used for quantitation of PCR products after the immobilization of a fluorescent complex. Reproduced from Vary (1992).

Incorporation of these primers into a duplex PCR product creates two separate sites for triplex formation. PCR products are captured via triplex formation on latex beads carrying an oligopyrimidine strand homologous to one end of the PCR product. Another, Texas Red-labeled, oligopyrimidine strand homologous to the second triplex-binding site is used to quantify the captured PCR product. After the immobilization of a fluorescent complex on latex microparticles, samples are transferred to a 96-well suction filtration plate for washing and filtration. The subsequent quantification is sensitive ($\sim 10^2$ copies) and reduces the background contributions.

Nonenzymatic Ligation of Double-Helical DNA Mediated by Triple Helix Formation

The binding of two triplex-forming oligonucleotides to adjacent sites on a double-stranded target sequence is a cooperative process due to stacking interactions between the bases at the oligonucleotide junction (Strobel and Dervan, 1989; Froehler et al., 1992a; Colocci et al., 1993; Colocci and Dervan, 1995). Therefore, a tandem PyPu tract has a high probability of being occupied by both oligonucleotides simultaneously. These oligonucleotides may then be ligated by either phosphodiester or pyrophosphate linkage (Luebke and Dervan, 1989; Dolinnaya et al., 1991). The 5′-terminal phosphate of one of the oligonucleotides activated with BrCN, imidazole, and $NiCl_2$ forms a covalent bond with the 3′-OH of the adjacent oligonucleotide (Luebke and Dervan, 1989). Oligomers may also be condensed using 1-(1-dimethylaminopropyl)-3-ethylcarbodiimide hydrochloride (Dolinnaya et al., 1991).

Triplex formation can also provide the basis for the nonenzymatic sequence-specific ligation of double-helical DNA fragments containing purine tracts at the juxtaposed blunt ends (Luebke and Dervan, 1991, 1992). A single-stranded template hybridizes simultaneously to both PyPu duplexes, and stacking interactions between the base pairs of two duplexes help stabilize such a complex. N-cyanoimidazole was used to ligate chemically these duplexes. The nature of the bonds was elucidated in the following experiment. The plasmid DNA, linearized by the restriction endonuclease *Stu*I producing blunt ends, was then nonenzymatically circularized. Sequential chemical ligation of two strands occurs at comparable reaction rates for the first and second strands. The susceptibility of ligated site to *Stu*I provided evidence that the chemical ligation produced phosphodiester linkage (Luebke and Dervan, 1992). Two variants allowed the formation of the sequences of the types 5′-$(\text{purine})_m(\text{purine})_n$-3′ and 5′-$(\text{purine})_m(\text{pyrimidine})_n$-3′. In the first sequence, a continuous homopyrimidine strand served as a template for the hybridization of two duplexes with purine strands on the same duplex side (Luebke and Dervan, 1991). In the second sequence, a homopyrimidine hybrid with two segments coupled 3′ to 3′ via abasic 1,2-dideoxyribose linker provided the formation of a triplex with a strand switch and purine strands on the opposite sides of the duplex (Luebke and Dervan, 1992).

This methodology seems promising for ligation of DNA fragments with blunt ends; however, accessible yields are still not high: 15% and 7% of the starting linearized plasmid was converted into a supercoiled form in the case of 5′-$(purine)_m(purine)_n$-3′ and 5′-$(purine)_m(pyrimidine)_n$-3′ products, respectively (Luebke and Dervan, 1991, 1992).

Triplex-Mediated Inhibition of Viral DNA Integration

The integration of a DNA copy of the viral RNA genome into the host cell DNA is one of the key steps in the replication cycle of retroviruses, including the immunodeficiency viruses HIV-1 and HIV-2. A viral protein, integrase, which binds to short sequences at the ends of viral DNA long terminal repeats (LTR), is especially important. Inhibition of protein function resulting from triplex formation at the protein-binding site was used to suppress viral DNA integration. The integrase-binding site located in the U3 site of LTR containing the purine sequence GGAAGGG was shown to form a stable intermolecular triplex with an oligonucleotide–intercalator conjugate (Mouscadet et al., 1994b). This triplex formation was further shown to prevent the catalytic functions of integrase in vitro, thereby sequence-specifically inhibiting integration at the U3 site of LTR (Mouscadet et al., 1994b). Thus, it is possible that in addition to the regulation of gene expression, triplex-forming oligonucleotides may be used to control other protein-mediated genetic functions.

Site-Directed Mutagenesis

There exist a number of oligonucleotide-based approaches to the creation of site-specific mutations. They are based on the use of single-stranded M13 vector (Derbyshire et al., 1986; Li et al., 1989), PCR (Higuichi et al., 1988; Ho et al., 1989; Kadowaki et al., 1986; Vallette et al., 1989) and double-stranded plasmids (DeChiara et al., 1986; Lai et al., 1993). These methods provide very useful models for the study of specific mutations at specific sites. However, they do not take into account the processes from which these particular mutations could result. Mutational mechanisms can be proposed as contributing to the activity of most human carcinogens, including various environmental chemical and radiation factors (see Barrett and Shelby, 1992; Pitot, 1993, for recent reviews). The susceptibility of DNA to the action of these factors depends on sequence and spatial structure. For example, various regions of DNA in chromatin are unequally accessible to chemical agents. Therefore, mechanisms of DNA damage and repair should be studied using localized modifications by potential carcinogens or irradiation. Yet the introduction of a specific mutation into a highly localized genome region has been hampered by the lack of highly specific DNA binding agents. New chemical building blocks have made it possible to introduce modified bases and interbase adducts (e.g., thymine dimers) into specific DNA sequences in order to study mutation consequences at the sequence level (Wang and Taylor, 1992; Hatahet et al., 1993; Jiang and Taylor, 1993). However, the number of possible mutation products can be quite large, and not all of them can be introduced into the desired gene region as chemical building blocks: for instance, pyrimidine hydrates usually resulting from UV irradiation are thermolabile, which poses definite limitations on reaction conditions. Therefore, chemical (Zarytova et al., 1981, 1992; Fedorova et al., 1988; Dikalov et al., 1991; Povsic et al., 1992) and photochemical (Le Doan et al., 1987, 1991; Lee et al., 1988; Praseuth et al., 1988a,b; Gasparro et al., 1991; Dobrikov et al., 1992a,b) methods can be used to obtain specific modifications at predetermined sites. Investigation of repair processes in DNA which were studied for the sequences with sufficiently arbitrarily distributed lesions (Friedberg, 1984; Ward, 1988; Sage, 1993) could be extended on sequences with the predetermined sites. In some of these studies, the extent of triplex-targeted modification was determined.

The first study of triplex-mediated photomutagenesis used a purine 10-mer linked to psoralen at its 5′ end that was targeted to a specific *supF* gene in an intact, double-stranded λ phage genome (Havre et al., 1993; Gasparro et al., 1994). In these experiments, the triplex-forming psoralen-conjugated oligonucleotide was incubated under complex-stabilizing conditions with the λ phage DNA. Long-wavelength (320–400 nm) UV irradiation was used to activate the psoralen, which formed adducts at the target site. The DNA was packaged in vitro into phage particles that were adsorbed to *E. coli.* Sequence analysis of the gene under study and the control gene, which did not contain a binding site for the psoralen-linked oligomer, showed a low background photomodification: the estimated frequency of unspecific mutations was 100–500 times lower than that of specific mutations. About 56% of the mutations were the TA to AT transversions consistent with the mutagenic action of psoralen. Psoralen forms adducts predominantly at pyrimidines, especially at thymidines (Sage, 1993). The utility of this approach may depend on the efficiency of targeted photomutagenesis. In the first attempt, a targeted mutation yield of 0.233% was achieved (Havre et al., 1993), which depends on the processing of the psoralen adducts by DNA repair and replication enzymes. Subsequent studies on targeted photomutagenesis used a simian virus 40 vector genome (Havre and Glazer, 1993; Wang, G. et al., 1995). The oligonucleotide–DNA complex was transfected into monkey COS-7 cells using cationic liposomes. There was a good agreement with the results obtained for a bacterial model with respect to the proportion of specific mutations (55% of TA to AT transversions). However, mutations were produced in the target gene in over 6% of the viral genomes. This 30-fold increase relative to similar experiments with I phage suggests that the repair system in mammalian cells was less efficient, and the triplex-directed photolesion was more often transformed into a mutation. The molecular details of psoralen-induced mutagenesis have not been studied enough. It is known that the major lesion resulting from the targeted psoralen photoreaction with DNA is a monoadduct, whereas the interstrand cross-link is a minor product (Gasparro et al., 1994). Human lymphoblastoid cells transfected with psoralen-modified DNA are able to repair both of these lesions; however, cross-link products seem to be repaired less efficiently and result in mutations (Sandor and Bredberg, 1994).

The ability to specifically mutate double-stranded DNA may provide tools for genetic engineering, gene therapy, and antiviral therapeutics. In combination with developed methods of repair studies (Boorstein et al., 1990; Ganguly et al., 1990; Hamilton et al., 1992; Kim and Sancar, 1993; May et al., 1993; Sancar and Tang, 1993; Svoboda et al., 1993), this could provide important information on survival mechanisms and gene evolution.

Detection of Mutations in Homopurine DNA Sequences

Since single-triad mismatches significantly destabilize the triple helix under appropriate stringent conditions, this phenomenon could be used to detect mutations in homopurine DNA sequences. This possibility was studied by Olivas and Maher (1994a), who used the human p53 tumor suppressor gene model. One of seven PyPu tracts in the p53 gene may contain the site of a clinically important 8-base-pair microdeletion (Figure 7.5). Oligonucleotide probes I and II should bind to the PyPu targets I and II in the wild-type p53 gene, respectively, whereas probe I should not bind to target I in mutant p53 gene. PCR was used to amplify 496 base-pair-DNA fragments from plasmid DNA containing wild-type and mutant p53 sequences. Samples of the resulting PCR products were directly mixed with labeled probe oligonucleotides in pH 5.0 buffer and analyzed by dimethylsulfate footprinting assay and by comigration in a 5% native polyacrylamide gel at the same pH in the presence of 1 m*M* $MgCl_2$. The data obtained confirm the expected binding specificities of the oligonucleotide probes. The above-described method does not require DNA denaturation and might facilitate certain DNA screening procedures. As for the necessity for homopurine sequences of approximately 10 base pairs and longer, the occurrence of such sequences in some mammalian genomes can be as frequent as 1 in every 250 base pairs (Behe, 1987). Our analysis of 25 randomly chosen human genes also showed a comparable value of 1 PyPu tract of 10 base pairs or longer in every 300-base-pair sequence (Potaman and Soyfer, unpublished results).

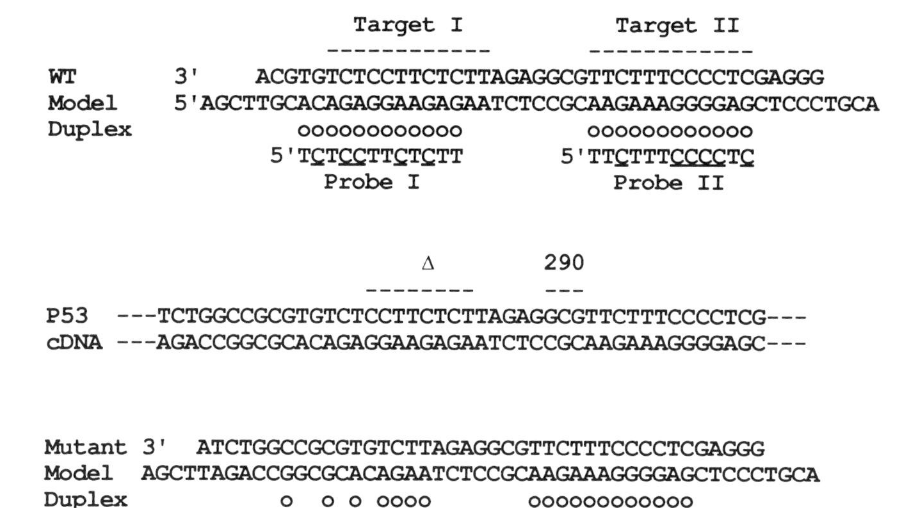

FIGURE 7.5. Experimental design for the detection of mutations in homopurine sequences of DNA. A portion of the nucleotide sequence of wild-type p53 cDNA with a reference codon 290 is shown in the middle. The position of a clinically important 8-base-pair microdeletion is indicated by Δ. Oligonucleotide probes I and II containing 5-methylcytosine (C) bind to the PyPu targets I and II in the wild-type sequence (top), and only probe II binds to its target in the mutant sequence. Circles between probes and duplex targets show stable triads. Experimentally, triplex formation is determined by dimethylsulfate footprinting assay and comigration of oligonucleotides and duplexes in 5% polyacrylamide gel under triplex-stabilizing conditions. From Olivas and Maher (1994a). Reprinted with permission.

Mapping of Genomic DNA

The primary goal of the Human Genome Program is to make a series of descriptive maps of each chromosome at increasingly finer resolutions. A single chromosome is cut by some rare-cutting agent into large pieces, which are then ordered and subdivided. The smaller pieces are then mapped further. Alternatives or additions to usual mapping of restriction nuclease fragments could include mapping repetitive sequences, promoters, thermally stable regions, specific DNA structural regions, etc. Site-specific binding of triplex-forming oligonucleotides may form the basis for rare cutting of chromosomal DNA and physical mapping of PyPu tracts which may have important functional roles in the cell (see Chapter 6).

Triple Helix Vector

Analysis of complex genomes requires extensive isolation of large DNA fragments, their cloning in cosmid vectors and yeast artificial chromosomes (YACs), with subsequent mapping and sequencing of these fragments. One of the first applications of the triplex methodology for mapping of genomic DNA was described in 1990 as a cosmid Triple Helix Vector (Moores, 1990). The Triple Helix Vector (Figure 7.6A) contains two different triplex-forming sequences external to the T3 and T7 promoters but internal to the *Not*I sites. Cosmid clones in the Triple Helix Vector are mapped by a complete digestion with *Not*I, followed by partial digestion with another enzyme of interest. The resulting fragments are then complexed at one or another PyPu site near the *Not*I ends with one of two labeled triplex-forming oligonucleotides. After triplex formation between duplex restriction fragments and labeled oligonucleotides, the complexes are separated on an agarose gel under triplex-stabilizing conditions. The restriction ladder obtained after gel drying and autoradiography can be converted into a restriction map of the clone (Figure 7.6B). Each band in the ladder represents the distance between one triplex site and a restriction enzyme site. Subtracting the size of one partial digest band from the size of the neighboring band reveals the distance separating the two restriction sites. This method has increased accuracy, since the sites are mapped from both ends of the clone. Thus, the region of lowest accuracy from one end (lowest-resolved large fragments) is the region of greatest accuracy from the other end. In comparison with alternative mapping techniques using hybridization of oligonucleotides complementary to the T3 and T7 promoters on a solid support, the strategy described above is about two times faster.

A related technique that combines the advantages of Triplex Helix Vector and TAC (Triplex Affinity Capturing) methodology was recently described (Ji et al., 1994). A modified cosmid was constructed from the SuperCos 1 vector by flanking the cloning site with two PyPu sequences, which are internal to both T3 and T7 promoters and *Not*I restriction sites. After *Not*I digestion, cosmid DNA is combined with a biotinylated triplex-forming oligonucleotide under complex-forming conditions. The triplex containing a cloned fragment is captured with streptaviridin-coated magnetic beads. Elution of the cloned fragment under triplex-destabilizing conditions results in a recovery of up to 95% and purity of >95%. The isolated insert DNA is directly digested with *Cvi*JI restriction endonuclease to generate random fragments for shotgun sequencing.

Artificial Endonucleases in Genome Mapping Strategies

Fragmentation of long chromosomal DNA for mapping and, in particular, sequencing purposes by naturally occurring restriction enzymes is sometimes inappropriate because of their too frequent cleavage. This is due

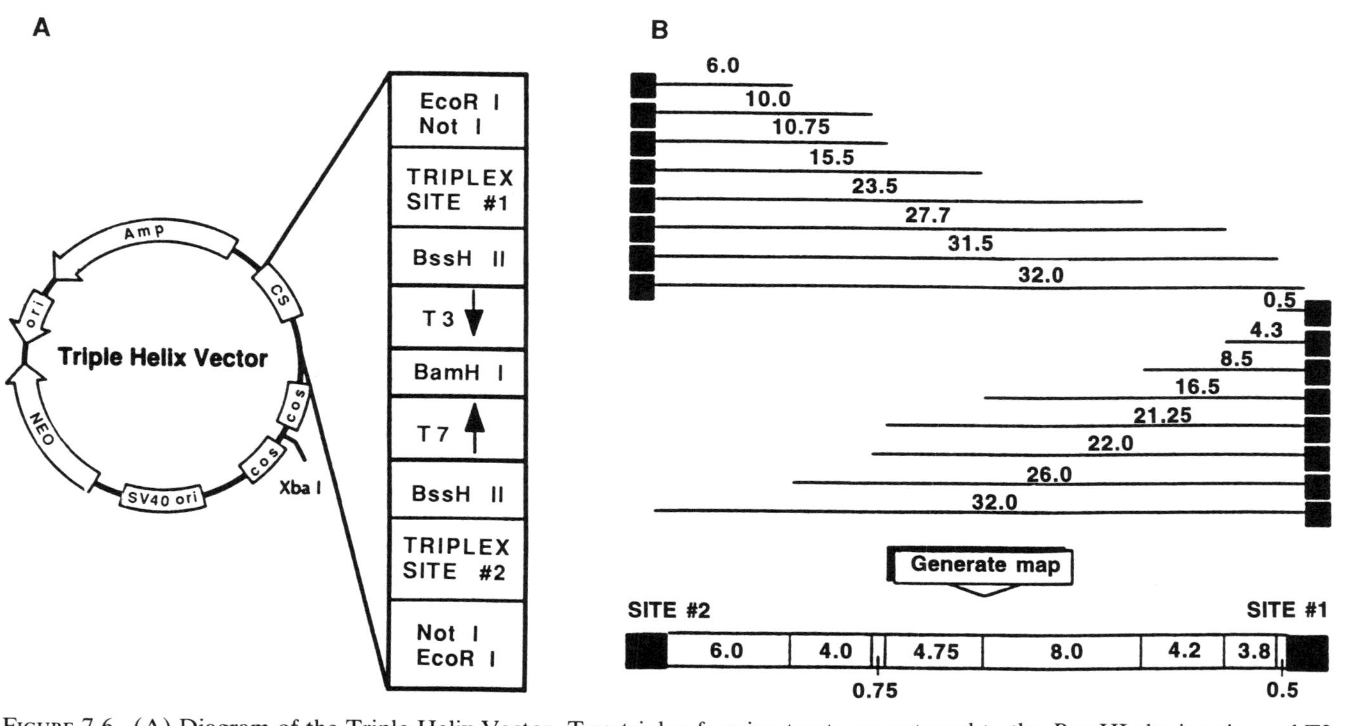

FIGURE 7.6. (A) Diagram of the Triple Helix Vector. Two triplex-forming tracts are external to the *Bam*HI cloning site and T3 and T7 promoters and are internal to the *Not*I restriction site. (B) Mapping using the Triple Helix Vector. The DNA with a cloned insert was digested to completion with *Not*I and then partially digested with *Eco*RI by removing aliquots from the reaction mixture at different points in time. Fragments in each set of the duplicate *Eco*RI-digested aliquots were hybridized with one of the labeled triplex-forming oligonucleotides. The triplex DNA was analyzed in a 1% agarose gel under triplex-stabilizing conditions. The resulting sets of fragments hybridized with oligonucleotides allow one to generate an *Eco*RI restriction map for the cloned DNA sequence. Adapted from Moores (1990).

to the relatively short recognition sites for most restriction nucleases. Therefore, alongside the discovery and application of new rare-cutting restriction endonucleases (e.g., *Not*I and *Mlu*I; Hanish and McCleland, 1990; see Billings et al., 1991, for review), efforts are being made to make some restriction sites more rare and to develop different kinds of synthetic endonucleases (see Sigman, 1990; Dervan, 1992; Povsic et al., 1992, for recent short reviews).

Transient site-specific protection from enzymatic methylation by a site-specific DNA binding protein leaves only rare recognition sites unmethylated for the restriction enzymes (Koob et al., 1988). When the protecting protein dissociates, only such unmethylated "Achilles heel" sites are susceptible to the action of corresponding endonucleases, thereby providing much rarer cleavage of DNA. The other recognition sites are not cleaved, due to methylation. In the oligonucleotide equivalent of the "Achilles heel" methodology, protection of restriction sites against methylation is accomplished by the triplex-forming oligonucleotide (Strobel and Dervan, 1991; Strobel et al., 1991; see Strobel and Dervan, 1992, for review). Subsequent triple-helix disruption and cleavage by a restriction enzyme yields larger DNA fragments (Figure 7.7). By preventing *Eco*RI methylase reaction at a specific site with a triplex-forming oligonucleotide, it was possible to cleave the yeast chromosome III (340-kb size) with the *Eco*RI endonucleases at a single site. As an extension of this methodology, Nehls et al. (1994) suggested to introduce triplex-forming tracts, which overlap *Eco*RI restriction site, into mammalian genome via homologous recombination. In such a way, large chromosomal domains may be created for cloning, sequencing and functional analysis. The triplex-mediated "Achilles heel" methodology is complementary to the protein-mediated variant, yet it has mainly been limited to the cases where homopurine–homopyrimidine tracts and restriction sites overlap (Strobel and Dervan, 1991, 1992; Strobel et al., 1991). In the only attempt to overcome this limitation, Kiyama & Oishi (1995) designed a DNA sequence containing two PyPu tracts separated by a short sequence with a restriction site. Long oligonucleotides capable of forming triplexes with both tracts simultaneously and wrapping around intermediary sequence without hydrogen bonding protected restriction site from the appropriate enzyme. Although the enzyme recognition site and the PyPu tract may not overlap, the requirement of the PyPu tracts still remains.

One of the promising directions in the development of artificial endonucleases is the use of triplex-forming oligonucleotides carrying DNA-cleaving groups. The ability to vary the length of the recognition site and gently control the complex stability can, in principle, provide targeting oligonucleotides at distant sites and produce DNA fragments of any desired size. Even a simple analysis of randomly chosen published sequences of 25 human genes shows that oligonucleotides of 9 bases and

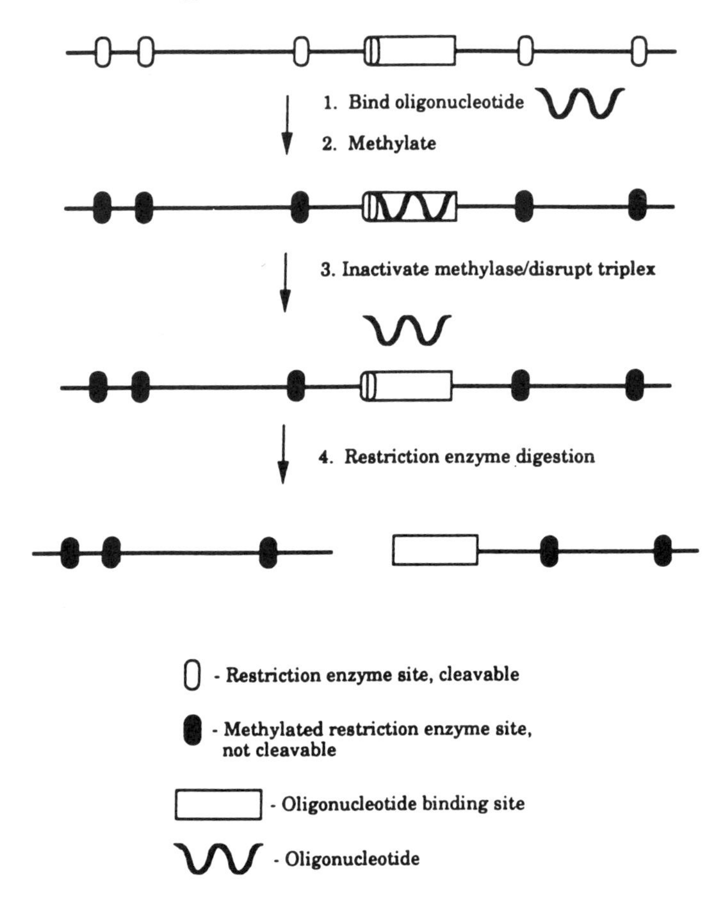

FIGURE 7.7. A generalized scheme for single-site restriction cleavage of genomic DNA by oligonucleotide-directed triplex formation. The triplex between the chromosomal DNA and the oligonucleotide is formed in a methylase-compatible buffer. *Eco*RI methylase methylates all adenines in DNA except the partially triplex protected site 5′-GAATTC-3′. After triplex disruption, all recognition sites for *Eco*RI restriction endonuclease are resistant to cleavage, with only the one temporarily triplex-protected site being susceptible. This strategy allows one to cut DNA rarely at restriction sites coinciding with the PyPu tract. From Strobel & Dervan (1992). Reprinted with permission.

longer can be used to produce chromosome fragments several dozen kilobases in length (Table 7.1; Potaman and Soyfer, unpublished results). For example, an arbitrary 9-mer consisting of the bases of four types is statistically expected to occur once in a sequence 32 kb in length, so this is the average size of fragments that can be produced with triplex-binding 9-mer bearing the DNA-cleaving group. Because of the nonrandom base distribution in the DNA sequences, for two-thirds of 9-mers found, the

TABLE 7.1. Repeatability of different PyPu tracts in 25 human genes (68,224 bases total length).

Length	Number found	Average occurrence of the same tract	Nonrepeated (unique) tracts	Per cent (unique)
8-mer	203	4.82	6	3
9-mer	111	2.10	37	33.3
10-mer	76	1.46	52	68.4
11-mer	57	1.12	48	84.2
12-mer	34	1.20	27	79.4
13-mer	19	1.00	19	100
14-mer	18	1.11	16	88.8
15-mer	11	1.00	11	100
16-mer	7	1.00	7	100
17-mer	2	1.00	2	100
18-mer	3	1.33	2	66.7
19-mer	2	1.00	2	100
20-mer	2	1.00	2	100

frequency of repeats was less than one in 68 kb, so they could be used to produce fragments longer than 68 kb. Apparently, for longer oligonucleotides, recognition sites will occur less frequently, but more extended search is required to find a relationship between the desired length of the DNA fragment and the length of the corresponding targeted oligonucleotide. Special attention should be paid to the sequences themselves, because among tracts of equal length the frequency of occurrence varies. For instance, the AAGGAGAA and AGAAGAGG sites occurred 8 times each in 68-kb sequences examined, whereas the GAGGAAAA and GAGGGAGA sites were unique.

Triplex-mediated artificial endonuclease strategies can be separated into several groups.

In the first (enzymatic) group, the oligonucleotide-directed recognition of DNA is coupled with an enzymatic cleavage of the latter. This can be accomplished via an attachment of a DNA-cleaving enzyme (e.g., staphylococcal nuclease) to the oligonucleotide (Pei et al., 1990). Specific double-stranded cleavage occurred predominantly at AT-rich sites at the 5′ end of the PyPu tract for both 5′-mono- and 5′,3′-diderivatized oligomers. Another method makes use of the specific feature of triple-helical complex of DNA with an oligomeric peptide nucleic acid (PNA) which forms with a PNA/DNA stochiometry 2:1 (see Figure 3.21). Formation of the PNA/DNA triplex results in a displacement of one DNA strand which is recognized and digested by a single-strand specific enzyme (e.g., S1 or mung bean nucleases) (Demidov et al., 1993). Although the yields of enzymatic reactions are greater than 75%, they do not result in point cleavage and the lengths of affected sites are 5 to 10 bp. This may be due to excessive freedom of targeted staphylococcal nuclease

(Pei et al., 1990) or the presence of a long looped part of the DNA (Demidov et al., 1993).

In the second (chemical) group, DNA-binding molecules are combined with reactive functionalities that result in the oxidation of the deoxyribose (Hertzberg and Dervan, 1982; Dervan, 1987; Moser and Dervan, 1987; Boidot-Forget et al., 1988; Strobel et al., 1988; François et al., 1989a,b; Strobel and Dervan, 1990) or the electrophilic modification of the bases (Baker and Dervan, 1985; Fedorova et al., 1988; Vlassov et al., 1988; Podust et al., 1989; Povsic and Dervan, 1990; Knorre and Vlassov, 1991; Shaw et al., 1991; Povsic et al., 1992). Among the best-known reagents are those capable of generating reactive species, such as the hydroxyl radical. For instance, metal complexes (e.g., Fe–EDTA and Cu–phenanthroline) are used to generate diffusible $OH^{\cdot}$ radicals which attack the deoxyriboses within one of the double helix grooves. Other examples of DNA-cleaving chemicals have been presented in Table 2.4. The DNA molecule may be chemically cleaved either in the course of the oxidative reaction or during additional chemical treatment. DNA at oxidized sites can be also cleaved enzymatically. As a result of a sufficient mobility of the cleaving moieties coupled to the oligonucleotide ends and a spatial diffusion of hydroxyl radicals, oxidative chemical cleavages occur at the length of several base pairs. Reaction yields are relatively low, 25% or less (Moser and Dervan, 1987; Strobel et al., 1988; Strobel and Dervan, 1990). For Cu–phenanthroline, however, 70% cleavage efficiency was reported (François et al., 1989a). The latest studies show that the cleavage efficiency can be further enhanced. Alkylation with the 4(N-methyl-N-2-chloroethylamino)benzyl group could result in 80% to 90% scission; however this has been demonstrated so far only for single-stranded DNA (Zarytova et al., 1992). Alkylation with N-bromoacetyl produced a single site cleavage with a yield of 85% to 90%, but two adjacent inverted PuPy tracts are required for two-strand cleavage (Povsic et al., 1992; see also Figure 2.24 in Chapter 2).

The third (photochemical) group includes DNA-binding molecules that carry photoactivatable moieties in order to produce either direct or indirect strand breaks (that is, appearing after chemical or enzymatic treatment) (Le Doan et al., 1987, 1991; Praseuth et al., 1988a,b; Perrouault et al., 1990; Gasparro et al., 1991; Dobrikov et al., 1992a,b). The photochemical reaction is initiated by long-wavelength ($\lambda > 300\,nm$) irradiation, which does not affect nucleic acids themselves. A number of photoactivatable groups were tested (Table 2.4). The photocrosslinks produced by azidoproflavine, azidophenacyl, and psoralen (Le Doan et al., 1987, 1990; Praseuth et al., 1988b; Takasugi et al., 1991) can be converted into strand breaks under alkaline conditions. The yield of the photoinduced crosslinking reaction can be quite high — about 80% (Takasugi et al., 1991; Dobrikov et al., 1992a). Ellipticine derivatives covalently attached to oligonucleotides were used to photoinduce cleavage of the two strands of a target PyPu sequence (Perroualt et al., 1990; Le Doan et al., 1991). The drawbacks of such photoinduced cleavages are that they are not 100% complete, and that due to "diffusible" active groups, photodamaged sites spanned several base pairs. In addition, some of the above-mentioned agents produce multiple effects: for instance, porphyrins induce crosslinking reactions and oxidation of guanine bases, whereas ellipticine induces both crosslinks and strand scissions (Le Doan et al., 1990, 1991).

Synthetic oligonucleotides that contain photosensitizing nucleobase analogs (e.g., halogen- or thiopurines and pyrimidines) were suggested for modifying DNA at specific sites (Demidov et al., 1992). Since halogenated pyrimidines have the tails of absorption spectra spanning much further into the long-wavelength UV region than common bases, and thiopurines and pyrimidines have absorption maxima at 330 to 340 nm, such base analogs (Figure 7.8) can be activated by a long-wavelength ($\lambda > 300$ nm)

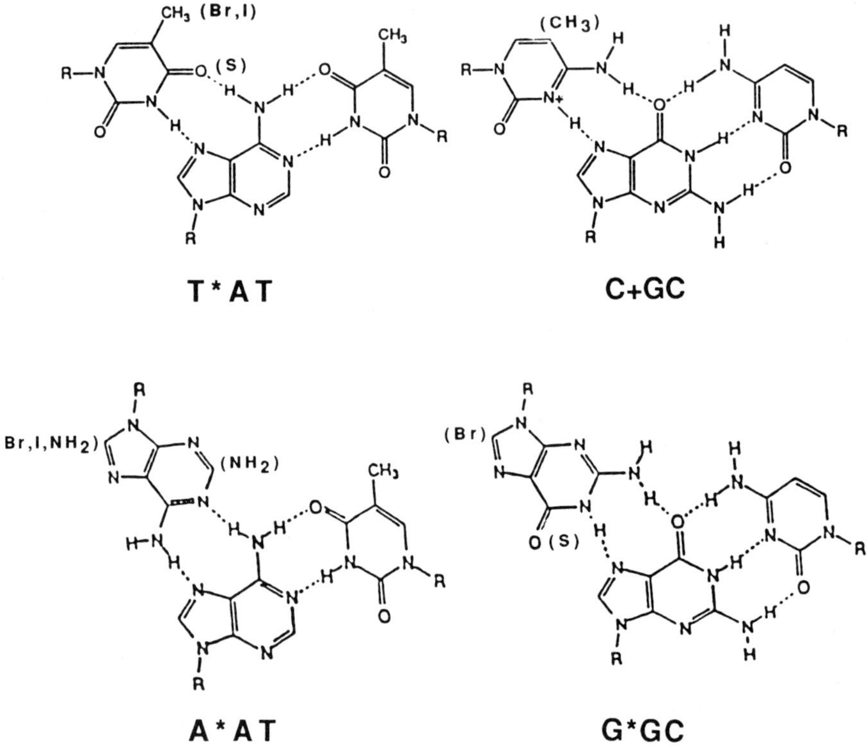

T*AT

C+GC

A*AT

G*GC

FIGURE 7.8. Some base triads for the PyPuPy and PyPuPu triplexes containing photoactivatable base analogs. Several of the indicated substitutions cause the bases to absorb light in the UV region ($\lambda > 300$ nm), where ordinary bases have negligible absorbance.

light (Kochetkov and Budovskii, 1972; Wang, 1976; Peak et al., 1984; Cadet and Vigny, 1990; Rahn, 1992). Once incorporated into the triplex-forming oligonucleotides, under UV irradiation these base analogs will induce photomodifications at specific DNA sites. Among these modifications are photodimers, strand breaks, and crosslinks with the target (Peak and Peak, 1987; Favre, 1990; Sugiyama et al., 1990, 1993; Rahn, 1992). This approach seems to have some important advantages over other photosensitizers. First, nucleobase analogs contribute to complex stability, as they form hydrogen bonds with the DNA target. Moreover, bromination of pyrimidines enhanced the affinities of oligonucleotides to their double-stranded target (Povsic and Dervan, 1989). Second, one or several preferable photoreactive base analogs can be introduced into desirable sites of the same oligonucleotide. Third, a nondiffusible photoactive moiety is expected to produce a highly localized effect (Sugiyama et al., 1990). Work is in progress to evaluate the specificities and accessible yields in the suggested photoreactions.

A recent report (Panyutin and Neumann, 1994) describes the formation of double-strand breaks (DSBs) induced by ^{125}I-labeled oligonucleotides targeted to a PyPu tract in the *nef* gene of the human immunodeficiency virus. The DSBs found only under triplex-forming conditions were produced with a high efficiency (0.8 DSB per decay) and were distributed within 10 base pairs of a maximum located just opposite the position of iodinated cytosine in the oligonucleotide.

Physical Genome Mapping by Electron Microscopy

Electron microscopic visualization of the triplex formed between the plasmid DNA and biotinylated oligonucleotides has been described (Cherny et al., 1993b). A bulky streptavidin molecule, which binds to the biotin moiety, was used as a marker of oligonucleotide on plasmid DNA. In such a way, the binding sites for a purine oligonucleotide targeted to the 17-base-pair-long PyPu sequence of the fragment from a human papilloma virus 16 inserted in plasmid DNA were determined. Under conditions favoring the formation of the PyPuPu triplex, the complex was detected by electron microscopy. Similarly, under conditions favorable for the PyPuPy triplex, the pyrimidine oligonucleotide formed a stable complex. More than 80% of DNA molecules bound the streptavidin marker in the correct position of the PyPu tract, and very few cases of nonspecific binding were detected. Recent results also show that biotinylated PNA, which forms 2:1 complexes with appropriate PyPu DNA targets, may be used for electron microscopic mapping of short PyPu sites in duplex DNA (Demidov et al., 1994a). Electron microscopic visualization of PyPu sites can be applied for physical genome mapping or for detection of triplexes as markers (e.g., for unambiguous identification of DNA ends). Perhaps the major limitation of this methodology is its limited applicability to only relatively short DNA fragments or plasmid DNA.

Control of Gene Expression

One of the most important applications of triple-stranded complexes is their potential to regulate gene expresion in vivo. Any disease caused by the expression of a gene can be treated at various stages of the cellular processes. Many traditional drugs are targeted against the functional proteins, and their design requires a lot of information about structure–activity relationships for these proteins. The oligonucleotide-based antisense approach is directed against the formation on the mRNA template of the secondary products, proteins. However, continuous gene expression may supply new mRNA molecules. The antigene strategy exploits the possibility of binding some exogenous compounds to DNA inside genes or regulatory regions in order to hamper mRNA synthesis on the DNA template. The antigene strategy is directed against the primary process of gene expression and, therefore, may be more efficient than the antisense strategy. Another advantage is that it avoids the need to determine the various specific mechanisms of drug–protein interaction. Triplex-based inhibition of gene expression is feasible due to a natural abundance of PyPu tracts in genes and their 5′ flanking regions (Szybalski et al., 1966; Birnboim et al., 1979; Larsen and Weintraub, 1982; Hoffman-Liebermann et al., 1986; Yavachev et al., 1986; Beasty and Behe, 1987; Behe, 1987, 1995; Hoffman et al., 1990; Kato et al., 1990; Bucher and Yagil, 1991; Carter et al., 1991; Tripathi and Brahmachari, 1991; Blume et al., 1992; see also Table 6.1).

Many years ago, Morgan and Wells (1968) showed that mRNA synthesis on a duplex polymer template is inhibited when the binding of a homopyrimidine third strand results in a triple-helical complex. During the past few years, triplex formation has been shown to inhibit transcrition and subsequent protein synthesis in a number of more complicated experimental cell-free systems. This inhibition is based on sufficiently high affinities of triplex-forming oligonucleotides for double-stranded DNA ($K_{diss} \sim 1{-}10\,nM$) and the lifetimes of triple-stranded complexes on the order of several hours (Maher et al., 1990; Durland et al., 1991; Rougée et al., 1992; Singleton and Dervan, 1993). These values approach those observed for many sequence-specific DNA-binding proteins, so the protein factors that take part in transcription cannot easily displace the oligomer bound to the PyPu tract. Specificity of triplex formation is high: even one mismatch in a 15-nucleotide-long triplex-forming oligonucleotide results in at least 10-fold decrease in affinity compared to a perfect triplex (Moser and Dervan, 1987).

Figure 7.9 schematically shows several mechanisms that are relevant to transcription inhibition by triplex-forming oligonucleotides (Thuong and

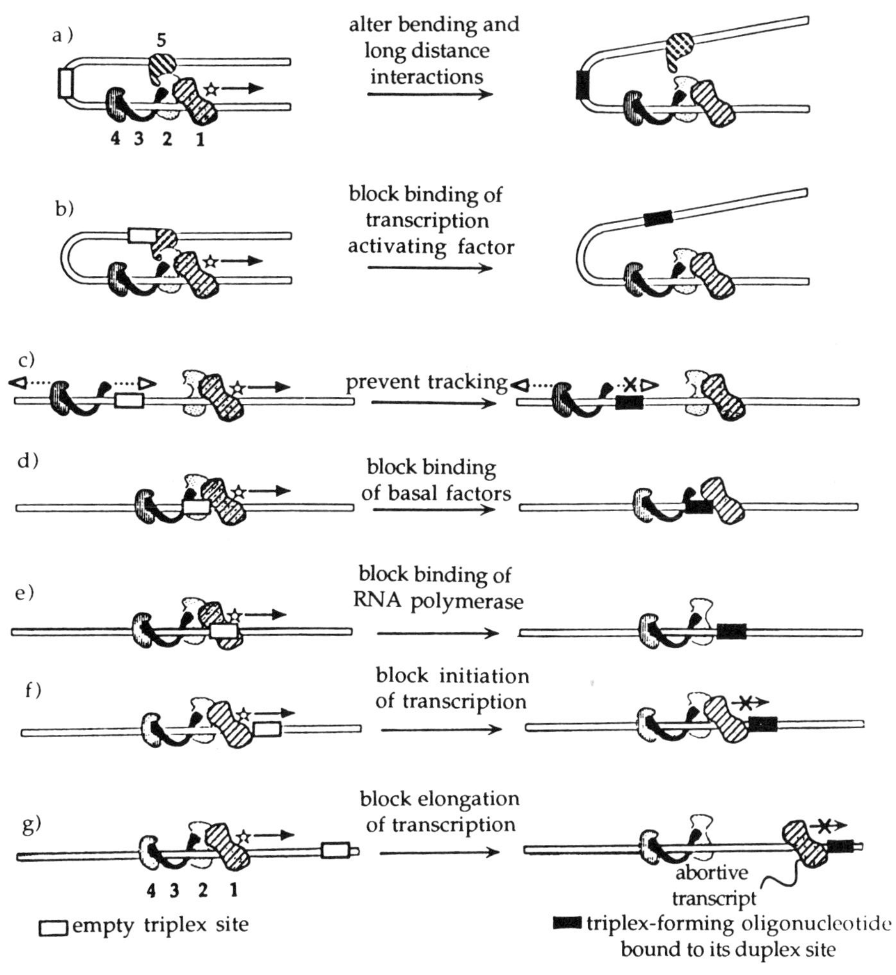

FIGURE 7.9. Hypothetical mechanisms by which triplex-forming oligonucleotides can inhibit transcription. The transcription process involves RNA polymerase and associated protein factors, activating factors that have binding sites in the promoter region, and activator proteins that bind to enhancer sequences at long distances upstream of the RNA polymerase binding site and require the double helix to fold. The star and arrow indicate the transcription start and the direction of transcription, respectively. An open box represents an unoccupied PyPu tract, whereas a filled box represents an intermolecular triplex formed on duplex DNA. Proteins: RNA polymerase (1), associated basal transcription factors (2), transcription activating factors (3,4,5). Triplex formation may result in: (a) disruption of long-distance interactions between enhancer-bound proteins and the transcription complex; (b) blockage of activator binding at the distant site; (c) blockage of trancription factor sliding to the promoter sequence from the enhancer sequence; (d) inhibition of binding of basal transcription factors; (e) a physical barrier to RNA polymerase binding; (f) inhibition of transcriptional initiation upon binding downstream but in contact with transcription complex; (g) transcription inhibition at the elongation step when the PyPu tract is located within the transcribed region of the gene. From Thuong and Hélène (1993). Reprinted with permission.

Hélène, 1993). The transcription machinery generally involves RNA polymerase and associated factors, activating proteins that bind upstream in the promoter region, and activators that bind to enhancer sequences at long distances from the RNA polymerase binding site and act via folding of the double helix. Triplex-forming oligonucleotides can (1) influence the local bending ability of DNA, thereby preventing long-distance interactions between enhancer-bound proteins and RNA polymerase and associated factors; (2) eliminate binding of activators, which exert their action either through long-distance interactions or through factors bound adjacent to the RNA polymerase (not shown); (3) block tracking of transcription factors initially bound to a distant enhancer sequence and sliding toward the RNA polymerase machinery bound around the promoter site; (4) repress binding and interactions of basal factors (i.e., polypeptides associated with the RNA polymerase); (5) inhibit the binding of an RNA polymerase to the promoter site; (6) block initiation of transcription upon binding downstream of promoter but in contact with the transcription machinery; (7) inhibit elongation of transcription when bound to the PyPu site within the transcribed gene. Some other mechanisms, such as the recruitment of inhibitory factors (e.g., proteins recognizing triple-stranded structure) or the alteration of chromatin assembly, may also be suggested (see also Maher et al., 1992). For more details, the reader is referred to recent reviews (Hélène, 1991; Hélène et al., 1992; Postel, 1992).

In addition to in vitro inhibition of gene expression, experiments have shown that the inhibiting activity is retained when the preformed DNA-triple-forming oligonucleotide complex is introduced into cultured cells (Grigoriev et al., 1992, 1993a,b; Lu and Ferl, 1992; Scaggiante et al., 1994). Moreover, incubation of the cells in the presence of triplex-forming oligonucleotides was shown to result in the uptake of the oligonucleotides and a subsequent inhibition of gene expression and protein synthesis (Orson et al., 1991; Postel et al., 1991; Ing et al., 1993; Ojwang et al., 1994; Roy, 1994). Table 7.2 lists some examples in which inhibition of transcription initiation or chain elongation by triplex-forming oligonucleotides was demonstrated.

In experiments in vitro, triplex-forming oligonucleotides produced significant inhibitory (up to 90%) effects on RNA polymerases when used in excess of their target duplexes (moles oligonucleotide/moles template >100) (Maher et al., 1992; Roy 1994; Xodo et al., 1994). In experiments in vivo, 50% inhibition of mRNA synthesis or cell proliferation may be obtained when the oligonucleotide concentration in the extracellular milieu is maintained at a level of up to 100 μM (Postel et al., 1991; Ing et al., 1993; Ojwang et al., 1994; Roy, 1994). Inhibition of transcription is determined by the lifetime of the oligonucleotides, which in enzyme-rich cellular media is on the order of a few dozen minutes (Akhtar et al., 1991; Orson et al., 1991); however, a decreased level of mRNA

TABLE 7.2. Inhibition of transcription of selected genes via intermolecular triplexes.

Gene	Oligomer	Inhibition	Conditions	Reference
Human *c-myc*	27-mer	Initiation	In vitro	Cooney et al., 1988
Human *c-myc*	27-mer	Initiation	HeLa cells (uptake)	Postel et al., 1991
G-free cassette plasmid	15-mer	Elongation	In vitro	Young et al., 1991
Mouse IL2R	28-mer	Initiation	Lymphocytes (uptake)	Orson et al., 1991
Mouse IL2R	15-mer-acridine	Initiation	HSB2 cells	Grigoriev et al., 1992
Mouse IL2R	15-mer-psoralen	Initiation	HSB3 cells (plasmid-oligo electroporation)	Grigoriev et al., 1993a
E. coli bla	13-mer	Initiation	In vitro	Duval-Valentin et al., 1992
Human dihydrofolate reductase	19-mer	Sp1 binding	In vitro	Gee et al., 1992
HIV-1	Various	Initiation	In vitro	Ojwang et al., 1994; Volkmann et al., 1993, 1995
	31 and 38-mers	Initiation	MT4 cells (uptake)	Mc Shan et al., 1992
T7 early promoter		Initiation	In vitro	Ross et al., 1992
Maize *Adh* 1-GUS	Various lengths	Initiation	Protoplasts (cotransformation)	Lu and Ferl, 1992
Human platelet-derived growth factor A-chain	24-mer	Initiation	In vitro	Wang et al., 1992
6-16 IRE	21-mer	Initiation	In vitro	Roy, 1993, 1994
Progesterone-responsive gene	38-mer	Initiation	In vivo (transfection)	Ing et al., 1993
HER-2/*neu*	various	Initiation	In vitro	Ebbinghaus et al., 1993
HER-2/*neu*	various	Transcription factor binding	In vitro	Noonberg et al., 1994c
Erythropoietin	various	Initiation	In vitro	Imagawa et al., 1994
Human *mdr1*	27-mer	Elongation	CEM-VLB 100 cells	Scaggiante et al., 1994

was noticed even 7 days after removal of natural phosphodiester oligonucleotides from the cell culture (Ojwang et al., 1994).

Triplex-stabilizing conditions may not be readily available in the cell. For example, the formation of PyPuPy triplexes with CG*C$^+$ triads requires lower pH, and that of PyPuPu triplexes with TA*A triads requires divalent metal cations or sufficient concentrations of polyamines. Therefore, some studies have concentrated on developing stably binding oligonucleotides containing T and G nucleotides (Cooney et al., 1988; Orson et al., 1991; Postel et al., 1991; Sun et al., 1991a; McShan et al., 1992; Ing et al., 1993; Roy, 1993, 1994; Cheng and Van Dyke, 1994; Durland et al., 1994; Fox, 1994; Ojwang et al., 1994; Olivas and Maher, 1995a). The idea behind these experiments is to delineate some principles of oligonucleotide composition and sequence that would allow formation of functional triplexes under ordinary conditions of the living cell. Although T and G containing oligonucleotides can form the PyPuPu type triplex at physiological pH in the presence of Mg^{2+}, K^+ ions destabilize such triplexes by promoting formation of oligonucleotide tetraplex stabilized by Hoogsteen hydrogen bonds (see Chapter 5). Using oligonucleotides containing 7-deazaxanthine and 6-thioguanine, which are not able to form Hoogsteen hydrogen bonds, results in efficient triplex formation (Gee et al., 1995; Olivas and Maher, 1995b).

The initiation of transcription or chain elongation can be inhibited in such a way when the bound oligonucleotide hampers RNA polymerase binding to its promoter site. It should be kept in mind, however, that the rate of triplex formation is slow. Therefore, for oligonucleotides competing with protein factors for binding to a specific sequence on DNA, kinetic phenomena might become a limiting factor. The effect caused by oligonucleotide binding can be enhanced by increasing its concentration, as demonstrated by Maher et al. (1989) in experiments on *Ava*I restriction nuclease protection assay.

It must be kept in mind that all claims of triplex-mediated inhibition of gene expression must have enough controls that confirm this specific mechanism of action. In one published work (Hobbs and Yoon, 1994), the differences in activity of reporter enzyme correlated with internalized plasmid DNA copy number rather than inhibition of transcription. The mechanism of oligonucleotide action in this case remains unclear. In other studies (Fedoseyeva et al., 1994; Ramanathan et al., 1994), an oligonucleotide designed to form a triplex exerted its effect by interrupting the cascade of events required for interferon-mediated induction of several genes, including the targeted one. Thus, an interpretation of oligonucleotide effects in vivo requires caution.

The growing knowledge of sequence information on the genes encoding the key proteins in the pathogenesis of various deseases has revealed many cases where the PyPu tracts are available and, therefore, triplex methodology is applicable. Table 7.3 shows examples of diseases that may

TABLE 7.3. Diseases that may potentially be treated with triplex-forming oligonucleotides and their analogs.

Virus-associated diseases
Adenovirus, herpes simplex viruses 1 and 2, herpes zoster, cytomegalovirus, Epstein-Barr virus, human papilloma virus, influenza A and B, parainfluenza, human T-cell lymphotropic virus, human immunodeficiency virus, hepatitis A and B
Oncologic diseases
Lymphoma, leukemia, melanoma, osteosarcoma, carcinoma (of colon, prostate, kidney, bladder, breast)
Other
Psoriasis, drug resistance, allergy, inflammation

be treated with triplex-forming oligonucleotides and their analogs (Chubb and Hogan, 1992). The least complicated and, therefore, most likely first therapeutic applications of triplex-forming oligonucleotides will be as antiviral agents. For many pathogenic viruses, the protein sequences and functions are relatively well understood. Many essential viral proteins whose genes should be suppressed have no human cellular analogs. This reduces the risk of undesirable suppression of normally functioning human genes.

The prospects for the treatment of cancer diseases with triplex-forming oligonucleotides are less clear. Although oncogenes have suggested to be involved in the development and progression of several human tumors, the understanding of the underlying processes is not clear enough. Besides, the difference between an oncogene and its normal cellular counterpart may be in only one base pair (Barbacid, 1987). Therefore, the risk of nonspecific inhibition of normal cellular genes should be carefully evaluated. A single base mismatch may reduce the efficiency of oligonucleotide binding by more than an order of magnitude when the tract is 15 base pairs long (Moser and Dervan, 1987); however, the optimum binding site may be in the range of 20 to 40 base pairs long (Chubb and Hogan, 1992). Thus, there is a danger of more significant nonspecific binding.

A number of other problems have arisen during the development of triplex-regulated gene expression. In some cases, the binding of short oligonucleotides may not be strong enough to stop the enzymatic machinery. For example, antisense oligonucleotides bound to the coding sequence of mRNA cannot prevent translation unless ribonuclease H cleaves the target mRNA at the binding site. Ribosomes moving along an mRNA template displace a bound oligonucleotide much like the unfolding of the secondary structures in mRNA.

On the other hand, repressor proteins inhibit the initiation of RNA polymerase action by preventing its binding to promoter DNA. Certain repressor–operator complexes can also act as transcriptional terminators.

However, the extent to which different protein–DNA complexes block transcription depends on the nature of the complex and its position relative to the promoter and the RNA polymerase. The *Lac* repressor protein–operator complex blocks transcription for *E. coli* RNA polymerase and eukaryotic RNA polymerase. In many other cases, only partial inhibition of transcription has been reported. These facts suggest a phenomenon of derepression by transcription across the operator sequences. Skoog and Maher (1993b) examined this problem as it applies to triplex-mediated transcription inhibition. They used a model system in which triplex-forming oligonucleotides inhibit a T7 RNA polymerase promoter by binding to an overlapping PyPu operator sequence (Maher, 1992; Skoog and Maher, 1993a). Inhibition in this system is due to triplex formation and occurs even in the case where the polymerase has prior access to the DNA template. However, efficient inhibition was demonstrated only at the level of transcription initiation. The triple-helical complexes located downstream from the promoter were unable to block T7 RNA polymerase elongation (Skoog and Maher, 1993a). Furthermore, the activity of the T7 promoter was restored when elongated phage polymerases transcribed across triple-helical complexes at the operator sequence (Skoog and Maher, 1993b).

These findings demonstrate that the examined artificial repressor–operator complex is subject to antagonism by *cis* elements (other promoters) acting at a distance. Although a high sensitivity of triple-helical complexes to disruption by phage polymerase elongation is likely to be an extreme case, less significantly pronounced effects, such as only transient pausing of eukaryotic RNA polymerase II at the triplex position, should be kept in mind when considering the inhibition of gene expression.

The triplex-disruptive transcription and the metabolic instability of the triplex-forming oligonucleotide can pose serious problems in controlling gene expression. To prolong the repressive action of triplex complexes, triplexes can be stabilized by the chemical modification of oligonucleotides. For example, the usefulness of phosphorothioate oligonucleotides for the inhibition of gene expression was recently demonstrated (Alunni-Fabbroni et al., 1994; Xodo et al., 1994b). The use of backbone-modified oligonucleotides could make oligomers resistant to nuclease action or enhance their binding to target duplex (see Chapter 4). Although backbone modifications were shown to generally increase the lifetimes of many oligonucleotide analogs in biological media (Akhtar et al., 1991; Crooke, 1992), the purposeful thermodynamic stabilization of triple-helical complexes is far from being understood. This is due to multiple effects of backbone substitutions and a lack of stereoselectivity during chemical synthesis of the analogs that results in a racemic mixture of oligomers with various biophysical and biochemical properties.

Attachment of intercalating molecules to the oligonucleotides, or chemical, or photochemical cross-linking of oligonucleotides, can increase the lifetime of triplex complexes. The triplex–duplex junction on the 5′ side of the third strand is a strong intercalation site for several intercalating drugs (Collier et al., 1991a; Sun et al., 1991c). For example, attachment of an acridine derivative to the 5′ end of a homopyrimidine oligonucleotide strongly stabilized the resulting triplex (Sun et al., 1989; Stonehouse and Fox, 1994). Such an acridine–oligonucleotide hybrid was used as a transcriptional repressor due to its inhibitory effect on the binding of nuclear proteins to the NFκB enhancer sequence (Grigoriev et al., 1992). Triple-stranded structures were also significantly stabilized by an intercalating benzo[e]pyridoindole derivative (Mergny et al., 1992). The use of oligonucleotide–intercalator conjugates allows one to use short triplex-forming oligomers. For example, a 7-mer purine oligomer-oxazolopyridocarbazolo conjugate formed a triplex with the U3 long terminal repeat sequence of HIV-1 which was thermostable up to 40°C (Mouscadet et al., 1994b).

The inhibition of gene expression can be made irreversible using chemical or photochemical cross-linking of an oligonucleotide to its duplex target. An example of appropriate chemical reaction is the cross-linking of an oligonucleotide containing the modified nucleoside N4,N4-ethano-5-methyldeoxycytidine to the N7 of a specific guanine in the double-stranded DNA target (Shaw et al., 1991). Such a reaction was shown to increase dramatically the relative amount of truncated transcripts in vitro (Young et al., 1991). A psoralen–oligonucleotide conjugate under long-wavelength UV irradiation is cross-linked to the target PyPu site. Such a stable complex irreversibly inhibits the in vitro expression of *E. coli bla* gene where the PyPu tract is inside the gene (Duval-Valentin et al., 1992). A similar strategy was shown to be effective for a psoralen–oligonucleotide conjugate targeted to the promoter of the α subunit of the interleukin-2 receptor gene (Grigoriev et al., 1993a). The PyPu site overlaps the binding site for the transcription factor NFκB, which activates transcription from the IL-2Rα promoter. Inhibition is observed when UV-induced cross-linking occurs both in vitro and in vivo (after transfection of a psoralen–oligonucleotide hybrid and reporter plasmid carrying the gene in question). Another way to block transcription is to produce a cross-link between two strands in DNA using a psoralen moiety triplex targeted to a specific PyPu tract (Takasugi et al., 1991; Giovannangéli et al., 1992b). The experiments with photochemical gene inactivation are very attractive because this process can be externally controlled; however, the transcription was inhibited at best down to 20% compared to the level of transcription in a control experiment. At long irradiation times, the inhibitory effect levels off, perhaps due to the light-induced destruction of the psoralen ring (Duval-Valentin et al., 1992).

Triplex regulation of DNA expression seems very promising; however, the question arises about the possibility of efficient delivery of polyanionic, hydrophilic compounds to their targets through the hydrophobic lipid membrane. According to the data of Zamecnik et al. (1986), HeLa and CEF cells incubated in 20 μ*M* solution of 20-mer oligonucleotide cells can take up the oligonucleotide and accumulate it up to an intracellular concentration of 1.5 μ*M* within 15 min. Several mechanisms of oligonucleotide delivery into the cell via liposome- (Leonetti et al., 1990) and receptor-mediated endocytosis (Loke et al., 1989), or directly in the hydrophobized form (Manoharan et al., 1992; Ing et al., 1993) have been demonstrated. When delivered into the cell, the full-length oligomer may persist for several hours (Orson et al., 1991). This is also the case for some nonionic analogs of triplex-forming oligonucleotides, e.g., methylphosphonates (Ts'o et al., 1992). A number of other analogs designed to improve triplex-forming ability and metabolic stability were described in Chapter 4. Recent experiments also show that after microinjection into the cell, oligonucleotides move rapidly to the nucleus by passive diffusion (Leonetti et al., 1991).

The pharmacokinetic and toxicological properties of oligonucleotides upon administration to animals have received only limited attention. After intravenous and intraperitoneal injections of 12-mer (Miller and Ts'o, 1987) and 38-mer (Zendegui et al., 1992), they were found to distribute rapidly (within several minutes) to all tissues except the brain. Their half-lives in serum, an enzyme-rich medium (Akhtar et al., 1991), are on the order of 10 min (Miller and Ts'o, 1987; Zendegui et al., 1992); however, they are accumulated intact in the tissues (Zendegui et al., 1992). Clearly, the issues of oligonucleotide dissipation in many tissues beyond the targeted one and the metabolic stability of oligonucleotides must receive more experimental effort.

A promising development in accumulating the desired oligonucleotides at their action sites has been recently described (Noonberg et al., 1994a). The oligomers of interest are generated from a vector containing promoter, capping, and termination sequences of the human small nuclear U6 gene, surrounding a synthetic sequence to be synthesized. In experiments in vivo, these oligonucleotides are produced without cell specificity and reach steady-state levels of expression ($5 \cdot 10^6$ copies per cell) within 9 hours posttransfection, and their synthesis continues for at least 7 days posttransfection.

To be efficient therapeutics, triplex-forming oligonucleotides should be nontoxic and nonimmunogenic. Little information is available on toxicities of oligonucleotides and their analogs. A few studies of oligonucleotide toxicity in mice showed that unmodified oligonucleotides and their methylphosphonate and phosphorothioate analogs were well tolerated at doses of 40 mg/kg body weight (Agrawal et al., 1988; Goodchild et al., 1988). At 160 mg/kg, the unmodified oligonucleotide resulted in 50% mortality (Agrawal et al., 1988), whereas the phosphorothioate analog gave 25% mortality (Goodchild et al., 1998). Similar results were also obtained for oligodeoxyribonucleotides and their derivatives bearing the $ClRCH_2NH$ alkylating group (Knorre et al., 1994).

During the development of genetic applications of triplex-forming oligonucleotides, the possibility of their unspecific binding to numerous target sites (e.g., transfer RNA, small nuclear RNA, accessible single-stranded regions in ribosomal RNA, aminoacyl-tRNA-synthetases, false binding to the initiation start, etc.) was also emphasized (Knorre et al., 1994). Experimental in vivo studies of transcription inhibition used several intracellular controls to determine the extent of nonspecific oligonucleotide action on DNA and proteins. Orson et al. (1991) showed a significant inhibitory effect of 28-mer oligonucleotide on the targeted gene of a α-subunit of an interleukin-2 receptor (IL-2Rα) without any effect on the control genes of *c-myc*, β-actin, IL-R2β, and IL-6. A 27-mer oligonucleotide triplex-targeted to the *c-myc* promoter P1 also produced a smaller inhibitory effect on *c-myc* P2 mRNA and no effect on β-actin mRNA in HeLa cells (Postel et al., 1991). Phosphorothioate oligonucleotide analogs are known to bind to proteins (Eckstein, 1985). This direct binding seems to be the primary cause of their inhibitory effects on DNA polymerase and reverse transcriptase activities (Gao et al., 1988; Majumdar et al., 1989). Nuclear extracts from murine and human cells contain polypeptides with molecular weights of about 1,500, 3,000, and 6,000 daltons, which interact with oligonucleotides (Svinarchuk et al., 1993). Thus, on the one hand, unspecific interactions of oligonucleotides may undesirably influence some cellular systems, yet on the other hand, they may effectively reduce available oligonucleotide concentrations. The range of limitations imposed by these side effects of oligonucleotides remains to be elucidated.

Conclusion

Besides the recognition of potential biological roles of H DNA, interest in triple-helical nucleic acids is stimulated by prospects for genetic applications of intermolecular triplexes, prospects that are much clearer than those raised by previously discovered structures (Z DNA, cruciforms). In some cases related to in vitro triplex formation (e.g., Triple Helix Vector methodology, artificial photoendonucleases, electron microscopic mapping, different variants of triplex affinity capture, quantitation of PCR products, detection of mutations in homopurine sequences), the above-described applications of triplex methodology can be accomplished using conditions most favorable to triplex formation. This allows the use of low-pH media, divalent metal cations, etc. When the triplexes must be formed in vivo (control of gene expression and, perhaps, more advanced versions of site-directed mutagenesis) and, in some cases, in vitro (enzymatic or chemical reactions proceeding at neutral pH and requiring strictly specified ionic conditions), triplex applications seem to be more limited. Hopefully, design of new recognition schemes will result in the application of triplex methodology to a broader set of sequence motifs capable of forming triplexes under physiological conditions. Knowledge of pharmacological consequences of oligonucleotide applications in vivo is still limited, but newly accumulated data are promising with respect to the absence of serious toxicity and immunogenicity problems. This gives definite hope for future widespread biotechnological and medicinal use of intermolecular triplex-based methodologies.

References

Adeniyi-Jones, S., and M. Zasloff (1985) Transcription, processing and nuclear transport of a B1 *Alu* RNA species complementary to an intron of the murine α-fetoprotein gene. Nature 317, 81–84.

Agazie, Y.M., J.S. Lee, and G.D. Burkholder (1994) Characterization of new monoclonal antibody to triplex DNA and immunofluorescent staining of mammalian chromosomes. J. Biol. Chem. 269, 7019–7023.

Agrawal, S., J. Goodchild, M.P. Civeira, A.H. Thornton, P.S. Sarin, and P.C. Zamecnik (1988) Oligonucleotide phosphoroamidates and phosphorothioates as inhibitors of human immunodeficiency virus. Proc. Natl. Acad. Sci. USA 85, 7079–7083.

Aharoni, A., N. Baran, and H. Manor (1993) Characterization of a multisubunit human protein which selectively binds single stranded $d(GA)_n$ and $d(GT)_n$ sequence repeats in DNA. Nucleic Acids Res. 21, 5221–5228.

Ahmed, S., A. Kintanar, and E. Henderson (1994) Human telomeric C-strand tetraplexes. Nature Struct. Biol. 1, 83–88.

Akhebat, A., C. Dagneaux, J. Liquier, and E. Taillandier (1992) Triple helical polynucleotidic structures: an FTIR study of the $C^+ \cdot G \cdot C$ triplet. J. Biomol. Struct. Dyn. 10, 577–588.

Akhtar, S., R. Kole, and R.L. Juliano (1991) Stability of antisense oligonucleotide analogs in cellular extracts and sera. Life Sci. 49, 1793–1801.

Akinrimisi, E.O., C. Sander, and P.O.P. Ts'o (1963) Properties of helical polycytidylic acid. Biochemistry 2, 340–344.

Alexeev, D.G., A.A. Lipanov, and I.Y. Skuratovskii (1987) Poly(dA) · poly(dT) is a B-type double helix with a distinctively narrow major groove. Nature 325, 821–823.

Alunni-Fabbroni, M., G. Manfioletti, G. Manzini, and L.E. Xodo (1994) Inhibition of T7 RNA polymerase transcription by phosphate and phosphorothioate triplex-forming oligonucleotides targeted to a RY site downstream from the promoter. Eur. J. Biochem. 226, 831–839.

Amirhaeri, S., F. Wohlrab, E.O. Major, and R.D. Wells (1988) Unusual DNA structure in the regulatory region of the human papovavirus JC virus. J. Virol. 62, 922–931.

Anshelevich, V.V., A.V. Vologodskii, A.V. Lukashin, and M.D. Frank-Kamenetskii (1984) Slow relaxational processes in the melting of linear biopolymers: a theory and its application to nucleic acids. Biopolymers 23, 39–58.

Antao, V.P., D.M. Gray, and R.L. Ratliff (1988) CD of six conformational rearrangements of poly[d(A-G)·d(C-T)] induced by low pH. Nucleic Acids Res. 16, 719–738.

Anwander, E.H.S., M.M. Probst, and B.M. Rode (1990) The influence of Li^+, Na^+, Ca^{2+}, and Zn^{2+} ions on the hydrogen bonds of the Watson-Crick base pairs. Biopolymers 30, 757–769.

Arnott, S., and P.J. Bond (1973) Structures for poly(U)·poly(dA)·poly(U) triple-stranded polynucleotides. Nature New Biol. 224, 99–101.

Arnott, S., P.J. Bond, E. Selsing, and P.J.C. Smith (1976) Models of triple-stranded polynucleotides with optimized stereochemistry. Nucleic Acids Res. 3, 2459–2470.

Arnott, S., and E. Selsing (1974) Structures for the polynucleotide complexes poly(dA)·poly(dT) and poly(dT)·poly(dA)·poly(dT). J. Mol. Biol. 88, 509–521.

Asseline, U., and N.T. Thuong (1993) Oligonucleotides tethered via nucleic bases — a potential new set of compounds for alternate strand triple-helix formation. Tetrahedron Lett. 34, 4173–4176.

Azhayeva, E., A Azhayev, A. Guzaev, J. Hovinen, and H. Lonnberg (1995) Looped oligonucleotides form stable hybrid complexes with a single-stranded DNA. Nucleic Acids Res. 23, 1170–1176.

Bacolla, A., and F.Y. Wu (1991) Mung bean nuclease cleavage pattern at a polypurine·polypyrimidine sequence upstream from the mouse metallothionein-I gene. Nucleic Acids Res. 19, 1639–1647.

Bagga, R., and S.K. Brahmachari (1993) Polypurine-polypyrimidine sequences adopt unwound structure in pBR322 form V DNA as probed by single-hit analysis of *Hpa*II sites. J. Biomol. Struct. Dyn. 10, 879–890.

Baker, B.F., and P.B. Dervan (1985) Sequence-specific cleavage of double-helical DNA. N-bromoacetyldistamicin. J. Am. Chem. Soc. 107, 8266–8268.

Bancroft, D., L.D. Williams, A. Rich, and M. Egli (1994) The low-temperature crystal structure of the pure-spermine form of Z-DNA reveals binding of a spermine molecule in the minor groove. Biochemistry 33, 1073–1086.

Baran, N., A. Lapidot, and H. Manor (1991) Formation of DNA triplexes accounts for arrests of DNA synthesis at $d(TC)_n$ tracts. Proc. Natl. Acad. Sci. USA 88, 507–511.

Baran, N., A. Neer, and H. Manor (1983) "Onion skin" replication of integrated polyoma virus DNA and flanking sequences in polyoma-transformed rat cells: termination within a specific cellular DNA segment. Proc. Natl. Acad. Sci. USA 80, 105–109.

Barawkar, D.A., V.A. Kumar, and K.N. Ganesh (1994) Triplex formation at physiological pH by oligonucleotides incorporating 5-Me-dC-(N4-spermine). Biochem. Biophys. Res. Commun. 205, 1665-1670.

Barbacid, M. (1987) *ras* genes. Annu. Rev. Biochem. 56, 779–827.

Barrett, J.C., and M.D. Shelby (1992) Mechanisms of human carcinogenesis. Prog. Clin. Biol. Res. 374, 415–434.

Basu, H.S., and L.J. Marton (1987) The interaction of spermine and pentamines with DNA. Biochem. J. 244, 243–246.

Beal, P.A., and P.B. Dervan (1991) Second structural motif for recognition of DNA by oligonucleotide directed triple-helix formation. Science 251, 1360–1363.

Beal, P.A., and P.B. Dervan (1992a) The influence of single base triplet changes on the stability of a Pur · Pur · Pyr triple helix determined by affinity cleaving. Nucleic Acids Res. 20, 2773–2776.

Beal, P.A., and P.B. Dervan (1992b) Recognition of double helical DNA by alternate strand triple helix formation. J. Am. Chem. Soc. 114, 4976–4982.

Beard, P., J.F. Morrow, and P. Berg (1973) Cleavage of circular, superhelical simian virus 40 DNA to a linear duplex by S1 nuclease. J. Virol. 12, 1303–1313.

Beasty, A.M., and M.J. Behe (1987) An oligopurine sequence bias occurs in eukaryotic viruses. Nucleic Acids Res. 16, 1517–1528.

Bebenek, K., J. Abbots, J.D. Roberts, S.H. Wilson, and T.A. Kunkel (1989) Specificity and mechanism of error-prone replication by human immunodeficiency virus-1 reverse transcriptase. J. Biol. Chem. 264, 16948–16956.

Becker, M.M., and G. Grossmann (1992) Photofootprinting DNA in vitro. Methods Enzymol. 212, 262–272.

Becker, M.M., and Z. Wang (1989) B → A transitions within a 5S ribosomal RNA gene are highly sequence-specific. J. Biol. Chem. 264, 4163–4167.

Behe, M.J. (1987) The DNA sequence of the human β-globin region is strongly biased in favor of long strings of contiguous purine and pyrimidine residues. Biochemistry 26, 7870–7875.

Behe, M.J. (1995) An overabundance of long oligopurine tracts occurs in the genome of simple and complex eukaryotes. Nucleic Acids Res. 23, 689–695.

Belintsev, B.N., A.V. Vologodskii, and M.D. Frank-Kamenetskii (1975) Influence of base sequence on the stability of the double helix of DNA. Mol. Biol. (Moscow) 10, 629–633.

Belland, R.J. (1991) H-DNA formation by the coding repeat elements of neisserial *opa* genes. Mol. Microbiol. 5, 2351–2360.

Belotserkovskii, B.P., M.M. Krasilnikova, A.G. Veselkov, and M.D. Frank-Kamenetskii (1992) Kinetic trapping of H-DNA by oligonucleotide binding. Nucleic Acids Res. 20, 1903–1908.

Belotserkovskii, B.P., A.G. Veselkov, S.A. Filippov, V.N. Dobrynin, S.M. Mirkin, and M.D. Frank-Kamenetskii (1990) Formation of intramolecular triplex in homopurine-homopyrimidine mirror repeats with point substitutions. Nucleic Acids Res. 18, 6621–6624.

Beltran, R., A. Martinez-Balbas, J. Bernues, R. Bowater, and F. Azorin (1993) Characterization of the zinc-induced structural transition at a $d(GA \cdot CT)_{22}$ sequence. J. Mol. Biol. 230, 966–978.

Berger, I., C.H. Kang, A. Fredian, R. Ratliff, R. Moyzis, and A. Rich (1995) Extension of the four stranded intercalated cytosine motif by adenine-adenine base pairing in the crystal structure of d(CCCAAT). Nature/Structural Biol. 2, 416–425.

Bernues, J., R. Beltran, J.M. Casasnovas, and F. Azorin (1989) Structural polymorphism of homopurine-homopyrimidine sequences: the secondary DNA structure adopted by a $d(GA \cdot CT)_{22}$ sequence in the presence of zinc ions. EMBO J. 8, 2087–2094.

Bernues, J., R. Beltran, J.M. Casasnovas, and F. Azorin (1990) DNA-sequence and metal-ion specificity of the formation of *H-DNA. Nucleic Acids Res. 18, 4067–4073.

Best, G.C., and P.B. Dervan (1995) Energetics of formation of sixteen triple helical complexes which vary at a single position within a pyrimidine motif. J. Am. Chem. Soc. 117, 1187–1193.

Bhattacharyya, R.G., K.K. Nayak, and A.N. Chakrabarty (1988) Interaction of (ATP)magnesate(2+) with DNA: assessment of metal binding sites and DNA conformations by spectroscopic and thermal denaturation studies. Inorg. Chim. Acta 153, 79–86.

Bianchi, A., R.D. Wells, N.H. Heintz, and M.S. Caddle (1990) Sequences near the origin of replication of the DHFR locus of Chinese hamster ovary cells adopt left-handed Z-DNA and triplex structures. J. Biol. Chem. 265, 21789–21796.

Biggin, M.D., and R. Tjian (1988) Transcription factors that activate the Ultrabithorax promoter in developmentally staged extracts. Cell 53, 699–711.

Billings, P.R., C.L. Smith, and C.R. Cantor (1991) New techniques for physical mapping of the human genome. FASEB J. 5, 28–34.

Birg, F., D. Praseuth, A. Zerial, N.T. Thuong, U. Asseline, T. Le Doan, and C. Hélène (1990) Inhibition of simian virus 40 DNA replication in CV-1 cells by an oligonucleotide covalently linked to an intercalating agent. Nucleic Acids Res. 18, 2901–2908.

Birnboim, H.C., R.M. Holford, and V.L. Seligy (1976) Random phasing of polypyrimidine/polypurine segments and nucleosome monomers in chromatin from mouse L cells. Cold Spring Harbor Symp. 39, 1161–1165.

Birnboim, H., R.R. Sederoff, and M.C. Paterson (1979) Distribution of polypyrimidine-polypurine segments in DNA from diverse organisms. Eur. J. Biochem. 98, 301–307.

Blackburn, E.H., and J.W. Szostak (1984) The molecular structure of centromeres and telomeres. Annu. Rev. Biochem. 53, 163–194.

Blaho, J.A., J.E. Larson, M.J. McLean, and R.D. Wells (1988) Multiple DNA secondary structures in perfect inverted repeat inserts in plasmids. J. Biol. Chem. 263, 14446–14455.

Blake, R.D. (1972) Thermodynamics of oligo$(A)_n \cdot 2$poly(U) from the dependence of the temperature of the helix-coil transition on oligomer concentration. Biopolymers 11, 913–933.

Blake, R.D., and J.R. Fresco (1966) Polynucleotides. VII. Spectrophotometric study of the kinetics of formation of the two-stranded helical complex resulting from the interaction of poly(riboadenylate) and poly(ribouridylate). J. Mol. Biol. 19, 509–521.

Blake, R.D., and J.R. Fresco (1973) Polynucleotides. XI. Thermodynamics of $(A)_N \cdot 2(U)$ from the dependence of $T_m{}^N$ (helix-coil transition temperature) on oligomer length. Biopolymers 12, 775–789.

Blake, R.D., L.C. Klotz, and J.R. Fresco (1968) Polynucleotides. IX. Temperature dependence of kinetics of complex formation in equimolar mixtures of polyriboadenylate and polyribouridylate. J. Am. Chem. Soc. 90, 3556–3562.

Blake, R.D., J. Massoulie, and J.R. Fresco (1967) Polynucleotides. VIII. A spectral approach to the equilibria between polyriboadenylate and polyribouridylate and their complexes. J. Mol. Biol. 30, 291–308.

Bloomfield, V.A. (1991) Condensation of DNA by multivalent cations: considerations on mechanism. Biopolymers 31, 1471–1481.

Blume, S.W., J.E. Gee, K. Shrestha, and D.M. Miller (1992) Triple helix formation by purine-rich oligonucleotides targeted to the human dihydrofolate reductase promoter. Nucleic Acids Res. 20, 1777–1784.

Boidot-Forget, M., M. Chassignol, M. Takasugi, N.T. Thuong, and C. Hélène (1988) Site-specific cleavage of single-stranded and double-stranded DNA sequences by oligodeoxyribonucleotides covalently linked to an intercalating agent and an EDTA-Fe chelate. Gene 72, 361–371.

Boles, T.C., and M.E. Hogan (1987) DNA structure in the human c-*myc* gene. Biochemistry 26, 367–376.

Bond, J.P., C.F. Anderson, and M.T. Record, Jr. (1994) Conformational transition of duplex and triplex nucleic acid helices: thermodynamic analysis of effects of salt concentration on stability using preferential interaction coefficients. Biophys. J. 67, 825–836.

Booher, M.A., S. Wang, and E.T. Kool (1994) Base pairing and steric interactions between pyrimidine strand bridging loops and the purine strand in DNA pyrimidine.purine.pyrimidine triplexes. Biochemistry 33, 4645–4651.

Boorstein, R.L., T.P. Hilbert, R.P. Cunningham, and G.W. Teebor (1990) Formation and stability of repairable pyrimidine photohydrates in DNA. Biochemistry 29, 10455–10460.

Bornet, O., G. Lancelot, L. Chanteloup, T.T. Nguyen, and J.M. Beau (1994) Selectively ^{13}C-enriched DNA: ^{13}C and ^{1}H assignments of a triple helix by two-dimensional relayed HMQC experiments. J. Biomol. NMR 4, 575–580.

Borowiec, J.A., and J.D. Gralla (1985) Supercoiling response of the lac p^s promoter in vitro. J. Mol. Biol. 184, 587–598.

Borowiec, J.A., L. Zhang, S. Sasse-Dwight, and J.D. Gralla (1987) DNA supercoiling promotes formation of a bent repression loop in lac DNA. J. Mol. Biol. 196, 101–111.

Bowater, R., F. Aboul-ela, and D.M.J. Lilley (1992) Two-dimensional gel electrophoresis of circular DNA topoisomers. Methods Enzymol. 212, 105–120.

Bowater, R., D. Chen, and D.M.J. Lilley (1994) Elevated unconstrained supercoiling of plasmid DNA generated by transcription and translation of the tetracycline resistance gene in eubacteria. Biochemistry 33, 9266–9275.

Brahms, J.G., O. Dargouge, S. Brahms, Y. Ohara, and V. Vagner (1985) Activation and inhibition of transcription by supercoiling. J. Mol. Biol. 181, 455–465.

Brahms, J., and W.F.H.M. Mommaerts (1964) A study of conformation of nucleic acids in solution by means of circular dichroism. J. Mol. Biol. 10, 73–88.

Brahms, S., J. Vergne, J.G. Brahms, E. Di Capua, P. Bucher, and T. Koller (1983) DNA formed by reassociation of complementary single-stranded circles from natural DNA is shown to contain left- and right-handed double helices. Cold Spring Harbor Symp. Quant. Biol. 47, 119–124.

Brereton, H.M., F.A. Firgaira, and D.R. Turner (1993) Origins of polymorphism at a polypurine hypervariable locus. Nucleic Acids Res. 21, 2563–2569.

Brinton, B.T., M.S. Caddle, and N.H. Heintz (1991) Position and orientation-dependent effects of a eukaryotic Z-triplex DNA motif on episomal DNA replication in COS-7 cells. J. Biol. Chem. 266, 5153–5161.

Broitman, S.L., D.D. Im, and J.R. Fresco (1987) Formation of the triple-stranded polynucleotide helix, poly(A · A · U). Proc. Natl. Acad. Sci. USA 84, 5120–5124.

Broom, A.D., V. Amarnath, R. Vince, and J. Brownell (1979) Poly(2-fluoroadenylic acid). The role of basicity in the stabilization of complementary helices. Biochim. Biophys. Acta 563, 508–517.

Brossalina E., E. Pascolo, and J.J. Toulmé (1993) The binding of an antisense oligonucleotide to a hairpin structure via triplex formation inhibits chemical and biological reactions. Nucleic Acids Res. 21, 5616–5622.

Brossalina E., and J.J. Toulmé (1993) A DNA hairpin as a target for antisense oligonucleotides. J. Am. Chem. Soc. 115, 796–797.

Bucher, P., and G. Yagil (1991) Occurence of oligopurine-oligopyrimidine tracts in eukaryotic and prokaryotic genes. DNA Sequence 1, 157–172.

Buckin, V.A., B.I. Kankiya, N.V. Bulichov, A.V. Lebedev, I.Y. Gukovsky, V.P. Chuprina, A.P. Sarvazyan, and A.R. Williams (1989a) Measurement of anomalously high hydration of $(dA)_n \cdot (dT)_n$ double helices in dilute solution. Nature 340, 321–322.

Buckin, V.A., B.I. Kankiya, D. Rentzeperis, and L.A. Marky (1994) Mg^{2+} recognizes the sequence of DNA through its hydration shell. J. Am. Chem. Soc. 116, 9423–9429.

Buckin, V.A., B.I. Kankiya, A.P. Sarvazyan, and H. Uedaira (1989b) Acoustical investigation of poly(dA)·poly(dT), poly[d(A-T)], poly(A)·poly(U) and DNA hydration in dilute aqueous solutions. Nucleic Acids Res. 17, 4189–4203.

Budarf, M., and E. Blackburn (1987) S1 nuclease sensitivity of a double-stranded telomeric sequence. Nucleic Acids Res. 15, 6273–6292.

Burkholder, G.D., L.J.P. Latimer, and J.S. Lee (1988) Immunofluorescent staining of mammalian nuclei and chromosomes with a monoclonal antibody to triplex DNA. Chromosoma 97, 185–192.

Caddle, M.S., R.H. Lussier, and N.H. Heintz (1990) Intramolecular DNA triplexes, bent DNA and DNA unwinding elements in the initiation region of an amplified dihydrofolate reductase replicon. J. Mol. Biol. 211, 19–33.

Cadet, J., and P. Vigny (1990) The Photochemistry of Nucleic Acids. In (H. Morrison, Ed.) Bioorganic Photochemistry, Vol. 1, Wiley, New York, pp. 1–272.

Callahan, D.E., T.L. Trapane, P.S. Miller, P.O.P. Ts'o, and L.S. Kan (1991) Comparative circular dichroism and fluorescence studies of oligodeoxyribonucleotide and oligodeoxyribonucleoside methylphosphonate pyrimidine strands in duplex and triplex formation. Biochemistry 30, 1650–1655.

Camerini-Otero, R.D., and P. Hsieh (1993) Parallel DNA triplexes, homologous recombination, and other homology-dependent DNA interactions. Cell 73, 217–223.

Campos, J.L., and J.A. Subirana (1987) The complex of poly(dG)·poly(dC) with arginine: stabilization of the B form and transition to multistranded structures. J. Biomol. Struct. Dyn. 5, 15–19.

Campos, J.L., and J.A. Subirana (1991) The influence of Mg^{++} and Zn^{++} on polypurine-polypyrimidine multistranded helices. J. Biomol. Struct. Dyn. 8, 793–800.

Cantor, C.R., and A. Efstratiadis (1984) Possible structures of homopurine homopyrimidine S1-hypersensitive sites. Nucleic Acids Res. 12, 8059–8072.

Cantor, C.R., and P.R. Schimmel (1980a) Biophysical Chemistry. Part I: The Conformation of Biological Macromolecules. Freeman, New York.

Cantor, C.R., and P.R. Schimmel (1980b) Biophysical Chemistry. Part II: Techniques for the Study of Biological Structure and Function. Freeman, New York.

Cantor, C.R., and P.R. Schimmel (1980c) Biophysical Chemistry. Part III: The Behavior of Biological Macromolecules. Freeman, New York.

Carter, P.E., C. Duponchel, M. Tosi, and J.E. Fothergill (1991) Complete nucleotide sequence of the gene for human C1 inhibitor with an unusually high density of *Alu* elements. Eur. J. Biochem. 197, 301–308.

Cassani, G.R., and F.J. Bollum (1969) Oligodeoxythymidilate: polydeoxyadenylate and oligodeoxyadenylate: polydeoxythymidilate interactions. Biochemistry 8, 3928–3936.

Cassidy, S.A., L. Strekowski, W.D. Wilson, and K.R. Fox (1994) Effect of triplex-binding ligand on parallel and antiparallel DNA triple helices using short unmodified and acridine-linked oligonucleotides. Biochemistry 33, 15338–15347.

Celano, P., C.M. Berchtold, D.L. Kizer, A. Weeraratna, B.D. Nelkin, S.B. Baylin, and R.A. Casero, Jr. (1992) Characterization of an endogeneous RNA transcript with homology to the antisense strand of the human c-*myc* gene. J. Biol. Chem. 267, 15092–15096.

Chalikian, T., G.E. Plum, A.P. Sarvazyan, and K.J. Breslauer (1994) Influence of drug binding on DNA hydration: acoustic and densimetric characterizations of netropsin binding to the poly(dAdT) · poly(dAdT) and poly(dA) · poly(dT) duplexes and the poly(dT) · poly(dA) · poly(dT) triplex at 25°C. Biochemistry 33, 8629–8640.

Chamberlin, M.J. (1965) Comparative properties of DNA, RNA and hybrid homopolymer pairs. Fed. Proc. 24, 1446–1457.

Champoux, J.J., and B.L. McConaughy (1975) Priming of superhelical SV40 DNA by *E. coli* RNA polymerase for in vitro DNA synthesis. Biochemistry 14, 307–316.

Chandler, S.P., and K.R. Fox (1993) Triple helix formation at $A_8XA_8 \cdot T_8YT_8$. FEBS Lett. 332, 189–192.

Chandler, S.P., and K.R. Fox (1995) Extension of DNA triple helix formation to a neighbouring $(AT)_n$ site. FEBS Lett. 360, 21–25.

Chastain, M., and I. Tinoco, Jr. (1991) Structural elements in RNA. In (W.E. Cohn and K. Moldave, Eds.) Progress in Nucleic Acids Research and Molecular Biology, Vol. 41. Academic Press, San Diego, California, pp. 131–177.

Chastain, M., and I. Tinoco, Jr. (1992) Poly(rA) binds poly(rG) · poly(rC) to form a triple helix. Nucleic Acids Res. 20, 315–318.

Chen, A.H., A. Reyes, and R. Akeson (1993) A homopurine-homopyrimidine sequence derived from the rat neuronal cell-adhesion molecule-encoding gene alters expression in transient transfections. Gene 128, 211–218.

Chen, C., D. Clark, K. Ueda, I. Pastan, M.M. Gottesman, and I.B. Roninson (1990) Genomic organization of the human multidrug resistance (MDR1) gene and origin of P-glycoproteins. J. Biol. Chem. 256, 506–514.

Chen, F.-M. (1991) Intramolecular triplex formation of the purine-purine-pyrimidine type. Biochemistry 30, 4472–4479.

Cheng, A.-J., and M.W. Van Dyke (1993) Monovalent cation effects on intermolecular purine-purine-pyrimidine triple-helix formation. Nucleic Acids Res. 21, 5630–5635.

Cheng, A.-J., and M.W. Van Dyke (1994) Oligodeoxyribonucleotide length and sequence effects on intermolecular purine-purine-pyrimidine triple-helix formation. Nucleic Acids Res. 22, 4742–4747.

Cheng, Y.K., and B.M. Pettitt (1992a) Stabilities of double- and triple-strand helical nucleic acids. Prog. Biophys. Mol. Biol. 58, 225–257.

Cheng, Y.K., and B.M. Pettitt (1992b) Hoogsteen versus reversed-Hoogsteen base pairing: DNA triple helices. J. Am. Chem. Soc. 114, 4465–4474.

Cheng, Y.K., and B.M. Pettitt (1995) Solvent effects on model $d(CG \cdot G)_7$ and $d(TA \cdot T)_7$ DNA triple helices. Biopolymers 35, 457–473.

Cherny, D.I., B.P. Belotserkovskii, M.D. Frank-Kamenetskii, M. Egholm, O. Buchardt, R.H. Berg, and P.E. Nielsen (1993a) DNA unwinding upon strand-displacement binding of a thymine-substituted polyamide to double-stranded DNA. Proc. Natl. Acad. Sci. USA 90, 1667–1670.

Cherny, D.I., V.A. Malkov, A.A. Volodin, and M.D. Frank-Kamenetskii (1993b) Electron microscopy visualization of oligonucleotide binding to duplex DNA via triplex formation. J. Mol. Biol. 230, 379–383.

Choi, O.R., and J.D. Engel (1988) Developmental regulation of beta-globin gene switching. Cell 55, 17–26.

Chomilier, J., J.S. Sun, D.A. Collier, T. Garestier, C. Hélène, and R. Lavery (1992) A computational and experimental study of the bending induced at a double-triple helix junction. Biophys. Chem. 45, 143–152.

Christophe, D., B. Cabrer, A. Bacolla, H. Targovnik, V. Pohl, and G. Vassart (1985) An unusually long poly(purine)-poly(pyrimidine) sequence is located upstream from the human thyroglobulin gene. Nucleic Acids Res. 13, 5127–5144.

Chubb, J.M., and M.E. Hogan (1992) Human therapeutics based on triple helix technology. Trends Biotechnol. 10, 132–136.

Chung, Y.T., and E.B. Keller (1990) Regulatory elements mediating transcription from the *Drosophila melanogaster* actin 5C proximal promoter. Mol. Cell. Biol. 10, 206–216.

Chuprina, V.P., W. Nerdal, E. Stetten, V.I. Poltev, and O.Y. Fedoroff (1993) Base dependence of B-DNA sugar conformation in solution and in the solid state. J. Biomol. Struct. Dyn. 11, 671–683.

Cimino, G.D., H.B. Gamper, S.T. Isaacs, and J.E. Hearst (1985) Psoralens as photoactive probes of nucleic acid structure and function: organic chemistry, photochemistry, and biochemistry. Annu. Rev. Biochem. 54, 1151–1193.

Clark, S.P., C.D. Lewis, and G. Felsenfeld (1990) Properties of BPG1, a poly(dG)-binding protein from chicken erythrocytes. Nucleic Acids Res. 18, 5119–5126.

Collado-Vides, J., B. Magasanik, and J.D. Gralla (1991) Control site location and transcriptional regulation in *Escherichia coli*. Microbiol. Rev. 55, 371–394.

Collick, A., M.G. Dunn, and A.J. Jeffreys (1991) Minisatellite binding protein Msbp-1 is a sequence-specific single-stranded DNA-binding protein. Nucleic Acids Res. 19, 6399–6404.

Collier, D.A., J.A. Griffin, and R.D. Wells (1988) Non-B right-handed DNA conformations of homopurine · homopurimidine sequences in the murine immunoglobulin C_α switch region. J. Biol. Chem. 263, 7397–7405.

Collier, D.A., J.L. Mergny, N.T. Thuong, and C. Hélène (1991a) Site specific intercalation at the triplex-duplex junction induces a conformational change which is detectable by hypersensitivity to diethypyrocarbonate. Nucleic Acids Res. 19, 4219–4224.

Collier, D.A., N.T. Thuong, and C. Hélène (1991b) Sequence-specific bifunctional DNA ligands based on triplex-forming oligonucleotides inhibit restriction enzyme cleavage under physiological conditions. J. Am. Chem. Soc. 113, 1457–1458.

Collier, D.A., and R.D. Wells (1990) Effect of length, supercoiling and pH on intramolecular triplex formation. Multiple conformers at Pur · Pyr mirror repeats. J. Biol. Chem. 265, 10652–10658.

Colocci, N., M.D. Distefano, and P.B. Dervan (1993) Cooperative oligonucleotide-directed triple helix formation at adjacent DNA sites. J. Am. Chem. Soc. 115, 4468–4473.

Colocci, N., and P.B. Dervan (1995) Cooperative triple helix formation at adjacent DNA sites: sequence composition at the junction. J. Am. Chem. Soc. 117, 4781–4787.

Cooney, M., G. Czernuszewicz, E.H. Postel, S.J. Flint, and M.E. Hogan (1988) Site-specific oligonucleotide binding represses transcription of the human c-*myc* gene in vitro. Science 241, 456–459.

Cozzarelli, N.R., T.C. Boles, and J.H. White (1990) Primer on the topology and geometry of DNA supercoiling. In (N.R. Cozzarelli and J.C. Wang, Eds.) Biological Effects of DNA Topology. Cold Spring Harbor Laboratory Press, Cold Spring Harbor, New York, pp. 139–184.

Craig, M.E., D.M. Crothers, and P. Doty (1971) Relaxation kinetics of dimer formation by self-complementary oligonucleotides. J. Mol. Biol. 62, 383–401.

CRC Handbood of Chemistry and Physics, 73d Edition (1992–1993).

Crooke, S.T. (1992) Therapeutic applications of oligonucleotides. Annu. Rev. Pharmacol. Toxicol. 32, 329–376.

Crothers, D.M., T.E. Haran, and J.G. Nadeu (1990) Intrinsically bent DNA. J. Biol. Chem. 265, 7093–7096.

Darnell, J., H. Lodish, and D. Baltimore (1986) Molecular Biology of the Cell. Scientific American Books, New York, p. 618.

Daube, S.S., and P.H. von Hippel (1992) Functional transcription elongation complexes from synthetic RNA-DNA bubble duplexes. Science 258, 1320-1324.

Davies, R.J.H., and N. Davidson (1971) Interaction of polyadenylic acid with methylated xanthines. Biopolymers 10, 21–33.

Davis, D.R. (1967) X-ray diffraction studies of macromolecules. Annu. Rev. Biochem. 36, 321–364.

Davis, R.H., D.R. Morris, and P. Coffino (1992) Sequestered end products and enzyme regulation: the case of ornithine decarboxylase. Microbiol. Rev. 56, 280–290.

Davis, T.L., A.B. Firulli, and A.J. Kinniburgh (1989) Ribonucleoprotein and protein factors bind to an H-DNA-forming c-*myc* DNA element: possible regulators ot the c-*myc* gene. Proc. Natl. Acad. Sci. USA 86, 9682–9686.

Davisson, E.C., and K. Johnsson (1993) Triple helix binding of oligodeoxyribonucleotides containing 8-oxo-2′-deoxyadenosine. Nucleosides Nucleotides 12, 237–243.

Dayn, A., S. Malkhosyan, and S.M. Mirkin (1992a) Transcriptionally driven cruciform formation in vivo. Nucleic Acids Res. 20, 5991–5997.

Dayn, A., G.M. Samadashwily, and S.M. Mirkin (1992b) Intramolecular DNA triplexes: unusual sequence requirements and influence on DNA polymerization. Proc. Natl. Acad. Sci. USA 89, 11406–11410.

Dean, W.W., and J. Lebowitz (1971) Partial alteration of secondary structure in native superhelical DNA. Nature (New Biol.) 231, 5–8.

Debart, F., B. Rayner, G. Degols, and J.L. Imbach (1992) Synthesis and base-pairing properties of the nuclease-resistant α-anomeric dodecaribonucleotide α-[r(UCUUAACCCACA)]. Nucleic Acids Res. 20, 1193–1200.

DeChiara. T.M., F. Erlitz, and S.J. Tarnowski (1986) Procedures for in vitro DNA mutagenesis of human leukocyte interferon sequences. Methods Enzymol. 119, 403–415.

De Clercq, E., P.F. Torrence, and B. Witkop (1976) Polynucleotide displacement reactions: detection by interferon induction. Biochemistry 15, 717–724.

Delcourt, S.G., and R.D. Blake (1991) Stacking energies in DNA. J. Biol.Chem. 266, 15160–15169.

de los Santos, C., M. Rosen, and D. Patel (1989) NMR studies of DNA $(R+)_n \cdot (Y-)_n \cdot (Y+)_n$ triple helices in solution: imino and amino proton markers of T·A·T and $C \cdot G \cdot C^+$ base-triple formation. Biochemistry 28, 7282–7289.

De Marky, H., and G.S. Manning (1975) Application of polyelectrolyte "limiting laws" to the helix-coil transition of DNA. III. Dependence of helix stability on excess univalent salt and on polynucleotide phosphate concentration for variable equivalent ratios of divalent metal ion to phosphate. Biopolymers 14, 1407–1422.

Demidov, V.V., D.I. Cherny, A.V. Kurakin, M.V. Yavnilovich, V.A. Malkov, M.D. Frank-Kamenetskii, S.H. Sonnichsen, and P.E. Nielsen (1994a) Electron microscopy mapping of oligopurine tracts in duplex DNA by peptide nucleic acid targeting. Nucleic Acids Res. 22, 5218–5222.

Demidov, V.V., V.A. Malkov, V.N. Soyfer, and M.D. Frank-Kamenetskii (1992) Triplex-forming pyrimidine brominated oligonucleotides as gene-targeting photosensitizers. Proc. Conference "Human Genome-91", Pereslavl-Zalesskii, USSR.

Demidov, V.V., M.D. Frank-Kamenetskii, M. Egholm, O. Buchardt, and P.E. Nielsen (1993) Sequence selective double strand DNA cleavage by Peptide Nucleic Acid (PNA) targeting using nuclease S1. Nucl. Acids Res. 21, 2103–2107.

Demidov, V.V., V.N. Potaman, M.D. Frank-Kamenetskii, M. Egholm, O. Buchardt S.H. Sonnichsen, and P.E. Nielsen (1994b) Stability of peptide nucleic acids in human serum and cellular extracts. Biochem. Pharmacol. 48, 1310–1313.

Demidov, V.V., M.V. Yavnilovich, B.P. Belotserkovskii, M.D. Frank-Kamenetskii, and P.E. Nielsen (1995) Kinetics and mechanism of polyamide ("peptide") nucleic acid binding to duplex DNA. Proc. Natl. Acad. Sci. USA 92, 2637–2641.

Derbyshire, K.M., J.J. Salvo, and N.D.F. Grindley (1986) A simple and efficient procedure for saturation mutagenesis using mixed oligodeoxynucleotides. Gene 46, 145–152.

Dervan, P.B. (1987) Sequence specific recognition and cleavage of double helical DNA. In Proceedings of the Robert A. Welch Foundation. XXXI. Design of enzymes and enzyme models. Houston, Texas, pp. 93–104.

Dervan, P.B. (1992) Reagents for the site-specific cleavage of megabase DNA. Nature 359, 87–88

Dickerson, R.E. (1992) DNA structure from A to Z. Methods Enzymol. 211, 67–111.

Dieter-Wurm, I., M. Sabat, and B. Lippert (1992) Model for a platinated DNA triplex: Watson-Crick and metal-modified Hoogsteen pairing. J. Am. Chem. Soc. 114, 357–359.

Dikalov, S.I., G.V. Rumyantseva, L.M. Weiner, D.S. Sergejev, E.I. Frolova, T.S. Godovikova, and V.F. Zarytova (1991) Hydroxyl radical generation by oligonucleotide derivatives of anthracycline antibiotic and synthetic quinone. Chem.-Biol. Interact. 77, 325–339.

Distefano, M.D., and P.B. Dervan (1993) Energetics of cooperative binding of oligonucleotides with discrete dimerization domains to DNA by triple helix formation. Proc. Natl. Acad. Sci. USA 90, 1179–1183.

Distefano, M.D., J.A. Shin, and P.B. Dervan (1991) Cooperative binding of oligonucleotides to DNA by triple helix formation. J. Am. Chem. Soc. 113, 5901–5902.

Distefano, M.D., and P.B. Dervan (1992) Ligand-promoted dimerization of oligonucleotides binding cooperatively to DNA. J. Am. Chem. Soc. 114, 11006–11007.

Dittrich, K., J. Gu, R. Tinder, M. Hogan, and X. Gao (1994) T · C · G triplet in an antiparallel purine · purine · pyrimidine DNA triplex. Conformational studies by NMR. Biochemistry 33, 4111–4120.

Dobrikov, M.I., V.V. Gorn, V.F. Zarytova, A.S. Levina, T.A. Prikhod'ko, G.V. Shishkin, D.R. Tatabadze, and M.M. Zaalishvili (1992a) Site-specific photomodification of nucleic acids with arylazide and perfluoroarylazide oligonucleotide derivatives. II. Specificity in relation to nucleosides. Bioorg. Khim. 18, 1190–1198.

Dobrikov, M.I., V.F. Zarytova, N.I. Komarova, A.S. Levina, S.A. Lokhov, T.A. Prikhod'ko, G.V. Shishkin, D.R. Tatabadze, and M.M. Zaalishvili (1992b) Effective complementarily-addressed photomodification of nucleic acids by oligonucleotide derivatives containing aromatic azido groups. Bioorg. Khim. 18, 540–549.

Dolinnaya, N.G., O.V. Pyatrauskene, and Z.A. Shabarova (1991) Probing DNA triple helix structure by chemical ligation. FEBS Lett. 284, 232–234.

Dorman, C.J., G.C. Barr, N.N. Bhriain, and C.F. Higgins (1988) DNA supercoiling and the anaerobic and growth phase regulation of *ton*B gene expression. J. Bacteriol. 170, 2816–2826.

Dove, W.H., and N. Davidson (1962) Cation effects on the denaturation of deoxyribonucleic acid (DNA). J. Mol. Biol. 5, 467–478.

Drew, H.R., and R.E. Dickerson (1981) Structure of a B-DNA dodecamer. III. Geometry of hydration. J. Mol. Biol. 151, 535–556.

Drew, H.R., T. Takano, S. Tanaka, K. Itakura, and R.E. Dickerson (1980) High-salt $d(C_pG_pC_pG)$, a left-handed Z′ DNA double helix. Nature 286, 567–573.

Drlica, K., and J. Rouviere-Yaniv (1987) Histone-like proteins of bacteria. Microbiol. Rev. 51, 301–319.

Droge, P. (1993) Transcription-driven site-specific DNA recombination in vitro. Proc. Natl. Acad. Sci. USA 90, 2759–2763.

D'Souza, D.J., and E.T. Kool (1992) Strong binding of single-stranded DNA by stem-loop oligonucleotides. J. Biomol. Struct. Dyn. 10, 141–152.

Duckett, D.R., A.I.H. Murchie, R.M. Clegg, A. Zechel, E. von Kitzing, S. Diekmann, and D.M.J. Lilley (1990) The structure of the Holliday junctions. In (R.H. Sarma and M.H. Sarma, Eds.) Structure and Methods, Vol. 1: Human Genome Initiative and DNA Recombination. Adenine Press, New York, pp. 157–181.

Duguid, J., V.A. Bloomfield, J. Benevides, and G.J. Thomas, Jr. (1993) Raman spectroscopy of DNA-metal complexes. I. Interactions and conformational effects of the divalent cations: Mg, Ca, Sr, Ba, Mn, Co, Ni, Cu, Pd and Cd. Biophys. J. 65, 1916–1928.

Duker, N.J. (1986) Rates of heat-induced pyrimidine alterations in synthetic polydeoxyribonucleotides. Chem. Biol. Interact. 60, 265–273.

Duker, N.J., T.L. Chao, and E.M. Resnick (1986) Rates of heat-induced DNA purine alterations in synthetic polydeoxyribonucleotides. Chem. Biol. Interact. 58, 241–251.

Durand, M., S. Peloille, N.T. Thuong, and J.C. Maurizot (1992a) Triple-helix formation by an oligonucleotide containing one $(dA)_{12}$ and two $(dT)_{12}$ sequences bridged by two hexaethylene glycol chains. Biochemistry 31, 9197–9204.

Durand, M., N.T. Thuong, and J.C. Maurizot (1992b) Binding of netropsin to a DNA triple helix. J. Biol. Chem. 267, 24394–24399.

Durand, M., N.T. Thuong, and J.C. Maurizot (1994a) Interaction of Hoechst 33258 with a DNA triple helix. Biochimie 76, 181–186.

Durand, M., N.T. Thuong, and J.C. Maurizot (1994b) Berenil complexation with a nucleic acid triple helix. J. Biomol. Struct. Dyn. 11, 1191–1202.

Durland, R.H., D.J. Kessler, S. Gunnell, M. Duvic, B.M. Pettitt, and M.E. Hogan (1991) Binding of triple helix forming oligonucleotides to sites in gene promoters. Biochemistry 30, 9246–9255.

Durland, R.H., T.S. Rao, G.R. Revankar, J.H. Tinsley, M.A. Myrick, D.M. Seth, J. Rayford, P. Singh, and K. Jayaraman (1994) Binding of T and T analogs to CG base pairs in antiparallel triplexes. Nucleic Acids Res. 22, 3233–3240.

Durland, R.H., T.S. Rao, V. Bodepudi, D.M. Seth, K. Jayaraman, and G.R. Revankar (1995) Azole substituted oligonucleotides promote antiparallel triplex formation at non-homopurine duplex targets. Nucleic Acids Res. 23, 647–653.

Duval-Valentin, G., N.T. Thuong, and C. Hélène (1992) Specific inhibition of transcription by triple helix-forming oligonucleotides. Proc. Natl. Acad. Sci. USA 89, 504–508.

Duval-Valentin, G., T. de Bizemont, M. Takasugi, J.-L. Mergny, E. Bisagni, and Hélène (1995) Triple-helix specific ligands stabilize H-DNA conformation. J. Mol. Biol. 247, 847–858.

Dybvig, K., C.D. Clark, G. Aliperti, and M.J. Schlesinger (1983) A chicken repetitive DNA sequence that is highly sensitive to single-strand specific endonucleases. Nucleic Acids Res. 11, 8495–8508.

Ebbinghaus, S.W., J.E. Gee, B. Rodu, C.A. Mayfield, G. Sanders, and D.M. Miller (1993) Triplex formation inhibits HER-2/*neu* transcription in vitro. J. Clin. Invest. 92, 2433–2439.

Eckstein, F. (1985) Nucleoside phosphorothioates. Annu. Rev. Biochem. 54, 367–402.

Egholm, M., O. Buchardt, P.E. Nielsen, and R.H. Berg (1992a) Peptide nucleic acids (PNA). Oligonucleotide analogues with an achiral peptide backbone. J. Am. Chem. Soc. 114, 1895–1897.

Egholm, M., O. Buchardt, P.E. Nielsen, and R.H. Berg (1992b) Recognition of guanine and adenine in DNA by cytosine and thymine containing peptide nucleic acids (PNA). J. Am. Chem. Soc. 114, 9677–9678.

Eichhorn, G.L. (1981) The effect of metal ions on the structure and function of nucleic acids. Adv. Inorg. Biochem. 3, 1–46.

Egli, M., L.D. Williams, Q. Gao, and A. Rich (1991) Structure of the pure-spermine form of Z-DNA (magnesium free) at a 1 Å resolution. Biochemistry 30, 11388–11402.

Escudé, C., J.C. François, J.S. Sun, G. Ott, M. Sprizl, T. Garestier, and C. Hélène (1993) Stability of triple helices containing RNA and DNA strands: experimental and molecular modeling studies. Nucleic Acids Res. 21, 5547–5553.

Escudé, C., J.S. Sun, M. Rougée, T. Garestier, and C. Hélène (1992) Stable triple helixes are formed upon binding of RNA oligonucleotides and their 2′-O-methyl derivatives to double-helical DNA. C.R. Acad. Sci., Ser. III 315, 521–525.

Evans, T., and A. Efstratiadis (1986) Sequence-dependent S1 nuclease hypersensitivity of a heteronomous DNA duplex. J. Biol. Chem. 261, 14771–14780.

Evans, T., E. Schon, G. Gora-Maslak, J. Patterson, and A. Efstratiadis (1984) S1-hypersensitive sites in eukaryotic promoter regions. Nucleic Acids Res. 12, 8043–8058.

Favre, A. (1990) 4-Thiouridine as an Intrinsic Photoaffinity Probe of Nucleic Acid Structure and Interaction. In (H. Morrison, Ed.) Bioorganic Photochemistry, Vol. 1, Wiley, New York, pp. 379–426.

Fedorova, O.S., D.G. Knorre, L.M. Podust, and V.F. Zarytova (1988) Complementary addressed modification of double-stranded DNA within a ternary complex. FEBS Lett. 228, 273–276.

Fedoseyeva, E.V., Y. Li, B. Huey, S. Tam, C.A. Hunt, G. Benichou, and M.R. Garovoy (1994) Inhibition of interferon-γ-mediated immune functions by oligonucleotides. Suppression of human T cell proliferation by downregulation of IFNγ-induced ICAM-1 and Fc-receptor on accessory cells. Transplantation 57, 606–612.

Feigon, J., V. Sklenar, E. Wang, D.E. Gilbert, R.F. Macaya, and P. Schultze (1992) ^{1}H NMR spectroscopy of DNA. Methods Enzymol. 211, 235–253.

Feigon, J., A.H.-J. Wang, G. van der Marel, J.H. van Boom, and A. Rich (1985) Z-DNA forms without an alternating purine-pyrimidine sequence sequence in solution. Science 230, 82–84.

Felsenfeld, G., D.R. Davies, and A. Rich (1957) Formation of a three-standed polynucleotide molecule. J. Am. Chem. Soc. 79, 2023–2024.

Felsenfeld, G., and H.T. Miles (1967) The physical and chemical properties of nucleic acids. Ann. Rev. Biochem. 36, 407–448.

Felsenfeld, G., and A. Rich (1957) Studies on the formation of two- and three-stranded polyribonucleotides. Biochim. Biophys. Acta 26, 457–468.

Feuerstein, B.G., N. Pattabiraman, and L.J. Marton (1986) Spermine-DNA interactions: a theoretical study. Proc. Natl. Acad. Sci. USA 83, 5948–5952.

Fiers, W., R.L. Sinsheimer (1962) The structure of the DNA of the bacteriophage ϕX174. I. The action of exopolynucleotidases. J. Mol. Biol. 5, 408–419.

Financsek, I., L. Tora, G. Kelemen, and E.J. Hedvegi (1986) Supercoil induced S1 hyper-sensitive sites in the rat and human ribosomal genes. Nucleic Acids Res. 14, 3263–3277.

Finer, M.H., S. Aho, L.C. Gerstenfeld, H. Boedtker, and P. Doty (1987) Unusual DNA sequences located within the promoter region and the first intron of the chicken pro-α1(I) collagen gene. J. Biol. Chem. 262, 13323–13332.

Firulli, A.B., D.C. Maibenco, and A.J. Kinniburgh (1992) The identification of a tandem H-DNA structure in the c-*myc* nuclease sensitive promoter element. Biochem. Biophys. Res. Commun. 185, 264–270.

Firulli, A.B., D.C. Maibenco, and A.J. Kinniburgh (1994) Triplex-forming ability of a c-*myc*-promoter element predicts promoter strength. Arch. Biochem. Biophys. 310, 236–242.

Fischer, K.D., and J. Nowock (1990) The T → C substitution at −198 of the Aγ-globin gene associated with the British form of HPFH generates overlapping recognition sites for two DNA-binding proteins. Nucleic Acids Res. 18, 5685–5693.

Fishel, R., P. Anziano, and A. Rich (1990) Z-DNA affinity chromatography. Methods Enzymol. 184, 328–340.

Flavell, R.A., and F.M. Van den Berg (1975) The isolation of duplex DNA containing (dA · dT) clusters by affinity chromatography on poly(U) Sephadex. FEBS Lett. 58, 90–93.

Fossella, J.A., Y.J. Kim, H. Shih, E.G. Richards, and J.R. Fresco (1993) Relative specificities in binding of Watson-Crick base pairs by third strand residues in a DNA pyrimidine triplex motif. Nucleic Acids Res. 21, 4511–4515.

Fowler, R.F., and D.M. Skinner (1986) Eukaryotic DNA diverges at a long and complex pyrimidine · purine tract than can adopt altered conformations. J. Biol. Chem. 261, 8994–9001.

Fox, K.R. (1990) Long $(dA)_n \cdot (dT)_n$ tracts can form intramolecular triplexes under superhelical stress. Nucleic Acids Res. 18, 5387–5391.

Fox, K.R. (1994) Formation of DNA triple helices incorporating blocks of G · GC and T · AT triplets using short acridine-linked olibonucleotides. Nucleic Acids Res. 22, 2016–2021.

Fox, K.R. (1995) Kinetic studies on the formation of acridine-linked DNA triple helices. FEBS Lett. 357, 312–316.

François, J.C., and C. Hélène (1995) Recognition and cleavage of single-stranded DNA containing hairpin structures by oligonucleotides forming both Watson–Crick and Hoogsteen hydrogen bonds. Biochemistry 34, 65–72.

François, J.C., T. Saison-Behmoaras, C. Barbier, M. Chassignol, N.T. Thuong, and C. Hélène (1989a) Sequence-specific recognition and cleavage of duplex DNA via triple-helix formation by oligonucleotides covalently linked to a phenanthroline-copper chelate. Proc. Natl. Acad. Sci. USA 86, 9702–9706.

François, J.C., T. Saison-Behmoaras, M. Chassignol, N.T. Thuong, and C. Hélène (1988a) Artificial nucleases: specific cleavage of the double helix of DNA by oligonucleotides linked to copper-phenanthroline complex. C.R. Acad. Sci. Ser. III 307, 849–854.

François, J.C., T. Saison-Behmoaras, M. Chassignol, N.T. Thuong, and C. Hélène (1989b) Sequnece-targeted cleavage of single- and double-stranded DNA by oligothymidilates covalently linked to 1,10-phenanthroline. J. Biol. Chem. 264, 5891–5898.

François, J.C., T. Saison-Behmoaras, and C. Hélène (1988b) Sequence-specific recognition of the major groove of DNA by oligodeoxynucleotides via triple helix formation. Footprinting studies. Nucleic Acids Res. 16, 11431–11440.

François, J.C., T. Saison-Behmoaras, N.T. Thuong, and C. Hélène (1989c) Inhibition of restriction endonuclease cleavage via triple helix formation by homopyrimidine oligonucleotides. Biochemistry 28, 9617–9619.
Frank-Kamenetskii, M.D. (1971) Simplification of the empirical relationship between melting temperature of DNA, its GC content and concentration of sodium ions in solution. Biopolymers 10, 2623–2624.
Frank-Kamenetskii, M.D. (1988) H-form DNA and the hairpin-triplex model. Nature 333, 214.
Frank-Kamenetskii, M.D. (1989) Waves of DNA supercoiling. Nature 337, 206.
Frank-Kamenetskii, M.D. (1990) DNA supercoiling and unusual structures. In (N.R. Cozzarelli, and J.C. Wang, Eds.) Biological Effects of DNA Topology. Cold Spring Harbor Laboratory Press, Cold Spring Harbor, New York, pp. 185–215.
Frank-Kamenetskii, M. (1992) Protonated DNA structures. Methods Enzymol. 211, 180–191.
Frank-Kamenetskii, M.D., V.A. Malkov, O.N. Voloshin, and V.N. Soyfer (1991) Stabilization of PyPuPu triplexes with bivalent cations. Nucleic Acids Res. Symp. Ser. 24, 159–162.
Frank-Kamenetskii, M.D., and S.M. Mirkin (1995) Triplex DNA structures. Annu. Rev. Biochem. 64, 65–95.
Freeman, L.A., and W.T. Garrard (1992) DNA supercoiling in chromatin structure and gene expression. Crit. Rev. Euk. Gene Exp. 2, 165–209.
Freifelder, D. (1987) Molecular Biology. Jones and Bartlett Publishers, Boston, Portola Valley, pp. 88–89.
Fresco, J.R. (1963) Some investigations of the secondary and tertiary structure of ribonucleic acids. In (Vogel. H.J., V. Bryson, and J.O. Lampen, eds.) Informational Macromolecules. Academic Press, New York, pp. 121–142.
Freier, S.M., R. Kierzek, J.A. Jaeger, N. Sugimoto, M.H. Caruthers, T. Neilson, and D.H. Turner (1986) Improved free-energy parameters for predictions of RNA duplex stability. Proc. Natl. Acad. Sci. USA 83, 9373–9377.
Friedberg, E. (1984) DNA Repair. Freeman, New York.
Froehler, B.C., and D.J. Ricca (1992) Triple-helix formation by oligonucleotides containing the carbocyclic analogs of thymidine and 5-methyl-2′-deoxycytidine. J. Am. Chem. Soc. 114, 8320–8322.
Froehler, B.C., T. Terhorst, J.P. Shaw, and S.N. McCurdy (1992a) Triple-helix formation and cooperative binding by oligonucleotides with a 3′–3′ internucleotide junction. Biochemistry 31, 1603–1609.
Froehler, B.C., S. Wadwani, T.J. Terhorst, and S.R. Gerrard (1992b) Oligonucleotides containing C-5 propyne analogs of 2′-deoxyuridine and 2′-deoxycytiding. Tetrahedron Lett. 33, 5307–5310.
Fuller, W., M.H.F. Wilkins, H.R. Wilson, and L.D. Hamilton (1965) The molecular configuration of deoxyribonucleic acid. IV. X-ray diffraction study of the A form. J. Mol. Biol. 12, 60–80.
Gama-Sosa, M.A., J.C. Hall, K.E. Schneider, G.C. Lukashewicz, and R.M. Ruprecht (1989) Unusual DNA structures at the integration site of HIV provirus. Biochem. Biophys. Res. Commun. 161, 134–142.
Ganguly, T., K.M. Weems, and N.J. Duker (1990) Ultraviolet-induced thymine hydrates in DNA are excised by bacterial and human DNA glycosylase activities. Biochemistry 29, 7222–7228.

Gao, W., C.A. Stein, J.S. Cohen, G.E. Dutschman, and Y. Cheng (1988) Effect of phosphorothioate homo-oligodeoxynucleotides on herpex simplex virus type 2-induced DNA polymerase. J. Biol. Chem. 264, 11521–11522.

Gao, Y.G., M. Sriram, and A.H.J. Wang (1993) Crystallographic studies of metal ion-DNA interactions: different binding modes of cobalt(II), copper(II) and barium(II) to N7 of guanines in Z-DNA and a drug-DNA complex. Nucleic Acids Res. 21, 4093–4101.

Gasparro, F.P., R.L. Edelson, M.E. O'Malley, S.J. Ugent, and H.H. Wong (1991) Photoactivatable antisense DNA: suppression of ampicillin resistance in normally resistant *Escherichia coli*. Antisense Res. Dev. 1, 117–140.

Gasparro, F.P., P.A. Havre, G.A. Olack, E.J. Gunther, and P.M. Glaser (1994) Site specific targeting of psoralen photoadducts with a triple helix-forming oligonucleotide: characterization of psoralen monoadduct and crosslink formation. Nucleic Acids Res. 22, 2845–2852.

Gautheret, D., S.H. Damberger, and R.R. Gutell (1995) Identification of base-triples in RNA using comparative sequence analysis. J. Mol. Biol. 248, 27–43.

Gee, J.E., S. Blume, R.C. Snyder, R. Ray, and D.M. Miller (1992) Triplex formation prevents Sp1 binding to the dihydrofolate reductase promoter. J. Biol. Chem. 267, 11163–11167.

Gee, J.E., R.-L. Yen, M.-C. Hung, and M.E. Hogan (1994) Triplex formation at the rat *neu* oncogene promoter. Gene 149, 109–114.

Gee, J.E., G.R. Revankar, T.S. Rao, and M.E. Hogan (1995) Triplex formation at the rat *neu* gene utilizing imidazole and 2′-deoxy-6-thioguanosine base substitutions. Biochemistry 34, 2042–2048.

Gehring, K., J.L. Leroy, and M. Gueron (1993) A tetrameric DNA structure with protonated cytosine-cytosine base pairs. Nature 363, 561–565.

Gellert, M. (1981) DNA topoisomerases. Annu. Rev. Biochem. 50, 879–910.

Germond, J.E., V.M. Vogt, and B. Hirt (1974) Characterization of the single-strand-specific nuclease S1 activity on double-strand supercoiled polyoma DNA, Eur. J. Biochem. 43, 591–600.

Gessner, R.V., G.J. Quigley, A.H.-J. Wang, G. Van der Marel, J.H. Van Boom, and A. Rich (1985) Structural basis for stabilization of Z-DNA of cobalt hexaammine and magnesium cations. Biochemistry 24, 237–240.

Gfrorer, A., M.E. Schnetter, J. Wolfrum, and K.O. Greulich (1993) Double and triple helices of nucleic-acid polymers, studied by UV-resonance Raman-spectroscopy. Ber. Bunsen Ges. Phys. Chem. (Int. J. Phys. Chem.) 97, 155–162.

Gilmour, D.S., G.H. Thomas, and S.C.R. Elgin (1989) *Drosophila* nuclear proteins bind to regions of alternating C and T residues in gene promoters. Science 245, 1487–1490.

Giovannangéli, C., T. Montenay-Garestier, M. Rougée, M. Chassignol, N.T. Thuong, and C. Hélène (1991) Single-stranded DNA as a target for triple-helix formation. J. Am. Chem. Soc. 113, 7775–7777.

Giovannangéli, C., M. Rougée, T. Garestier, N.T. Thuong, and C. Hélène (1992a) Triple-helix formation by oligonucleotides containing the three bases thymine, cytosine, and adenine. Proc. Natl. Acad. Sci. USA 89, 8631–8635.

Giovannangéli, C., N.T. Thuong, and C. Hélène (1992b) Oligonucleotide-directed photo-induced cross-linking of HIV proviral DNA via triple-helix formation. Nucleic Acids Res. 20, 4275–4281.

Giovannangéli, C., N.T. Thuong, and C. Hélène (1993) Oligonucleotide clamps arrest DNA synthesis on a single-stranded DNA target. Proc. Natl. Acad. Sci. USA 90, 10013–10017.

Glaser, R., and E.J. Gabbay (1968) Topography of nucleic acid helices in solutions. III. Interactions of spermine and spermidine derivatives with polyadenylic-polyuridylic and polyinosinic-polycytidylic acid helices. Biopolymers 6, 243–254.

Glaser, R.L., G.H. Thomas, E.S. Siegfried, S.C.R. Elgin, and J.T. Lis (1990) Optimal heat-induced expression of the *Drosophila hsp26* gene requires a promoter sequence containing $(CT)_n \cdot (GA)_n$ repeats. J. Mol. Biol. 211, 751–761.

Glikin, G.C., G. Gargiulo, L. Rena-Descalzi, and A. Worcel (1983) *Escherichia coli* single-strand binding protein stabilizes specific denatured sites in superhelical DNA. Nature 303, 770–774.

Glover, J.N.M., C.S. Farah, and D.E. Pulleyblank (1990) Structural chracterization of separated DNA conformers. Biochemistry 29, 11110–11115.

Glover, J.N.M., D.B. Haniford, and D.E. Pulleyblank (1988) Intermediate range effects in DNA. I: Low pH/stress induced conformational changes in the vicinity of an extruded $d(AT)_n \cdot d(AT)_n$ in cruciform. Nucleic Acids Res. 16, 5473–5490.

Goding, C.R., and W.C. Russell (1983) S1 sensitive sites in adenovirus DNA. Nucleic Acids Res. 11, 21–36.

Golas, T., M. Fikus, Z. Kazimierczuk, and D. Shugar (1976) Preparation and properties of an analogue of poly(A) and poly(G) : poly(isoguanylic acid). Eur. J. Biochem. 65, 183–192.

Goller, M., B. Funke, C. Gehe-Becker, B. Kroger, F. Lottspeich, and I. Horak (1994) Murine protein which binds preferentially to oligo-C-rich single-stranded nucleic acids. Nucleic Acids Res. 22, 1885–1889.

Goodchild, J., S. Agrawal, S., M.P. Civeira, P.S. Sarin, D. Sun, and P.C. Zamecnik (1988) Inhibition of human immunodeficiency virus replication by antisense oligonucleotides. Proc. Natl. Acad. Sci. USA 85, 7079–7083.

Goss, T.A., M. Bard, and H.W. Jarrett (1990) High-performance affinity chromatography of DNA. J. Chromatogr. 508, 279–287.

Goss, T.A., M. Bard, and H.W. Jarrett (1991) High-performance affinity chromatography of messenger RNA. J. Chromatogr. 588, 157–164.

Grabczyk, E., and M.C. Fishman (1995) A long purine-pyrimidine homopolymer acts as a transcriptional diode. J. Biol. Chem. 270, 1791–1797.

Gralla, J., and D.M. Crothers (1973) Free energy of imperfect nucleic acid helices. III. Small internal loops resulting from mismatches. J. Mol. Biol. 78, 301–319.

Granot, J.H., J. Feigon, and D.R. Kearns (1982) Interactions of DNA with divalent metal cations. I. ^{31}P-NMR studies. Biopolymers 21, 181–201.

Gray, D.M., R.L. Ratliff, V.P. Antao, and C.W. Gray (1988) CD spectroscopy of acid induced structures of polydeoxyribonucleotides: importance of $C \cdot C^+$ base pairs. In (Sarma, R.H., and M.H. Sarma, Eds.) Structure and Expression, Vol. 2: DNA and Its Drug Complexes. Adenine Press, New York, pp. 147–166.

Green, P.J., O. Pines, and M. Inouye (1986) The role of antisense RNA in gene regulation. Annu. Rev. Biochem. 55, 569–597.

Greenberg, W.A., and P.B. Dervan (1995) Energetics of formation of sixteen triple helical complexes which vary at a single position within a purine motif. J. Am. Chem. Soc. 117, 5016–5022.

Griffin, B.E., W.J. Haslam, and C.B. Reese (1964) Synthesis and properties of some methylated polyadenylic acids. J. Mol. Biol. 10, 353–356.

Griffin, L.C., and P.B. Dervan (1989) Recognition of thymine · adenine base pairs by guanine in a pyrimidine triple helix motif. Science 245, 967–971.

Griffin, L.C., L.L. Kiessling, P.A. Beal, P. Gillespie, and P.B. Dervan (1992) Recognition of all four base pairs of double-helical DNA by triple-helix formation: design of nonnatural deoxy-ribonucleotides for pyrimiding · purine base pair binding. J. Am. Chem. Soc. 114, 7976–7982.

Grigoriev, M., D. Praseuth, P. Robin, A. Hemar, T. Saison-Behmoaras, A. Dautry-Varsat, N.T. Thuong, C. Hélène, and A. Harrel-Ballan (1992) A triple helix-forming oligonucleotide-intercalator conjugate acts as a transcriptional repressor via inhibition of NFκB binding to interleukin-2 receptor α-regulatory sequence. J. Biol. Chem. 267, 3389–3395.

Grigoriev, M., D. Praseuth, A.L. Guyesse, P. Robin, N.T. Thuong, C. Hélène, and A. Harrel-Bellan (1993a) Inhibition of gene expression by triple helix-directed DNA cross-linking at specific genes. Proc. Natl. Acad. Sci. USA 90, 3501–3505.

Grigoriev, M., D. Praseuth, A.L. Guyesse, P. Robin, N.T. Thuong, C. Hélène, and A. Harrel-Bellan (1993b) Inhibition of interleukin-2 receptor alpha-subunit gene expression by oligonucleotide-directed triple-helix formation. C. R. Acad. Sci. III 316, 492–495.

Gross, D.S., and W.T. Garrard (1988) Nuclease hypersensitive sites in chromatin. Annu. Rev. Biochem. 57, 159–197.

Gryaznov, S.M., and D.H. Lloyd (1993) Modulation of oligonucleotide duplex and triplex stability via hydrophobic interactions. Nucleic Acids Res. 21, 5909–5915.

Guarente, L. (1988) UACs and enhancers: common mechanism of transcriptional activation in yeast and mammals. Cell 52, 303–305.

Haas, B.L., and W. Guschlbauer (1976) Protonated polynucleotide structures. 18. Interaction of oligocytidylates with poly (G). Nucleic Acids Res. 3, 205–218.

Haas, B.L., M.T. Sarocchi, and W. Guschlbauer (1976) Protonated polynucleotide structures. 20. Interaction between poly(dG)-poly(dC) and poly(rC). Nucleic Acids Res. 3, 1549–1559.

Hacia, J.G., P.B. Dervan, and B.J. Wold (1994a) Inhibition of Klenow fragment DNA polymerase on double-helical templates by oligonucleotide-directed triple-helix formation. Biochemistry 33, 6192–6200.

Hacia, J.G., B.J. Wold, and P.B. Dervan (1994b) Phosphorothioate oligonucleotide-directed triple helix formation. Biochemistry 33, 5367–5369.

Hagerman, P.J. (1990) Sequence-directed curvature of DNA. Annu. Rev. Biochem. 59, 755–781.

Hamilton, K.K., P.M.H. Kim, and P.W. Doetsch (1992) A eukaryotic DNA glycosylase/lyase recognizing ultraviolet-light induced pyrimidine dimers. Nature 356, 725–728.

Hampel, K.J., C. Ashley, and J.S. Lee (1994) Kilobase-range communication between polypurine · polypyrimidine tracts in linear plasmids mediated by triplex formation: a braided knot between two linear duplexes. Biochemistry 33, 5674–5681.

Hampel, K.J., G.D. Burkholder, and J.S. Lee (1993) Plasmid dimerization mediated by triplex formation between polypyrimidine-polypurine repeats. Biochemistry 32, 1072–1077.

Hampel, K.J., P. Crosson, and J.S. Lee (1991) Polyamines favor DNA triplex formation at neutral pH. Biochemistry 30, 4455–4459.

Hampel, K.J., and J.S. Lee (1993) Two-dimensional pulsed-field electrophoresis of yeast chromosomes: evidence for triplex-mediated DNA condensation. Biochem. Cell Biol. 71, 190–196.

Han, H., and P.B. Dervan (1993) Sequence-specific recognition of double helical RNA and RNA · DNA by triple helix formation. Proc. Natl. Acad. Sci. USA 90, 3806–3810.

Han, H., and P.B. Dervan (1994) Different conformational families of pyrimidine · purine · pyrimidine triple helices depending on backbone composition. Nucleic Acids Res. 22, 2837–2844.

Haner, R., and P.B. Dervan (1990) Single-strand DNA triplex formation. Biochemistry 29, 9761–9765.

Hanish, J., and M. McClelland (1990) Methylase-limited partial *Not*I cleavage for physical mapping of genomic DNA. Nucleic Acids Res. 18, 3287–3291.

Hanvey, J.C., J. Klysik, and R.D. Wells (1988a) Influence of DNA sequence on the formation of non-B right-handed helices in oligopurine · oligopyrimidine inserts in plasmids. J. Biol. Chem. 263, 7386–7396.

Hanvey, J.C., M. Shimizu, and R.D. Wells (1988b) Intramolecular DNA triplexes in supercoiled plasmids. Proc. Natl. Acad. Sci. USA 85, 6292–6296.

Hanvey, J.C., M. Shimizu, and R.D. Wells (1989) Intramolecular DNA triplexes in supercoiled plasmids. II. Effect of base composition and noncentral interruptions on formation and stability. J. Biol. Chem. 264, 5950–5956.

Hanvey, J.C., M. Shimizu, and R.D. Wells (1990) Site-specific inhibition of *Eco*RI restriction/modification enzymes by a DNA triple helix. Nucleic Acids Res. 18, 157–161.

Hanvey, J.C., E.M. Williams, and J.M. Besterman (1991) DNA triple-helix formation at physiologic pH and temperature. Antisense Res. Dev. 1, 307–317.

Harder, M.E., and W.C. Johnson, Jr. (1990) Stabilization of the Z′ form of poly(dGdC):poly(dGdC) in solution by multivalent ions relates to the ZII form in crystals. Nucleic Acids Res. 18, 2141–2148.

Hartman, D.A., S.R. Kuo, T.R. Broker, L.T. Chow, and R.D. Wells (1992) Intermolecular triplex formation distorts the DNA duplex in the regulatory region of human papillomavirus type-11. J. Biol. Chem. 267, 5488–5494.

Harvey, S.C., J. Luo, and R. Lavery (1988) DNA stem-loop structures in oligopurine-oligopyrimidine triplexes. Nucleic Acids Res. 16, 11795–11809.

Hashimoto, H., M.B. Nelson, and C. Switzer (1993) Zwitterionic DNA. J. Am. Chem. Soc. 115, 7128–7134.

Hatahet, Z., A.A. Purmal, and S.S. Wallace (1993) A novel method for site specific introduction of single model oxidative DNA lesions into oligodeoxyribonucleotides. Nucleic Acids Res. 21, 1563–1568.

Hattori, M., J. Frazier, and H.T. Miles (1975) Poly(8-aminoguanylic acid): formation of ordered self-structures and interaction with poly(cytidylic acid). Biochemistry 18, 5033–5045.

Hattori, M., J. Frazier, and H.T. Miles (1976) The structure of triple-stranded G-2C polynucleotide helices. Biopolymers 15, 523–531.

Hattori, M., M. Ikehara, and H.T. Miles (1974) Poly(2-methyl-N6-methyladenylic acid): synthesis, properties, and interaction with poly(uridylic acid). Biochemistry 13, 2754–2761.

Hausheer, F.H., U.C. Singh, J.D. Saxe, O.M. Colvin, and P.O. Ts'o (1990) Can oligonucleotide methylphosphonates form a stable triplet with a double DNA helix? Anticancer Drug Des. 5, 159–167.

Hausheer, F.H., U.C. Singh, J.D. Saxe, J.P. Flory, and K.B. Tufto (1992) Thermodynamic and conformational characterization of 5-methylcytosine-versus cytosine-substituted oligomers in DNA triple helices: *ab initio* quantum mechanical and free energy perturbation studies. J. Am. Chem. Soc. 114, 5356–5362.

Havre, P.A., and P.M. Glazer (1993) Targeted mutagenesis of simian virus 40 DNA mediated by a triple helix-forming oligonucleotide. J. Virol. 67, 7324–7331.

Havre, P.A., E.J. Gunther, F.P. Gasparro, and P.M. Glazer (1993) Targeted mutagenesis of DNA using triple helix-forming obligonucleotides linked to psoralen. Proc. Natl. Acad. Sci. USA 90, 7879–7883.

Hayes, J.J., J. Bashkin, T.D. Tullius, and A.P. Wolffe (1991) The histone core exerts a dominant constraint on the structure of DNA in a nucleosome. Biochemistry 30, 8434–8440.

Hélène, C. (1991) The antigene strategy: control of gene expression by triple-helix-forming oligonucleotides. Anticancer Drug Des. 6, 569–584.

Hélène, C., and G. Lancelot (1982) Interactions between functional groups in protein-nucleic acid associations. Prog. Biophys. Mol. Biol. 39, 1–68.

Hélène, C., N.T. Thuong, and A. Harrel-Bellan (1992) Control of gene expression by triple helix-forming oligonucleotides. The antigene strategy. Ann. NY Acad. Sci. 660, 27–36.

Hélène, C., and J.J. Toulmé (1990) Specific regulation of gene expression by antisense, sense and antigene nucleic acids. Biochim. Biophys. Acta 1049, 99–125.

Helm, C.W., K. Shrestha, S. Thomas, H.M. Shingleton, and D.M. Miller (1993) A unique c-*myc*-targeted triplex-forming oligonucleotide inhibits the growth of ovarian and cervical carcinomas in vitro. Gynecol. Oncol. 49, 339–343.

Henderson, E., C.C. Hardin, S.K. Walk, I. Tinoco, and E.H. Blackburn (1987) Telomeric DNA oligonucleotides form novel intramolecular structures containing guanine · guanine base pairs. Cell 51, 899–908.

Hentschel, C.C. (1982) Homocopolymer sequences in the spacer of a sea urchin histone gene repeat are sensitive to S1 nuclease. Nature 295, 714–716.

Herbeck, R., T.J. Yu, and W.L. Peticolas (1976) Effect of cross linkiing on the secondary structure of DNA. I. Cross linking by photodimerization. Biochemistry 15, 2656–2660.

Herr, W. (1985) Diethyl pyrocarbonate: a chemical probe for secondary structure in a negatively supercoiled DNA. Proc. Natl. Acad. Sci. USA 82, 8009–8013.

Hertzberg, R.P., and P.B. Dervan (1982) Cleavage of double helical DNA by methydiumpropyl-EDTA-iron(II). J. Am. Chem. Soc. 104, 313–315.

Higuichi, R., B. Krummel, and R.K. Saiki (1988) A general method of in vitro preparation and specific mutagenesis of DNA fragments: study of protein and DNA interactions. Nucleic Acids Res. 16, 7351–7367.

Hill, R.J., and B.D. Stollar (1983) Dependence of Z-DNA antibody binding to polythene chromosomes on acid fixation and DNA torsional strain. Nature 305, 338–340.

Hillen, W., and H.G. Gassen (1975) Physical and coding properties of poly(5-aminouridylic) acid and of 5-aminouridine-containing trinucleotides. Biochim. Biophys. Acta 407, 347–356.

Hillen, W., and H.G. Gassen (1979) Physical and coding properties of poly(5-methoxyuridylic) acid. Biochim. Biophys. Acta 562, 207–213.

Ho, S., H.D. Hunt, R.M. Horton, J.K. Pullen, and L.R. Pease (1989) Site-directed mutagenesis by overlap extension using the polymerase chain reaction. Gene 77, 51–59.

Hobbs, C.A., and K. Yoon (1994) Differential regulation of gene expression in vivo by triple helix-forming oligonucleotides as detected by a reporter enzymes. Antisense Res. Dev. 4, 1–8.

Hoffman, E.K., S.P. Trusko, M. Murphy, and D.L. George (1990) An S1 nuclease-sensitive homopurine/homophyrimidine domain in the c-Ki-*ras* promoter interacts with a nuclear factor. Proc. Natl. Acad. Sci. USA 87, 2705–2709.

Hoffman-Liebermann, B., D. Liebermann, A. Troutt, L.H. Kedes, and S.N. Cohen (1986) Human homologs of TU transposon sequences: polypurine/polypyrimidine sequence elements that can alter DNA conformation in vitro and in vivo. Mol. Cell Biol. 6, 3632–3642.

Hogeland, J.S., and D.D. Weller (1993) Investigations of oligodeoxyinosine for triple helix formation. Antisense Res. Dev. 3, 285–290.

Hollingsworth, M.A., C. Closken, A. Harris, C.D. McDonald, G.S. Pahwa, and L.J. Maher, III (1994) A nuclear factor that binds purine-rich, single-standed oligonucleotides derived from S1-sensitive elements upstream of the CFTR gene and the MUC1 gene. Nucleic Acids Res. 22, 1138–1146.

Honigberg, S.M., B.J. Rao, and C.M. Radding (1986) Ability of RecA protein to promote a search for rare sequences in duplex DNA. Proc. Natl. Acad. Sci. USA 83, 9586–9590.

Hoogsteen, K. (1959) The structure of crystals containing a hydrogen-bonded complex of 1-methylthymine and 9-methyladenine. Acta Crystallogr. 12, 822–823.

Hoogsteen, K. (1963) The crystal and molecular structure of a hydrogen-bonded complex between 1-methylthymine and 9-methyladenine. Acta Crytallogr. 16, 907–916.

Hopkins, H.P., D.D. Hamilton, W.D. Wilson, and G. Zon (1993) Duplex and triple-helix formation with dA_{19} and dT_{19} — thermodynamic parameters from calorimetric, NMR, and circular dichroism studies. J. Phys. Chem. 97, 6555–6563.

Horne, D.A., and P.B. Dervan (1990) Recognition of mixed-sequence duplex DNA by alternate-strand triple-helix formation. J. Am. Chem. Soc. 112, 2435–2437.

Horne, D.A., and P.B. Dervan (1991) Effects of an abasic site on triple helix formation characterized by affinity cleaving. Nucleic Acids Res. 19, 4963–4965.

Horwitz, E.M., K.A. Maloney, and T.J. Ley (1994) A human protein containing a "cold shock" domain binds specifically to H-DNA upstream from the human γ-globin genes. J. Biol. Chem. 269, 14130–14139.

Howard, F.B., J. Frazier, M.N. Lipsett, and H.T. Miles (1964) Infrared demonstration of two- and three-strand helix formation between poly C and guanosine mononucleotides and oligonucleotides. Biochem. Biophys. Res. Commun. 17, 93–102.

Howard, F.B., J. Frazier, and H.T. Miles (1966a) A new polynucleotide complex stabilized by 3 interbase hydrogen bonds, poly-2-aminoadenylic acid + polyuridylic acid. J. Biol. Chem. 241, 4293–4295.

Howard, F.B., J. Frazier, and H.T. Miles (1974) Poly-8-bromoadenylic acid. A helical, all-*syn* homopolynucleotide. J. Biol. Chem. 249, 2987–2990.

Howard, F.B., J. Frazier, and H.T. Miles (1975) Poly(8-bromoadenylic acid): synthesis and characterization of all-*syn* polynucleotide. J. Biol. Chem. 250, 3951–3959.

Howard, F.B., J. Frazier and H.T. Miles (1976) Poly(2-aminoadenylic acid): interaction with poly(uridylic acid). Biochemistry 15, 3783–3795.

Howard, F.B., J. Frazier, M.F. Singer and H.T. Miles (1966b) Helix formation between polyribonucleotides and purines, purine nucleosides and nucleotides. II. J. Mol. Biol. 16, 415–439.

Howard, F.B., W. Limn, and H.T. Miles (1985) Poly(2-amino-8-methyladenylic acid). Competing structural and energetic effects of substituents. Biochemistry 24, 5033–5039.

Howard, F.B. and H.T. Miles (1984) $2NH_2A \times T$ helices in the ribo- and deoxypolynucleotide series. Structural and energetic consequences of $2NH_2A$ substitution. Biochemistry 23, 6723–6732.

Howard, F.B., H.T. Miles, K. Liu, J. Frazier, G. Raghunathan, and V. Sasisekharan (1992) Structure of $d(T)_n \cdot d(A)_n \cdot d(T)_n$: the DNA triple helix has B-form geometry with C2′-endo sugar pucker. Biochemistry 31, 10671–10677.

Htun, H. and J.E. Dahlberg (1988) Single strands, triple strands, and kinks in H-DNA. Science 241, 1791–1796.

Htun, H. and J.E. Dahlberg (1989) Topology and formation of triple-stranded H-DNA. Science 243, 1571–1576.

Htun, H. and B.J. Johnston (1992) Mapping adducts of DNA structural probes using transcription and primer extension approaches. Methods Enzymol. 212, 272–294.

Htun, H., E. Lund, and J.E. Dahlberg (1984) Human U1 RNA genes contain an unusually sensitive nuclease S1 cleavage site within the conserved 3′ flanking region. Proc. Natl. Acad. Sci. USA 81, 7288–7292.

Huang, C.C., D. Nguen, R. Martinez, and C.A. Edwards (1992) Triple-helix formation is compatible with an adjacent DNA-protein complex. Biochemistry 31, 993–998.

Huang, C.-Y. and P.S. Miller (1993) Triplex formation by an oligoribonucleotide containing N^4-(6-aminopyridinyl)-2′-deoxycytidine. J. Am. Chem. Soc. 115. 10456–10457.

Huang, W.M. and P.O.P. Ts'o (1966) Physico-chemical basis of the recognition process in nucleic acid interactions. I. Interactions of polyuridylic acid and nucleosides. J. Mol. Biol. 16, 523–543.

Hudson, R.H.E. and M.J. Damha (1993) Association of branched nucleic acids. Nucleic Acids Symp. Ser. 29, 97–99.

Hung, S.H., Q. Yu, D.M. Gray, and R.L. Ratliff (1994) Evidence from CD spectra that d(purine) · r(pyrimidine) and r(purine) · d(pyrimidine) hybrids are in different structural classes. Nucleic Acids Res. 22, 4326–4334.

Husler, P.L. and H.H. Klump (1994) Unfolding of a branched double-helical DNA three-way junction with triple-helical ends. Arch. Biochem. Biophys. 313, 29–38.

Ikeda, K., J. Frazier, and H.T. Miles (1970) Poly 2-amino-6-N-methyladenylic acid: synthesis, characterization and interaction with polyuridylic acid. J. Mol. Biol. 54, 59–84.

Ikehara, M., I. Tazawa, and T. Fukui (1969) Polynucleotides. VII. Synthesis of ribopolynucleotides containing 8-substituted purine nucleotides by polynucleotide phosphorylase. Biochemistry 8, 736–743.

Imagawa, S., T. Izumi and Y. Miura (1994) Positive and negative regulation of the erythropoietin gene. J. Biol. Chem. 269, 9038–9044.

Ing, N.H., J.M. Beekman, D.J. Kessler, M. Murphy, K. Jayaraman, J.G. Zendegui, M.E. Hogan, B.W.O'Malley, and M.J. Tsai (1993) In vivo transcription of a progesterone-responsive gene is specifically inhibited by a triple-forming oligonucleotide. Nucleic Acids Res. 21, 2789–2796.

Inman, R.B. (1964a) Transitions of DNA homopolymers. J. Mol. Biol. 9, 624–637.

Inman, R.B. (1964b) Multistranded DNA homopolymer interactions. J. Mol. Biol. 10, 137–146.

Inman, R.B. and R.L. Baldwin (1964) Helix-random coil transition in DNA homopolymer pairs. J. Mol. Biol. 8, 452–469.

Irvine, D., C. Tuerk, and L. Gold (1991) SELEXION. Systematic evolution of ligands by exponential enrichment with integrated optimization by non-linear analysis. J. Mol. Biol. 222, 739–761.

Ishikawa, F., J. Frazier, and H.T. Miles (1973) Poly(2-dimethylaminoadenylic acid). Synthesis and characterization of the homopolymer. Biochemistry 12, 4790–4798.

Ito, T., C.L. Smith, and C.R. Cantor (1992a) Sequence-specific DNA purification by triplex affinity capture. Proc. Natl. Acad. Sci. USA 89, 495–498.

Ito, T., C.L. Smith, and C.R. Cantor (1992b) Affinity capture electrophoresis for sequence-specific DNA purification. Genet. Anal. Tech. Appl. 9, 96–99.

Ito, T., C.L. Smith, and C.R. Cantor (1992c) Triplex affinity capture of a single copy clone from a yeast genomic library. Nucleic Acids Res. 20, 3524.

Ivanov, V.I. and D.Y. Krylov (1992) A-DNA in solution as studied by diverse approaches. Methods Enzymol. 211, 111–126.

Ivanov, V.I., L.E. Minchenkova, A.K. Schyolkina, and A.I. Poletayev (1973) Different conformations of double-stranded nucleic acid in solution as revealed by circular dichroism. Biopolymers 12, 89–110.

Ivanov, V.L., L.E. Minchenkova, E.E. Minyat, M.D. Frank-Kamenetskii, and A.K. Schyolkina (1974) The B to A transition of DNA in solution. J. Mol. Biol. 87, 817–833.

Jacob, F. and J. Monod (1961) Genetic regulatory mechanisms in the synthesis of proteins. J. Mol. Biol. 3, 318–356.

Jain, S., G. Zon and M. Sundaralingam (1989) Base only binding of spermine in the deep groove of the A-DNA octamer d(GTGTACAC). Biochemistry 28, 2360–2364.

Jaishree, T.N. and A.H.-J. Wang (1993) NMR studies of pH-dependent conformational polymorphism of alternating $(C\text{-}T)_n$ sequences. Nucleic Acids Res. 21, 3839–3844.

Jane, S.M., D.L. Gumucio, P.A. Ney, J.M. Cunningham, and A.W. Nienhuis (1993) Methylation-enhanced binding of Sp1 to the stage selector element of the human γ-globin gene promoter may regulate developmental specificity of expression. Mol. Cell. Biol. 13, 3272–3281.

Jaworski, A., J.A. Blaho, J.E. Larson, M. Shimizu, and R.D. Wells (1989) Tetracycline promoter mutations decrease non-B DNA structural transitions, negative linking differences and deletions in recombinant plasmids in *Escherichia coli*. J. Mol. Biol. 207, 513–526.

Jayasena, S.D. and M.J. Behe (1989) Nucleosome reconstitution of core-length poly(dG) · poly(dC) and poly(rG-dG) · poly(rG-dC). Biochemistry 28, 975–980.

Jayasena, S.D. and B.H. Johnston (1992a) Intramolecular triple-helix formation at $(Pu_nPy_n) \cdot (Pu_nPy_n)$ tracts: recognition of alternate strands via Pu · PuPy and Py · PuPy base triplets. Biochemistry 31, 320–327.

Jayasena, S.D. and B.H. Johnston (1992b) Oligonucleotide-directed triple helix formation at adjacent oligopurine and oligopyrimidine DNA tracts by alternate strand recognition. Nucleic Acids Res. 20, 5279–5288.

Jayasena, S.D. and B.H. Johnston (1993) Sequence limitations of triplex formation by alternate-strand recognition. Biochemistry 32, 2800–2807.

Jean, Y.C., Y.G. Gao, and A.H.J. Wang (1993) Z-DNA structure of a modified DNA hexamer at 1.4-Å resolution: aminohexyl-5′-d(pCpGp[br5C]pGpCpG). Biochemistry 32, 381–388.

Jeffrey, G.A. and W. Saenger (1991) Hydrogen Bonding in Biological Structures. Springer-Verlag, New York.

Jetter, M.C. and F.W. Hobbs (1993) 7,8-dihydro-8-oxoadenine as a replacement for cytosine in the third strand of triple helices. Triplex formation without hypochromicity. Biochemistry 32, 3249–3254.

Ji, H. and L.M. Smith (1993) Rapid purification of double-stranded DNA by triple-helix-mediated affinity capture. Anal. Chem. 65, 1323–1328.

Ji, H., L.M. Smith, and R.A. Guilfoyle (1994) Rapid isolation of cosmid insert DNA by triple-helix-mediated affinity capture. Genet. Anal. Techn. Appl. 11, 43–47.

Jia, X. and L.G. Marzilli (1991) Zinc ion-DNA polymer interaction. Biopolymers 31, 23–44.

Jiang, N. and J.-S. Taylor (1993) In vivo evidence that UV-induced C → T mutations at dipyrimidine sites could result from the replicative bypass of *cis-syn* cyclobutane dimers or their deamination products. Biochemistry 32, 472–481.

Jin, R., W.H. Chapman, Jr., A.R. Srinivasan, W.K. Olson, R. Breslow, and K.J. Breslauer (1993) Comparative spectroscopic, calorimetric, and computational studies of nucleic acid complexes with 2′,5″- versus 3′,5″-phosphodiester linkages. Proc. Natl. Acad. Sci. USA 90, 10568–10572.

Johnson, C.A., Y. Jinno, and G.T. Merlino (1988) Modulation of epidermal growth factor receptor proto-oncogene transcription by a promoter site sensitive to S1 nuclease. Mol. Cell. Biol. 8, 4174–4184.

Johnson, D. and A.R. Morgan (1978) Unique structures formed by pyrimidine-purine DNAs which may be four-stranded. Proc. Natl. Acad. Sci. USA 75, 1637–1641.

Johnson, K.H., D.M. Gray, and J.C. Sutherland (1991) Vacuum UV CD spectra of homopolymer duplexes and triplexes containing A · T or A · U base pairs. Nucleic Acids Res. 19, 2275–2280.

Johnson, K.H., R.H. Durland, and M.E. Hogan (1992) The vacuum UV CD spectra of G · G · C triplexes. Nucleic Acids Res. 20, 3859–3864.

Johnston, B.H. (1988) The S1-sensitive form of $d(C\text{-}T)_n \cdot d(A\text{-}G)_n$: chemical evidence for a three-stranded structure in plasmids. Science 241, 1800–1804.

Johnston, B.H. (1992a) Generation and detection of Z-DNA. Methods Enzymol. 211, 127–158.

Johnston, B.H. (1992b) Hydroxylamine and methoxylamine as probes of DNA structure. Methods Enzymol. 212, 180–194.

Johnston, B.H. and A. Rich (1985) Chemical probes of DNA conformation: detection of Z-DNA at nucleotide resolution. Cell 42, 713–724.

Jolley, M.E., C.H.J. Wang, S.J. Ekenberg, M.S. Zuelke, and D.M. Kelso (1984) Particle concentration fluorescence immunoassay (PCFIA): a new rapid immunoassay technique with high sensitivity. J. Immunol. Methods 67, 21–35.

Jones, R.J., K.Y. Lin, J.F. Milligan, S. Wadwani, and M.D. Matteucci (1993a) Synthesis and binding properties of pyrimidine oligodeoxyribonucleotide analogs containing neutral phosphodiester replacements-the formacetal and 3′-thioformacetal internucleoside linkages. J. Org. Chem. 58, 2983–2991.

Jones, R.J., S. Swaminathan, J.F. Milligan, S. Wadwani, B.C. Froehler, and M.D. Matteucci (1993b) Oligonucleotides containing a covalent conformationally restricted phosphodiester analog for high-affinity triple helix formation: the riboacetal internucleotide linkage. J. Am. Chem. Soc. 115, 9816–9817.

Joshi, R.R. and K.N. Ganesh (1994) Duplex and triplex directed DNA cleavage by oligonucleotide-Cu(II)/Cu(III) metallodesferal conjugates. Biochim. Biophys. Acta 1201, 454–460.

Jovin, T.M., K. Rippe, N. Bramsing, R. Klement, W. Elhorst, and M. Vorlickova (1990) Parallel-stranded DNA. In (R.H. Sarma and M.H. Sarma, Eds.) Structure and Methods: Vol. 3: DNA and RNA. Adenine Press, New York, pp. 155–174.

Kabakov, A.E. and A.M. Poverenny (1993) Immunochemical probing of DNA structure with monoclonal antibody to OsO_4/2,2′-bipyridine adduct. Anal. Biochem. 211, 224–232.

Kadowaki, H., T. Kadowaki, F.E. Wondisford, and S.I. Taylor (1989) Use of polymerase chain reaction catalyzed by *Taq* DNA polymerase for site-specific mutagenesis. Gene 76, 161–166.

Kan, L.S., D.E. Callahan, T.L. Trapane, P.S. Miller, P.O.P.Ts'o and D.H. Huang (1991) Proton NMR and optical spectroscopic studies on the DNA triplex formed by d-A-$(G\text{-}A)_7$-G and d-C-$(T\text{-}C)_7$-T. J. Biomol. Struct. Dyn. 8, 911–933.

Kanaya, E.N., F.B. Howard, J. Frazier, and H.T. Miles (1987) Poly(2-amino-8-methyldeoxyadenylic acid): contrasting effects in deoxy- and ribopolynucleotides of 2-amino and 8-methyl substituents. Biochemistry 26, 7159–7165.

Kandimalla, E.R. and S. Agrawal (1994) Single-strand-targeted triplex formation: stability, specificity and RNase H activation properties. Gene 149, 115–121.

Kandpal, R.P., D.C. Ward, and S.M. Weisman (1992) Chromosome fishing: an affinity capture method for selective enrichment of large genomic DNA fragments. Methods Enzymol. 216, 39–54.

Kang, S. and R.D. Wells (1992) Central non-Pur · Pyr sequences in oligo(dG · dC) tracts and metal ions influence the formation of intramolecular DNA triplex isomers. J. Biol. Chem. 267, 20887–20891.

Kang, S., F. Wohlrab, and R.D. Wells (1992a) Metal ions cause the isomerization of certain intramolecular triplexes, J. Biol. Chem. 267, 1259–1264.

Kang, S., F. Wohlrab, and R.D. Wells (1992b) GC-rich flanking tracts decrease the kinetics of intramolecular DNA triplex formation. J. Biol. Chem. 267, 19435–19442.

Karlovsky, P., P. Pecinka, M. Vojtiskova, E. Makarturova, and E. Palecek (1990) Protonated triplex DNA in *E. coli* cells as detected by chemical probing. FEBS Lett. 274, 39–42.

Kato, M. (1993) Polypyrimidine/polypurine sequence in plasmid DNA enhances formation of dimer molecules in *Escherichia coli*. Mol. Biol. Rep. 18, 183–187.

Kato, M., J. Kudoh, and N. Shimizu (1990) The pyrimidine/purine-biased region of the epidermal growth factor receptor gene is sensitive to S1 nuclease and may form an intramolecular triplex. Biochem. J. 268, 175–180.

Kato, M. and N. Shimizu (1992) Effect of potential triplex DNA region on the in vitro expression of bacterial β-lactamase gene in superhelical recombinant plasmids. J. Biochem. 112, 492–494.

Kessler, D.J., B.M. Pettitt, Y.K. Cheng, S.R. Smith, K. Jayaraman, H.M. Vu, and M.E. Hogan (1993) Triple helix formation at distant sites: hybrid oligonucleotides containing a polymeric linker. Nucleic Acids Res. 21, 4810–4815.

Kibler-Herzog, L., B. Kell, G. Zon, K. Shinozuka, S. Mizan, and W.D. Wilson (1990) Sequence dependent effects in methylphosphonate deoxyribonucleotide double, and triple helical complexes. Nucleic Acid Res. 18, 3545–3555.

Kibler-Herzog, L., G. Zon, G. Whittler, S. Mizan, and W.D. Wilson (1993) Stabilities of duplexes, and triplexes of dA_{19}, and dT_{19} with alternating methylphosphonate, and phosphodiester linkages. Anticancer Drug Des. 8, 65–79.

Kiessling, L.L., L.C. Griffin, and P.B. Dervan (1992) Flanking sequence effects within the pyrimidine triple-helix motif characterized by affinity cleaving. Biochemistry 31, 2829–2834.

Kilpatrick, M.W., A. Torri, D.S. Kang, J.A. Engler, and R.D. Wells (1986) Unusual DNA structures in the adenovirus genome. J. Biol. Chem. 261, 11350–11354.

Kim, R.A., and J.C. Wang (1989) A subthreshold level of DNA topoisomerases leads to the excision of yeast rDNA as extrachromosomal rings. Cell 57, 975–985.

Kim, S.G., S. Tsukahara, S. Yokoyama, and H. Takaku (1992) The influence of oligodeoxyribonucleotide phosphorothioate pyrimidine strands on triplex formation. FEBS Lett. 314, 29–32.

Kim, S.K., P.E. Nielsen, M. Egholm, O. Buchardt, R.H. Berg, and B. Norden (1993) Right-handed triplex formed between peptide nucleic acid PNA-T_8, and poly(dA) shown by linear, and circular dichroism spectroscopy. J. Am. Chem. Soc. 115, 6477–6481.

Kim, S.K., and B. Norden (1993) Methyl green. A DNA major-groove binding drug. FEBS Lett. 315, 61–64.

Kim, S.-T., and A. Sancar (1993) Photochemistry, photophysics, and mechanism of pyrimidine dimer repair by DNA photolyase. Photochem. Photobiol. 57, 895–904.

Kinniburgh, A.J. (1989) A cis-acting transcription element of the c-*myc* gene can assume an H-DNA conformation. Nucleic Acid Res. 17, 7771–7778.

Kinniburgh, A.J., A.B. Firulli, and R. Kolluri (1994) DNA triplexes, and regulation of the c-*myc* gene. Gene 149, 93–100.

Kitagawa, Y., and E. Okuhara (1987) Formation of triple-helical nucleic acids studied by using antibodies specific for poly(A) · poly(U) · poly(U). J. Biochem. 102, 1203–1212.

Kiyama, R., and R.D. Camerini-Otero (1991) A triplex DNA-binding protein from human cells: purification, and characterization. Proc. Natl. Acad. Sci. USA 88, 10450–10454.

Kiyama, R., N. Nishikawa, and M. Oishi (1994) Enrichment of human DNAs that flank poly(dA) · poly(dT) tracts by triplex DNA formation. J. Mol. Biol. 237, 193–200.

Kiyama, R., and M. Oishi (1995) Protection of DNA sequences by triplex-bridge formation. Nucleic Acids Res. 23, 452–458.

Klinck, R., E. Guittet, J. Liquier, E. Taillandier, C. Gouette, and T. Huynh-Dinh (1994) Spectroscopic evidence for an intramolecular RNA triple helix. FEBS Lett. 355, 297–300.

Klysik, J. (1992) Cruciform extrusion facilitates intramolecular triplex formation between distal oligopurine · oligopyrimidine tracts: long range effects. J. Biol. Chem. 267, 17430–17437.

Klysik, J. (1995) An intramolecular triplex structure from non-mirror repeated sequence containing both Py:Pu · Pu, and Pu:Pu · Py triads. J. Mol. Biol. 245, 499–507.

Klysik, J., K. Rippe, and T.M. Jovin (1991) Parallel-stranded DNA under topological stress: rearrangement to $(dA)_{15} \cdot (dT)_{15}$ to a $d(A \cdot A \cdot T)_n$ triplex. Nucleic Acids Res. 19, 7145–7154.

Klysik, J., and M. Shimizu (1993) *Escherichia coli* single-stranded DNA-binding protein alters the structure of intramolecular triplexes in plasmids. FEBS Lett. 333, 261–267.

Klysik, J., S.M. Stirdivant, J.E. Larson, P.A. Hart, and R.D. Wells (1981) Left-handed DNA in restriction fragments, and a recombinant plasmid. Nature 290, 672–677.

Knorre, D.G., and V.V. Vlassov (1991) Reactive oligonucleotide derivatives as gene-targeted biologically-active compounds, and affinity probles. Genetica 85, 53–63.

Knorre, D.G., V. V. Vlassov, V.F. Zarytova, A.V. Lebedev, and O.S. Fedorova (1994) Design, and targeted reactions of oligonucleotide derivatives. CRC Press, Boca Raton, Florida.

Kochetkov, N.K., and E.I. Budovskii, Eds. (1972) Organic Chemistry of Nucleic Acids, Part B. Plenum Press, London, and New York, pp. 543–618.

Kochevar, I.E., and D.A. Dunn (1990) Photosensitized reactions of DNA: cleavage, and addition. In (H. Morrison, Ed.) Bioorganic Photochemistry, Vol. 1. Wiley, New York, pp. 273–315.

Koh, J.S., and P.B. Dervan (1992) Design of a nonnatural deoxyribonucleoside for recognition of GC base pairs by oligonucleotide-directed triple helix formation. J. Am. Chem. Soc. 114, 1470–1478.

Kohwi, Y. (1989a) Non-B DNA structure: preferential target for the chemical carcinogen glycidaldehyde. Carcinogenesis 10, 2035–2042.

Kohwi, Y. (1989b) Cationic metal-specific structures adopted by the poly (dG) region, and the direct repeats in the chicken adult β^A globin gene promoter. Nucleic Acids Res. 17, 4493–4502.

Kohwi, Y., and T. Kohwi-Shigematsu (1988) Magnesium ion-dependent triple-helix structure formed by homopurine-homopyrimidine sequences in supercoiled plasmid DNA. Proc. Natl. Acad. Sci. USA 85, 3781–3785.

Kohwi, Y., and T. Kohwi-Shigematsu (1991) Altered gene expression correlates with DNA structure. Genes Dev. 5, 2547–2554.

Kohwi, Y., and T. Kohwi-Shigematsu (1993) Structural polymorphism of homopurine-homopyrimidine sequences at neutral pH. J. Mol. Biol. 231, 1090–1101.

Kohwi, Y., S.R. Malkhosyan, and T. Kohwi-Shigematsu (1992) Intramolecular dG · dG · dC triplex detected in *Escherichia coli* cells. J. Mol. Biol. 223, 817–822.

Kohwi, Y., and Y. Panchenko (1993) Transcription-dependent recombination induced by triple-helix formation. Genes Dev. 7, 1766–1778.

Kohwi-Shigematsu, T., R. Gelinas, and H. Weintraub (1983) Detection of an altered DNA conformation at specific sites in chromatin, and supercoiled DNA. Proc. Natl. Acad. Sci. USA 80, 4389–4393.

Kohwi-Shigematsu, T., and Kohwi, Y. (1991) Detection of triple-helix related structures adopted by poly(dG)-poly(dC) sequences in supercoiled plasmid DNA. Nucleic Acids Res. 19, 4267–4271.

Kohwi-Shigematsu, T., and Kohwi, Y. (1992) Detection of non-B-DNA structures at specific sites in supercoiled plasmid DNA, and chromatin with haloacetaldehyde, and diethylpyrocarbonate. Methods Enzymol. 212, 155–180.

Koller, E., A.R. Hayman, and B. Trueb (1991) The promoter of the chicken α 2(VI) collagen gene has features characteristic of house-keeping genes, and of proto-oncogenes. Nucleic Acids Res. 19, 485–491.

Kolluri, R., T.A. Torrey, and A.J. Kinniburgh (1992) A CT promoter element binding protein: definition of a double-strand, and a novel single-strand DNA-binding motif. Nucleic Acids Res. 20, 111–116.

Koob, M., E. Grimes, and W. Szybalski (1988) Conferring operator specificity of restriction endonucleases. Science 241, 1084–1086.

Kool, E.T. (1991) Molecular recognition by circular oligonucleotides: increasing the selectivity of DNA binding. J. Am. Chem. Soc. 113, 6265–6266.

Koshlap, K.M., P. Gillespie, P.B. Dervan, and J. Feigon (1993) Nonnatural deoxyribonucleoside-D(3) incorporated in an intramolecular DNA triplex binds sequence-specifically by intercalation. J. Am. Chem. Soc. 115, 7908–7909.

Kovalsky, O.I., G.A. Dvoryankin, and E.I. Budowsky (1993) Laser Photofootprinting of d(C) · d(G) · d(G) intramolecular triplex. J. Biomol. Struct. Dyn. 10, 933–943.

Krakauer, H. (1971) Birding of Mg^{2+} ions to polyadenylate, polyuridylate, and their complexes. Biopolymers 10, 2459–2490.

Krakauer, H. (1974) A thermodynamic analysis of the influence of simple mono, and divalent cations on the conformational transitions of polynucleotide complexes. Biochemistry 13, 2579–2589.

Krakauer, H., and J.M. Sturtevant (1968) Heats of the helix-coil transitions of the poly A–poly U complexes. Biopolymers 6, 491–512.

Krauss, G., H. Sinderman, U. Schomburg, and G. Maass (1981) *Escherichia coli* single-strand deoxyribonucleic acid binding protein: stability, specificity, and kinetics of complexes with oligonucleotides, and deoxyribonucleic acid. Biochemistry 20, 5346–5352.

Krawczyk, S.H., J.F. Milligan, S. Wadwani, C. Moulds, B. Froehler, and M.D. Matteucci (1992) Oligonucleotide-mediated triple helix formation using an N^3-protonated deoxycytidine analog exhibiting pH-independent binding within the physiological range Proc. Natl. Acad. Sci. USA 89, 3761–3764.

Krystal, G.W., B.C. Armstrong, and J. Battey (1990) N-*myc* mRNA forms an RNA-RNA duplex with endogenous antisense transcripts. Mol. Cell. Biol. 10, 4180–4191.

Kuderova-Krejcova, A., A.M. Poverenny, and E. Palecek (1991) Probing of DNA structure with osmium tetroxide-2,2′-bipyridine adduct-specific antibodies. Nucleic Acids Res. 19, 6811–6817.

Kurfurst, R., V. Roig, M. Chassignol, U. Asseline, and N.T. Thuong (1993) Oligo-α-deoxyribonucleotides with a modified nucleic acid base, and covalently linked to reactive agents. Tetrahedron 49, 6975–6990.

Kyogoku, Y., R.C. Lord, and A. Rich (1967) The effect of substituents on the hydrogen bonding of adenine, and uracil derivatives. Proc. Natl. Acad. Sci. USA 57, 250–257.

Lacour, F., E. Nahon-Merlin, and M. Michelson (1973) Immunological recognition of polynucleotide structure. Curr. Top. Microbiol. Immunol. 62, 1–39.

Lafyatis. R., F. Denhez, T. Williams, M. Sporn, and A. Roberts (1991) Sequence-specific protein binding to, and activation of the TGF-β3 promoter through a repeated TCCC motif. Nucleic Acids Res. 19, 6419–6425.

Lai, D., X. Zhu, and S. Pestka (1993) A simple, and efficient method for site-directed mutagenesis with double-stranded plasmid DNA. Nucleic Acids Res. 21, 3977–3980.

Langlais, M, H.A. Tajmir-Riahi, and R. Savoie (1990) Raman spectroscopic study of the effects of Ca^{2+}, Mg^{2+}, Zn^{2+}, Cd^{2+} ions on calf thymus DNA: binding sites, and conformational changes. Biopolymers 30, 743–752.

Langowski, J., A.S. Benight, B.S. Fujimoto, J.M. Schurr, and U. Schomburg (1985) Change of conformation, and internal dynamics of supercoiled DNA upon binding of *Escherichia coli* single-strand binding protein. Biochemistry 24, 4022–4028.

Langridge, R., H.R. Wilson, C.W. Hooper, M.H.F. Wilkins, and L.D. Hamilton (1961) The molecular configuration of deoxyribonucleic acid. I. X-ray diffraction study of a crystalline form of the lithium salt. J. Mol. Biol. 2, 19–37.

Lapidot, A., N. Baran, and H. Manor (1989) $(dT\text{-}dC)_n$, and $(dG\text{-}dA)_n$ tracts arrest single stranded DNA replication in vitro. Nucleic Acids Res. 17, 883–900.

Larsen, A., and H. Weintraub (1982) An altered DNA conformation detected by S1 nuclease occurs at specific regions in active chick globin chromatin. Cell 29, 609–622.

Laskowski, M., Sr. (1980) Purification, and properties of the mung bean nuclease. Methods Enzymol. 65, 263–276.

Latimer, L.J.P., K. Hampel, and J.S. Lee (1989) Synthetic repeating sequence DNAs containing phosphorothioates: nuclease sensitivity, and triplex formation. Nucleic Acids Res. 17, 1549–1561.

Laughton, C.A., and S. Neidle (1991) DNA triple helices–a molecular dynamics study. J. Chim. Phys. Phys.-Chim. Biol. 88, 2597–2603.

Laughton, C.A., and S. Neidle (1992a) Molecular dynamics simulation of the DNA triplex $d(TC)_5 \cdot d(GA)_5 \cdot d(C^{+}T)_5$. J. Mol. Biol. 223, 519–529.

Laughton, C.A., and S. Neidle (1992b) Prediction of the structure of the Y+.R–.R(+)-type DNA triple helix by molecular modelling. Nucleic Acids Res. 20, 6535–6541.

Le Doan, T., L. Perrouault, D. Praseuth, N. Habhoub, J.L. Decoult, N.T. Thuong, J. Lhomme, and C. Hélène (1987) Sequence-specific recognition, photocrosslinking, and cleavage of the DNA double-helix by an oligo-[α]-thymidilate covalently linked to an azidoproflavine derivative. Nucleic Acids Res. 15, 7749–7760.

Le Doan, T., L. Perrouault, U. Asseline, N.T. Thuong, C. Rivalle, E. Bisagni, and C. Hélène (1991) Recognition, and photo-induced cleavage, and photo-crosslinking of nucleic acids by oligonucleotides covalently linked to ellipticine. Antisense Res. Dev. 1, 43–54.

Le Doan, T., D. Praseuth, L. Perrouault, M. Chassignol, N.T. Thuong, and C. Hélène (1990) Sequence-targeted photochemical modification of nucleic acids by complementary oligonucleotides covalently linked to porphyrins. Bioconjugate Chem. 1, 108–113.

Lee, B.L., A. Murakami, K.R. Blake, S.-B. Lin, and P.S. Miller (1988) Interaction of psoralen-derivatized oligodeoxynucleotide methylphosphonates with single-stranded DNA. Biochemistry 27, 3197–3203.

Lee, J.S., and A.R. Morgan (1982) Novel aspects of the structure of the *Escherichia coli* nucleoid investigated by a rapid sedimentation assay. Can. J.Biochem. 60, 952–961.

Lee, J.S., D.A. Johnson, and A.R. Morgan (1979) Complexes formed by $(pyrimidine)_n \cdot (purine)_n$ DNAs on lowering the pH are three-stranded. Nucleic Acids Res. 6, 3073–3091.

Lee, J.S., M.L. Woodsworth, L.J. Latimer, and A.R. Morgan (1984) Poly(pyrimidine) · poly(purine) synthetic DNAs containing 5-methylcytosine form stable triplexes at neutral pH. Nucleic Acids Res. 12, 6603–6614.

Lee, J.S., G.D. Burkholder, L.J.P. Latimer, B.L. Haug, and R.P. Braun (1987) A monoclonal antibody to triplex DNA binds to eukaryotic chromosomes. Nucleic Acids Res. 15, 1047–1061.

Lee, J.S., L.J.P. Latimer, B.L. Haug, D.E. Pulleyblank, D.M. Skinner, and G.D. Burkholder (1989) Triplex DNA in plasmids, and chromosomes. Gene 82, 191–199.

Lehninger A.L., D.L. Nelson, and M.M. Cox (1994) Principles of Biochemistry. Second Edition. Worth, New York, p. 338.

Lehrman, E., and D.M. Crothers (1977) An ethidium induces double helix of poly(rA) · poly(rU). Nucleic Acids Res. 4, 1381–1392.

Leonetti, J.-P., P. Machy, G. Degols, B. Lebleu, and L. Leserman (1990) Antibody-targeted liposomes containing oligodeoxyribonucleotides complementary to viral RNA selectively inhibit viral replication. Proc. Natl. Acad. Sci. USA 87, 2448–2451.

Leonetti, J.-P., N. Mechti, G. Degols, C. Gagnor, and B. Lebleu (1991) Intracellular distribution of microinjected antisense oligonucleotides. Proc. Natl. Acad. Sci. USA 88, 2702–2706.

Le Pecq, J.B. (1971) Use of ethidium bromide for separation, and determination of nucleic acids of various conformational forms, and measurement of their associated enzymes. Methods Biochem. Anal. 20, 41–86.

Le Pecq, J.B., and C. Paoletti (1965) Displacement between polyribonucleotides as studied by ethidium hydrobromide–displacement of poly(A-2I) by poly U. C.R. Acad. Sci. 260, 7033–7036.

Letai, A.G., M.A. Palladino, E. From, V. Rizzo, and J.R. Fresco (1988) Specificity in formation of triple-stranded nucleic acid helical complexes: studies with agarose-linked polyribonucleotide affinity columns. Biochemistry 27, 9108–9112.

Letsinger R.L., G. Zhang, D.K. Sun, T. Ikeuchi, and P.S. Sarin (1989) Cholesteryl-conjugated oligonucleotides: synthesis, properties, and activity as inhibitors of replication of human immunodeficiency virus in cell culture. Proc. Natl. Acad. Sci. USA 86, 6553–6556.

Li, B., J.A. Langer, B. Schwartz, and S. Pestka (1989) Creation of phosphorylation sites in proteins: construction of phosphorylable human interferon α. Proc. Natl. Acad. Sci. USA 86, 558–562.

Lilley, D.M.J. (1980) The inverted repeat as a recognizable structural feature in supercoiled DNA molecules. Proc. Natl. Acad. Sci. USA 77, 6468–6472.

Lilley, D.M.J. (1992) Probes of DNA structure. Methods Enzymol. 212, 133–139.

Limn, W., S. Uesugi, M. Ikehara, and H.T. Miles (1983) Poly (8-methyladenylic acid): a single-stranded regular structure with alternating *syn-anti* conformations. Biochemistry 22, 4217–4222.

Lin, C.H., and D.J. Patel (1992) Site-specific covalent duocarmycin A–intramolecular DNA triplex complex. J. Am. Chem. Soc. 114, 10658–10660.

Lin, S.-B., C.-F. Kao, S.-C. Lee, and L.-S. Kan (1994) DNA triplex formed by d-A-(G-A)$_7$-G, and d-^{m}C-(T-^{m}C)$_7$-T in aqueous solution at neutral pH. Anti-Cancer Drug Des. 9, 1–8.

Lipsett, M.N. (1963) The interactions of poly C, and guanine trinucleotide. Biochem. Biophys. Res. Commun. 11, 224–231.

Lipsett, M.N. (1964) Complex formation between polycytidylic acid, and guanine oligonucleotides. J. Biol. Chem. 239, 1256–1260.

Lipsett, M.N., L.A. Heppel, and D.F. Bradley (1960) Complex formation between adenine oligonucleotides, and polyuridylic acid. Biochim. Biophys. Acta 41, 175–177.

Lipsett, M.N., L.A. Heppel, and D.F. Bradley (1961) Complex formation between oligonucleotides, and polymers. J. Biol. Chem. 236, 857–863.

Liquier, J., P. Coffinier, M. Firon, and E. Taillandier (1991) Triple helical polynucleotidic structures: sugar conformations determined by FTIR spectroscopy. J. Biomol. Struct. Dyn. 9,437–435.

Liquier, J., R. Letellier, C. Dagneaux, M. Ouali, F. Morvan, B. Raynier, J.L. Imbach, and E. Taillandier (1993) Triple helix formation by α-oligonucleotides: a vibrational spectroscopy, and molecular modeling study. Biochemistry 32, 10591–10598.

Liquori, A.M., L. Constantino, V. Crescenzi, V. Elia, E. Giglio, R. Puliti, M. De Santis Savino, and V. Vitagliano (1967) Complexes between DNA, and polyamines: a molecular model. J. Mol. Biol. 24, 113–122.

Liu, Q.R., and P.K. Chan (1990) Identification of a long stretch of homopurine · homopyrimidine sequence in a cluster of retroposons in the human genome. J. Mol. Biol. 212, 453–459.

Liu, K., H.T. Miles, J. Frazier, and V. Sasisekharan (1993) A novel DNA duplex. A parallel stranded DNA helix with Hoogsteen base pairing. Biochemistry 32, 11802–11809.

Liu, K., H.T. Miles, K.D. Parris, and V. Sasisekharan (1994) Fibre-type X-ray diffraction patterns from single crystals of triple helical DNA. Nature Struct. Biol. 1, 11–12.

Liu, L.F., and J.C. Wang (1987) Supercoiling of the DNA template during transcription. Proc. Natl. Acad. Sci. USA 84, 7024–7027.

Live, D.H., I. Radhakrishnan, V. Misra, and D.J. Patel (1991) Characterization of protonated cytidine in oligonucleotides by ^{15}N NMR studies at natural abundance. J. Am. Chem. Soc. 113, 4687–4688.

Loke, S.L., C.A. Stein, X.H. Zhang, K. Mori, M. Nakanishi, C. Subasinghe, J.S. Cohen, and L.M. Neckers (1989) Characterization of oligonucleotide transport into living cells. Proc. Natl. Acad. Sci. USA 86, 3874–3878.

Lopez, S., and G. Lancelot (1992) 2D NMR-studies of a triple helix. J. Chim. Phys. Phys.-Chim. Biol. 89, 157–165.

Loprete, D.M., and K.A. Hartman (1993) Conditions for the stability of the B, C, and Z structural forms of poly (dG-dC) in the presence of lithium, potassium, magnesium, calcium, and zinc cations. Biochemistry 32, 4077–4082.

Lu, G., and R.J. Ferl (1992) Site-specific oligodeoxynucleotide binding to maize *Adh*1 gene promoter represses *Adh*1-GUS gene expression in vivo. Plant Mol. Biol. 19, 715–723.

Lu, G., and R.J. Ferl (1993) Homopurine/homopyrimidine sequences as potential regulatory elements in eukaryotic cells. Int. J. Biochem. 25, 1529–1537.

Lu, M., N. Zhang, S. Raimondi, and A.D. Ho (1992) S1 nuclease hypersensitive sites in an oligopurine/oligopyrimidine DNA from the t(10; 14) breakpoint cluster region. Nucleic Acids Res. 20, 263–266.

Lu, Q., L.L. Wallrath, H. Granok, and S.C.R. Elgin (1993) $(CT)_n \cdot (GA)_n$ repeats, and heat shock elements have distinct roles in chromatin structure, and transcriptional activation of the *Drosophila hsp*26 gene. Mol. Cell Biol. 13, 2802–2814.

Luebke, K.J., and P.B. Dervan (1989) Nonenzymatic ligation of oligodeoxyribonucleotides on a duplex DNA template by triple-helix formation. J. Am. Chem. Soc. 111, 8733–8735.

Luebke, K.J., and P.B. Dervan (1991) Nonenzymatic sequence-specific ligation of double-helical DNA. J. Am. Chem. Soc. 113, 7447–7448.

Luebke, K.J., and P.B. Dervan (1992) Nonenzymatic ligation of double-helical DNA by alternate-strand triple helix formation. Nucleic Acids Res. 20, 3005–3009.

Lukashin, A.V., A.V. Vologodskii, M.D. Frank-Kamenetskii, and Y.L. Lyubchenko (1976) Fluctuational opening of the double helix as revealed by theoretical, and experimental study of DNA interaction with formaldehyde. J. Mol. Biol. 108, 665–682.

Lyamichev, V. (1991) Unusual confomration of $(dA)_n \cdot (dT)_n$-tracts as revealed by cyclobutane thymine-thymine dimer formation. Nucleic Acids Res. 19, 4491–4496.

Lyamichev, V.I., M.D. Frank-Kamenetskii, and V.N. Soyfer (1990) Protection against UV-induced pyrimidine dimerization in DNA by triplex formation. Nature 344, 568–570.

Lyamichev, V.I., S.M. Mirkin, O.N. Danilevskaya, O.N. Voloshin, S.V. Balatskaya, V.N. Dobrynin, S.A. Filippov, and M.D. Frank-Kamenetskii (1989a) An unusual DNA structure detected in a telomeric sequence under superhelical stress, and at low pH. Nature 339, 634–636.

Lyamichev, V.I., S.M. Mirkin, and M.D. Frank-Kamenetskii (1985) A pH-dependent structural transition in the homopurine-homopyrimidine tract in superhelical DNA. J. Biomol. Struct. Dyn. 3, 327–338.

Lyamichev, V.I., S.M. Mirkin, and M.D. Frank-Kamenetskii (1986) Structures of homopurine-homopyrimidine tract in superhelical DNA. J. Biomol. Struct. Dyn. 3, 667–669.

Lyamichev, V.I., S.M. Mirkin, and M.D. Frank-Kamenetskii (1987) Structure of $(dG)_n \cdot (dC)_n$ under superhelical stress, and acid pH. J. Biomol. Struct. Dyn. 5, 275–282.

Lyamichev, V.I., S.M. Mirkin, M.D. Frank-Kamenetskii, and C.R. Cantor (1988) A stable complex between homopyrimidine oligomers, and homologous regions of duplex DNAs. Nucleic Acids Res. 16, 2165–2178.

Lyamichev, V.I., S.M. Mirkin, V.P. Kumarev, L.V. Baranova, A.V. Vologodskii, and M.D. Frank-Kamenetskii (1989b) Energetics of B-H transition in supercoiled DNA carrying $d(CT)_x \cdot d(AG)_x$, and $d(C)_n \cdot d(G)_n$ inserts. Nucleic Acids Res. 17, 9417–9423.

Lyamichev, V.I., O.N. Voloshin, M.D. Frank-Kamenetskii, and V.N. Soyfer (1991) Photofootprinting of DNA triplexes. Nucleic Acids Res. 19, 1633–1638.

Macaya, R., D.E. Gilbert, S. Malek, J.S. Sinsheimer, and J. Feigon (1991) Structure, and stability of X · G · C mismatches in the third strand of intramolecular triplexes. Science 254, 270–274.

Macaya, R., P. Schultze,, and J. Feigon (1992a) Sugar conformations in intramolecular DNA triplexes determined by coupling constants obtained by automated simulation of P. COSY cross peaks. J. Am. Chem. Soc. 114, 781–783.

Macaya, R., E. Wang, P. Schultze, V. Sklenar, and J. Feigon (1992b) Proton nuclear magnetic resonance assignments, and structural characterization of an intramolecular DNA triples. J. Mol. Biol. 225, 755–773.

McCarthy, J.G., and S.M. Heywood (1987) A long polypyrimidine/polypurine tract induces an altered DNA conformation on the 3′ coding region of the adjacent myosin heavy chain gene. Nucleic Acids Res. 15, 8069–8085.

Mace, H.A.F., H.R.B. Pelham, and A.A. Travers (1983) Association of an S1 nuclease-sensitive structure with short direct repeats 5′ of *Drosophila* heat shock genes. Nature 304, 555–557.

McCurdy, S., C. Moulds, and B. Froehler (1991) Deoxyoligonucleotides with inverted polarity: synthesis, and use in triple-helix formation. Nucleosides Nucleotides 10, 287–290.

McCall, M., T. Brown, and O. Kennard (1985) The crystal structure of d(G-G-G-G-C-C-C-C). A model for poly(dG) · poly(dC). J. Mol. Biol. 183, 385–396.

McDonald, C.D., and L.J. Maher, III (1995) Recognition of duplex DNA by RNA polynucleotides. Nucleic Acids Res. 23, 500–506.

McGhee, J.D., and P.H. von Hippel (1974) Theoretical aspects of DNA-protein interactions: co-operative, and non-co-operative binding of large ligands to a one-dimensional homogeneous lattice. J. Mol. Biol. 86, 469–489.

McKeon, C., A. Schmidt, and B. de Crombrugghe (1984) A sequence conserved in both the chicken, and mouse α2(I) collagen promoter contains sites sensitive to S1 nuclease. J. Biol. Chem. 259, 6636–6640.

McShan, W.M., R.D. Rossen, A.H. Laughter, J. Trial, D.J. Kessler, J.G. Zendegui, M.E. Hogan, and F.M. Orson (1992) Inhibition of transcription of HIV-1 in infected human cells by oligodeoxynucleotides designed to form DNA triple helices. J. Biol. Chem. 267, 5712–5721.

Maher, L.J., III (1992) Inhibition of T7 RNA polymerase initiation by triple-helical DNA complexes: a model for artificial gene repression. Biochemistry 31, 7587–7594.

Maher, L.J., III, P.B. Dervan, and B. Wold (1990) Kinetic analysis of oligodeoxyribonucleotide-directed triple-helix formation on DNA. Biochemistry 29, 8820–8826.

Maher, L.J., III, P.B. Dervan, and B. Wold (1992) Analysis of promoter-specific repression by triple-helical DNA complexes in a eukaryotic cell-free transcription system. Biochemistry 31, 70–81.

Maher, L.J., III, B. Wold, and P.B. Dervan (1989) Inhibition of DNA binding proteins by oligonucleotide-directed triple helix formation. Science 245, 725–730.

Maine, I.P., and T. Kodadek (1994) Efficient unwinding of triplex DNA by a DNA helicase. Biochem. Biophys. Res. Commun. 204, 1119–1124.

Majumdar, C., C.A. Stein, J.S. Cohen, S. Brooder, and S.H. Wilson (1989) Stepwise mechanism of HIV reverse transcriptase: primer function of phosphorothioate oligodeoxynucleotide. Biochemistry 28, 1340–1346.

Malenkov, G., L. Minchenkova, E. Minyat, A. Schyolkina, and V. Ivanov (1975) On the nature of the B-A transition of DNA in solution. FEBS Lett. 51, 38–42.

Malkov, V.A., V.N. Soyfer, and M.D. Frank-Kamenetskii (1992) Effect of intermolecular triplex formation on the yield of cyclobutane photodimers in DNA. Nucleic Acids Res. 20, 4889–4895.

Malkov, V.A., O.N. Voloshin, V.M. Rostapshov, I. Jansen, V.N. Soyfer, and M.D. Frank-Kamenetskii (1993a) Protonated pyrimidine-purine-purine triplex. Nucleic Acids Res. 21, 105–111.

Malkov, V.A., O.N. Voloshin, V.N. Soyfer, and M.D. Frank-Kamenetskii (1993b) Cation, and sequence effects on stability of intermolecular pyrimidine-purine-purine triplex. Nucleic Acids Res. 21, 585–591.

Mangel, W.F., and M.J. Chamberlin (1974) Studies of ribonucleic acid chain initiation by *Escherichia coli* ribonucleic acid polymerase bound to T7 deoxyribonucleic acid. Assay for the rate, and extent of ribonucleic acid chain initiation. J. Biol. Chem. 249, 3007–3013.

Manning, G.S. (1975) Application of polyelectrolyte "limiting laws" to the helix-coil transition of DNA. II. Effect of Mg^{2+} counterions. Biopolymers 11, 951–955.

Manning, G.S. (1978) The molecular theory of polyelectrolyte solutions with applications to the electrostatic properties of polynucleotides. Q. Rev. Biophys. 11, 179–246.

Manoharan, M., L.K. Johnson, D.P.C. McGee, C.J. Guinosso, K. Ramasamy, R.H. Springer, C.F. Bennett, D.J. Ecker, T. Vickers, L. Cowsert, and P.D. Cook (1992) Chemical modifications to improve uptake, and bioavailability of antisense oligonucleotides. Ann. NY Acad. Sci. 660, 306–309.

Manor, H., B.S. Rao, and R.G. Martin (1988) Abundance, and degree of dispersion of genomic $d(GA)_n \cdot d(CT)_n$ sequences. J. Mol. Evol. 27, 96–101.

Manzini, G., L.E. Xodo, and D. Gasparotto (1990) Triple helix formation by oligopyrine-oligopyrimidine DNA fragments — electrophoretic, and thermodynamic behaviour. J. Mol. Biol. 213, 833–843.

Marck, C., and D. Thiele (1978) Poly(dG) · poly(dC) at neutral, and alkaline pH: the formation of triple stranded poly(dG) · poly(dG) · poly(dG). Nucleic Acids Res. 5, 1017–1028.

Margot, J.B., and R.C. Hardison (1985) DNase I, and nuclease S1 sensitivity of the rabbit β1 globin gene in nuclei, and supercoiled plasmids. J. Mol. Biol. 184, 195–210.

Marky, L.A., and K.J. Breslauer (1987) Calculating thermodynamic data for transitions of any molecularity from equilibrium melting curves. Biopolymers 26, 1601–1620.

Marky, L., and R.B. Macgregor, Jr. (1990) Hydration of dA · dT polymers: role of water in the thermodynamics of ethidium, and propidium intercalation. Biochemistry 29, 4805–4811.

Marshall, W.S., and M.H. Caruthers (1993) Phosphorodithioate DNA as a potential therapeutic drug. Science 259, 1564–1570.

Martin, R.B. (1985) Nucleoside sites for transition metal ion binding. Acc. Chem. Res. 18, 32–38.

Martin, R.B., and Y.H. Mariam (1979) Interactions between metal ions, and nucleic bases, nucleosides, and nucleotides in solution. Metal Ions Biol. Syst. 8, 57–124.

Martinez-Balbas, A., and F. Azorin (1993) The effect of zinc on the secondary structure of $d(GA \cdot TC)_n$ DNA sequences of different length: a model for the formation *H-DNA. Nucleic Acids Res. 21, 2557–2562.

Massom, L.R., and H.W. Jarrett (1992) High-performance affinity chromatography of DNA. II. Porosity effects. J. Chromatogr. 600, 221–228.

Massoulie, J. (1964) Formation of complexes between poly(adenylic acids), and poly(uridylic acids). C.R. Acad. Sci. Paris 259, 3392–3393.

Massoulie, J., R. Blake, L.C. Klotz, and J.R. Fresco (1964a) Spectrophotometric method permitting the complexes of polyriboadenylic acid, and polyribouridylic acid in double, and triple helix forms to be studied separately. C.R. Acad. Sci. Paris 259, 3104–3107.

Massoulie, J., W. Guschlbauer, L.C. Klotz, and J.R. Fresco (1964b) Complexes formed by polyadenylic, and polyuridylic acids; characterization of the complexes by difference spectra. C.R. Acad. Sci. Paris 260, 1284–1285.

Massoulie, J., A.M. Michelson, and F. Pochon (1966) Polynucleotide analogues. VI. Physical studies on 5-substituted pyrimidine polynucleotides. Biochim. Biophys. Acta 114, 16–26.

Matteucci, M., K.Y. Lin, S. Butcher, and C. Moulds (1991) Deoxyoligonucleotides bearing neutral analogues of phosphodiester linkages recognize duplex DNA via triple-helix formation. J. Am. Chem. Soc. 113, 7767–7768.

Mavrothalassitis, G.J., D.K. Watson, and T.S. Papas (1990) Molecular, and functional characterization of promoter of ETS2, the human c-*ets*-2 gene. Proc. Natl. Acad. Sci. USA 87, 1047–1051.

Maxam, A.W., and W. Gilbert (1980) Sequencing end-labeled DNA with base-specific chemical cleavages. Methods Enzymol. 65, 499–560.

May, A., R.S. Nairn, D.S. Okumoto, K. Wassermann, T. Stevsner, J.C. Jones, and V.A. Bohr (1993) Repair of individual DNA strands in the hamster dihydrofolate reductase gene after treatment with ultraviolet light, alkylating agents, and cisplatin. J. Biol. Chem, 268, 1650–1657.

Mayfield, C., M. Squibb, and D. Miller (1994) Inhibition of nuclear protein binding to the human Ki-*ras* promoter by triplex-forming oligonucleotides. Biochemistry 33, 3358–3363.

Menzel, R., and M. Gellert (1983) Regulation of the genes for *E. coli* DNA gyrase: homeostatic control of DNA supercoiling. Cell 34, 105–113.

Mcrgny, J.L., D. Collier, M. Rougée, T. Montenay-Garestier, and C. Hélène (1991a) Intercalation of ethidium bromide into a triple-stranded oligonucleotide. Nucleic Acids Res. 19, 1521–1526.

Mergny, J.L., G. Duval-Valentin, C.H. Nguyen, L. Perrouault, B. Faucon, M. Rougée, T. Montenay-Garestier, E. Bisagni, and C. Hélène (1992) Triple-helix specific ligands. Science 256, 1681–1684.

Mergny, J.L., T. Garestier, M. Rougée, A.V. Lebedev, M. Chassignol, N.T. Thuong, and C. Hélène (1994) Fluorescence energy transfer between two triple helix-forming oligonucleotides bound to duplex DNA. Biochemistry 33, 15321–15328.

Mergny, J.L., J.S. Sun, M. Rougée, T. Montenay-Garestier, F. Barcelo, J. Chomilier, and C. Hélène (1991b) Sequence specificity in triple-helix formation: experimental, and theoretical studies of the effect of mismatches on triplex stability. Biochemistry 30, 9791–9798.

Michel, D., G. Chatelain, Y. Herault, and G. Brun (1992) The long repetitive polypurine/polypyrimidine sequence $(TTCCC)_{48}$ forms DNA triplex with Pu-Pu-Py base triplets in vivo. Nucl. Acids Res. 20, 439–443.

Michel, D., G. Ghatelain, Y. Herault, F. Harper, and G. Brun (1993) H-DNA can act as a transciptional insulator. Cell. Mol. Biol. Res. 39, 131–140.

Michel, F., A.D. Ellington, S. Couture, and J.W. Szostak (1990) Phylogenetic, and genetic evidence for base-triples in the catalytic domain of group I introns. Nature 347, 578–580.

Michel, F., M. Hanna, R. Green, D.P. Bartel, and J.W. Szostak (1989) The guanosine binding site of the Tetrahymena ribozyme. Nature 342, 391–395.

Michelson, A.M., J. Dondon, and M. Grunberg-Manago (1962) The action of polynucleotide phosphorylase on 5-halogenuridine-5′ pyrophosphates. Biochim. Biophys. Acta 55, 529–540.

Michelson, A.M., J. Massoulie, and W. Guschlbauer (1967) Synthetic polynucleotides. Progr. Nucleic Acids Res. Mol. Biol. 6, 83–141.

Michelson, A.M., and F. Pochon (1966) Polynucleotide analogs. VII. Methylation of polynucleotides. Biochim. Biophys. Acta 114, 469–480.

Miles, H.T. (1964) The structure of the three-stranded helix, poly(A+2U). Proc. Natl. Acad. Sci. USA 51, 1104–1109.
Miles, H.T., and J. Frazier (1964a) Infrared study of helix strandedness in the poly A-poly U system. Biochem. Biophys. Res. Commun. 14, 21–28.
Miles, H.T., and J. Frazier (1964b) A strand disproportionation reaction in a helical polynucleotide system. Biochem. Biophys. Res. Commun. 14, 129–136.
Millar, D.B. (1974) The quaternary structure of lactate dehydrogenase. II. The mechanisms, kinetics, and thermodynamics of dissociation, denaturation, and hybridization in ethylene glycol. Biochim. Biophys. Acta 359, 152–176.
Miller, J.H., and H.M. Sobell (1966) A molecular model for gene repression. Proc. Natl. Acad. Sci. USA 55, 1201–1205.
Miller, P.S., P. Bhan, C.D. Cushman, and T.L. Trapane (1992) Recognition of guanine-cytosine base pair by 8-oxoadenine. Biochemistry 31, 6788–6793.
Miller, P.S., and C.D. Cushman (1993) Triplex formation by oligodeoxyribonucleotides involving the formation of X · U · A triads. Biochemistry 31, 2999–3004.
Miller, P.S., and P.O.P. Ts'o (1987) A new approach to chemotherapy based on molecular biology, and nucleic acid chemistry: Matagen (masking tape for gene expression). Anti-Cancer Drug Des. 2, 117–128.
Miller, S.K., D.G. VanDerveer, and L.G. Marzilli (1985) Models for the interaction of Zn^{2+} with DNA. The synthesis, and X-ray structural characterization of two octahedral Zn complexes with mono-methyl phosphate esters of 6-oxopurine 5′-monophosphate nucleotides. J. Am. Chem. Soc. 107, 1048–1055.
Milligan, J.F., S.H. Krawczyk, S. Wadwani, and M.D. Matteucci (1993) An antiparallel triple helix motif with oligo-deoxynucleotides containing 2′-deoxyguanosine, and 7-deaza-2′-deoxyxanthosine. Nucleic Acids Res. 21, 327–333.
Minchenkova, L.E., and V.I. Ivanov (1967) Influence of reductants upon optical characteristics of the DNA-Cu^{2+} complex. Biopolymers 5, 615–625.
Minton, K.W. (1985) The triple helix: a potential mechanism for gene regulation. J. Exp. Pathol. 2, 135–148.
Minyat, E.E., V.I. Ivanov, A.M. Krizyn, L.E. Minchenkova, and A.K. Schyolkina (1978) Spermine, and spermidine-induced B to A transition of DNA in solution. J. Mol. Biol. 128, 397–409.
Mirkin, S.M., and M.D. Frank-Kamenetskii (1994) H-DNA, and related structures. Annu. Rev. Biophys. Biomol. Struct. 23, 541–576.
Mirkin, S.M., V.I. Lyamichev, K.N. Drushlyak, V.M. Dobrynin, and M.D. Frank-Kamenetskii (1987) DNA H form requires a homopurine-homopyrimidine mirror repeat. Nature 330, 495–497.
Mizuno, T., M.-Y. Cou, and M. Inoue (1984) A unique mechanism regulating gene expression: translational inhibition by complementary RNA transcript (micRNA). Proc. Natl. Acad. Sci. USA 81, 1966–1970.
Mohan, V., Y.K. Cheng, G.E. Marlow, and B.M. Pettitt (1993a) Molecular recognition of Watson-Crick base-pair reversals in triple-helix formation: use of nonnatural oligonucleotide bases. Biopolymers 33, 1317–1325.
Mohan, V., P.E. Smith, and B.M. Pettitt (1993b) Evidence for a new spine of hydration: solvation of DNA triple helices. J. Am. Chem. Soc. 115, 9297–9298.

Mollegaard, N.E., O. Buchardt, M. Egholm, and P.E. Nielsen (1994) Peptide nucleic acid · DNA strand displacement loops as artificial transcription promoters. Proc. Natl. Acad. Sci. USA 91, 3892–3895.

Mooren, M.M.W., D.E. Pulleyblank, S.S. Wijmenga, M.J.J. Blommers, and C. Hilbers (1990) Polypurine/polypyrimidine hairpins form a triple helix structure at low pH. Nucleic Acids Res. 18, 6523–6529.

Moores, J.C. (1990) A new vector for rapid mapping of genomic DNA. Strategies 3, 23–29.

Morgan, A.R. (1994) Three-stranded (triplex) DNAs (RNAs): do they have a role in biology? Ind. J. Biochem. Biophys. 31, 83–87.

Morgan, A.R., J.S. Lee, D.E. Pulleyblank, N.L. Murray, and D.E. Evans (1979) Ethidium bromide fluorescence assays. Part I. Physicochemical studies. Nucleic Acids Res. 7, 547–569.

Morgan, A.R., and R.D. Wells (1968) Specificity of the three-stranded complex formation between double-stranded DNA, and single-stranded RNA containing repeating nucleotide sequences. J. Mol. Biol. 37, 63–80.

Moser, H., and P.B. Dervan (1987) Sequence-specific cleavage of double helical DNA by triplex helix formation. Science 238, 645–650.

Mouscadet, J.-F., S. Carteau, H. Goulaouic, F. Subra, and C. Auclair (1994a) Triplex-mediated inhibition of HIV DNA integration in vitro. J. Biol. Chem. 269, 21635–21638.

Mouscadet, J.-F., C. Ketterlé, H. Goulaouic, S. Carteau, F. Subra, M. Le Bret, and C. Auclair (1994b) Triple helix formation with short oligonucleotide-intercalator conjugates matching the HIV-1 U3 LTR end sequence. Biochemistry 33, 4187–4196.

Mullis, K., F. Faloona, S. Scharf, R. Saiki, G. Horn, and H. Erlich (1986) Specific enzymatic amplification of DNA in vitro: the polymerase chain reaction. Cold Spring Harbor Symp. Quant. Biol. 51, 263–273.

Mullis, K.B., and F.A. Faloona (1989) Specific synthesis of DNA in vitro via a polymerase-catalyzed chain reaction. Methods Enzymol. 155, 2423–2427.

Muniyappa, K., S.L. Shaner, S.S. Tsang, and C.M. Radding (1984) Mechanism of the concerted action of recA protein, and helix-destabilizing proteins in homologous recombination. Proc. Natl. Acad. Sci. USA 81, 2757–2761.

Muraiso, T., S. Nomoto, H. Yamazaki, Y. Mishima, and R. Kominami (1992) A single-stranded DNA binding protein from mouse tumor cells specifically recognizes the C-rich strand of the $(AGG:CCT)_n$ repeats that can alter DNA conformation. Nucleic Acids Res. 20, 6631–6635.

Murchie, A.I.H., R. Bowater, F. Aboul-ela, and D.M.J. Lilley (1992) Helix opening transitions in supercoiled DNA. Biochim. Biophys. Acta 1131, 1–15.

Murray, N.L., and A.R. Morgan (1973) Enzymic, and physical studies on the triplex $dT_n \cdot dA_n \cdot rU_n$ [(deoxyribosylthymine)$_n$ · (deoxyadenosine)$_n$ · (ribouridine)$_n$)]. Can. J. Biochem. 51, 436–449.

Musti, A.M., V.M. Ursini, E.V. Avvedimento, V. Zimarino, and R. Di Lauro. (1987) A cell type specific factor recognizes the rat thyroglobulin promoter. Nucleic Acids Res. 15, 8149–8167.

Naylor. R., and P.T. Gilham (1966) Studies on some interactions, and reactions of oligonucleotides in aqueous solution. Biochemistry 5, 2722–2728.

Nehls, M.C., S. Krause, and T. Boehm (1994) Neomycin- and hygromycin-resistance expression cassettes containing an artificial triple-helix site, and a synthetic lac operator facilitate restriction endonuclease cleavage at predefined sites, and recovery of specific fragments from mammalian genomes. Mammalian Genome 5, 183–186.

Nelson, H.C.M., J.T. Finch, F.L. Bonaventura, and A. Klug (1987) The structure of an oligo(dA) · oligo(dT) tract, and its biological implications. Nature, 330, 221–226.

Neumann, E., and T. Ackermann (1969) Thermodynamic investigation of the helix-coil transitions of a polyribonucleotide system. J. Phys. Chem. 73, 2170–2178.

Nickol, J.M., and G. Felsenfeld (1983) DNA conformation at the 5′ end of the chicken adult β-globin gene. Cell 35, 467–477.

Nielsen, P.E. (1991) Sequence-selective DNA recognition by synthetic ligands. Bioconjugate Chem. 2, 1–12.

Nielsen, P.E. (1992) Uranyl photofootprinting of triple helical DNA. Nucleic Acids Res. 20, 2735–2739.

Nielsen, P.E., M. Egholm, and O. Buchardt (1994) Sequence-specific transcription arrest by peptide nucleic acid bound to the DNA template strand. Gene 149, 139–145.

Nielsen, P.E., M. Egholm, O. Buchardt, and R.H. Berg (1991) Sequence-selective recognition of DNA by strand displacement with a thymine-substituted polyamide. Science 254, 1497–1500.

Nielsen, P.E., C. Jeppesen, and O. Buchardt (1988) Uranyl salts as photochemical agents for cleavage of DNA, and probing of protein-DNA contacts. FEBS Lett. 235, 122–124.

Nishikawa, N., M. Oishi, and R. Kiyama (1995) Construction of a human genomic library of clones containing poly(dG-dA) · poly(dT-dC) tracts by Mg^{2+}-dependent triplex affinity capture. J. Biol. Chem. 270, 9258–9264.

Noonberg, S.B., G.K. Scott, M.R. Garovoy, C.C. Benz, and C.A. Hunt (1994a) In vivo generation of highly abundant sequence-specific oligonucleotides for antisense, and triplex gene regulation. Nucleic Acids Res. 22, 2830–2836.

Noonberg, S.B., G.K. Scott, C.A. Hunt, and C.C. Benz (1994b) Detection of triplex-forming RNA oligonucleotides by triplex blotting. Biotechniques 16, 1070–1073.

Noonberg, S.B., G.K. Scott, C.A. Hunt, M.E. Hogan, and C.C. Benz (1994c) Inhibition of transcription factor binding to the HER2 promoter by triplex-forming oligodeoxyribonucleotides. Gene 149, 123–126.

Ohms, J., and T. Ackermann (1990) Thermodynamics of double- and triple-helical aggregates formed by self-complementary oligoribonucleotides of the type rA_xU_y. Biochemistry 29, 5237–5244.

Ojwang, J., A. Elbaggari, H.B. Marshall, K. Jayaraman, M.S. McGrath, and R.F. Rando (1994) Inhibition of human immunodeficiency virus type 1 activity in vitro by oligonucleotides composed entirely of guanosine, and thymidine. J. Acquired Immune Deficiency Syndromes 7, 560–570.

Olivas, W.M., and L.J. Maher, III (1994a) Analysis of duplex DNA by triple helix formation: application to detection of a p53 microdeletion. BioTechniques 16, 128–132.

Olivas, W.M., and L.J. Maher, III (1994b) DNA recognition by alternate strand triple helix formation: affinities of oligonucleotides for a site in the human p53 gene. Biochemistry 33, 983–991.

Olivas, W.M., and L.J. Maher, III (1995a) Competitive triplex/quadruplex equilibria involving guanine-rich oligonucleotides. Biochemistry 34, 278–284.

Olivas, W.M., and L.J. Maher, III (1995b) Overcoming potassium-mediated triplex inhibition. Nucleic Acids Res. 23, 1936–1941.

Olmsted, J., III, and D.R. Kearns (1977) Mechanism of ethidium bromide fluorescence enhancement on binding to nucleic acids. Biochemistry 16, 3647–3654.

O'Neill, D., K. Bornschlegel, M. Flamm, M. Castle, and A. Bank (1991) A DNA-binding factor in adult hematopoietic cells interacts with a pyrimidine-rich domain upstream from the human delta-globin gene. Proc. Natl. Acad. Sci. USA 88, 8953–8957.

O'Neill, D., J. Kaysen, M. Donovan-Peluso, M. Castle, and A. Bank (1990) Protein-DNA interactions upstream from the Human A gamma globin gene. Nucleic Acids Res. 18, 1977–1982.

Ono, A., C.-N. Chen, and L.-S. Kan (1991a) DNA triplex formation of oligonucleotide analogues consisting of linker groups, and octamer segments that have opposite sugar-phosphate backbone polarities. Biochemistry 30, 9914–9921.

Ono, A., P.O.P. Ts'o, and L.-S. Kan (1991b) Triplex formation of oligonucleotides containing 2′-O-methyl-pseudoisocytidine in substitution for 2′-deoxycytidine. J. Am. Chem. Soc. 113, 4032–4033.

Orson, F.M., B.M. Kinsey, and W.M. McShan (1994) Linkage structures strongly influence the binding cooperativity of DNA intercalators conjugated to triplex forming oligonucleotides. Nucleic Acid Res. 22, 479–484.

Orson, F.M., D.W. Thomas, W.M. McShan, D.J. Kessler, and M.E. Hogan (1991) Oligonucleotide inhibition of IL2Rα mRNA transcription by promoter region collinear triplex formation in lymphocytes. Nucleic Acids Res. 19, 3435–3441.

Ouali, M., R. Letellier, F. Adnet, J. Liquier, J.S. Sun, R. Lavery, and E. Taillandier (1993) A possible family of B-like triple helix structures: comparison with the Arnott A-like triple helix. Biochemistry 32, 2098–2103.

Ouali, M., R. Letellier, J.S. Sun, A. Ahkebat, F. Adnet, J. Liquier, and E. Taillandier (1993) Determination of G*G · C triple-helix structure by molecular modeling, and vibrational spectroscopy. J. Am. Chem. Soc. 115, 4264–4270.

Palecek. E. (1991) Local supercoil-stabilized DNA structures. Crit. Rev. Biochem. Mol. Biol. 26, 151–226.

Palecek. E. (1992a) Probing DNA structure with osmium tetroxide complexes in vitro. Methods Enzymol. 212, 139–155.

Palecek. E. (1992b) Probing of DNA structure in cells with osmium tetroxide-2,2′ bipyridine. Methods Enzymol. 212, 305–318.

Panayotatos, N., and R.D. Wells (1981) Cruciform structures in supercoiled DNA. Nature 289, 466–470.

Paner, T.M., F.J. Gallo, M.J. Doktycz, and A.S. Benight (1993) Studies of DNA dumbbells. V. A DNA triplex formed between a 28 base-pair DNA dumbbell substrate, and a 16 base linear single strand. Biopolymers 33, 1779–1789.

Panyutin, I.G., O.I. Kovalsky, and E.I. Budowsky, (1989) Magnesium-dependent supercoiling-induced transition in $(dG)_n \cdot (dC)_n$ stretches, and formation of a new G-structure by $(dG)_n$ strand. Nucleic Acids Res. 17, 8257–8271.

Panyutin, I.G., and R.D. Neumann (1994) Sequence-specific DNA double-strand breaks induced by triplex forming ^{125}I labeled oligonucleotides. Nucleic Acids Res. 22, 4979–4982.

Panyutin, I.G., and R.D. Wells (1992) Nodule DNA in the $(GA)_{37} \cdot (CT)_{37}$ insert in superhelical plasmids. J. Biol. Chem. 267, 5495–5501.

Park, Y.W., and K.J. Breslauer (1992) Drug binding to higher ordered DNA structures: netropsin complexation with a nucleic acid triple helix. Proc. Natl. Acad. Sci. USA 89, 6653–6657.

Parniewski, P., G. Galazka, A. Wilk, and J. Klysik (1989) Complex structural behavior of oligopurine-oligopyrimidine sequence cloned within the supercoiled plasmid. Nucleic Acids Res. 17, 617–629.

Parniewski, P., M. Kwinkowski, A. Wilk, and J. Klysik (1990) Dam methyltransferase sites located within the loop region of the oligopurine-oligopyrimidine sequences capable of forming H-DNA are undermethylated in vivo. Nucleic Acids Res. 18, 605–611.

Parvin, J.D., and P.A. Sharp (1993) DNA topology, and a minimal set of basal factors for transcription by RNA polymerase II. Cell 73, 533–540.

Pattabiraman, N. (1986) Can the double helix be parallel? Biopolymers 25, 1603–1606.

Pauling. L., and R.B. Corey (1953) A proposed structure for the nucleic acids. Proc. Natl. Acad. Sci. USA 39, 84–97.

Peak, M.J., and J.G. Peak (1987) Photosensitized DNA damages. Photochem. Photobiol. 45, 57S.

Peak, J.G., M.J. Peak, and M. MacCoss (1984) DNA breakage caused by 334-nm UV light is enhanced by naturally occurring nucleic acid components, and nucleotide coenzymes. Photochem. Photobiol. 39, 713–716.

Peck, L.J., A. Nordheim, A. Rich, and J.C. Wang (1982) Flipping of cloned $d(pCpG)_n \cdot d(pCpG)_n$ DNA sequences from right- to left-handed helical structure by salt, Co(III), or negative supercoiling. Proc. Natl. Acad. Sci. USA 79. 4560–4564.

Pei, D., D.R. Corey, and P.G. Schultz (1990) Site-specific cleavage of duplex DNA by a semisynthetic nuclease via triple-helix formation. Proc. Natl. Acad. Sci. USA 87, 9858–9862.

Pei, D., H.D. Ulrich, and P.G. Schultz (1991) A combinatorial approach toward DNA recognition. Science 253, 1408–1411.

Peleg, M., V. Kopel, J.A. Borowiec, and H. Manor (1995) Formation of DNA triple helices inhibits DNA unwinding by the SV40 large T-antigen helicase. Nucleic Acids Res. 23, 1292–1299.

Perrouault, L., U. Asseline, C. Rivalle, N.T. Thuong, E. Bisagni, C. Giovannangéli, T. Le Doan, and C. Hélène (1990) Sequence-specific artificial photo-induced endonuclease based on triple helix-forming oligonucleotides. Nature 344, 358–360.

Pestov, D.G., A. Dayn, E.Y. Siyanova, D.L. George, and S.M. Mirkin (1991) H-DNA, and Z-DNA in the mouse c-Ki-*ras* promoter. Nucleic Acids Res. 19, 6527–6532.

Pettijohn, D.E., and R. Hecht (1973) RNA molecules bound to the folded bacterial genome sabilize DNA folds, and segregate domains of supercoiling. Cold Spring Harbor Symp. Quant. Biol. 38, 31–41.

Pettitt, B.M., and P.J. Rossky (1990) Modeling of solvation effects in biopolymer solution. In (D.L. Beveridge, Ed.) Theoretical Biochemistry, and Molecular Biophysics. Adenine Press, New York, pp. 223–229.

Pilch, D.S., and K.J. Breslauer (1994) Ligand-induced formation of nucleic acid triple helices. Proc. Natl. Acad. Sci. USA 91, 9332–9336.

Pilch, D.S., R. Brousseau, and R.H. Shafer (1990a) Thermodynamics of triple helix formation: spectrophotometric studies on the $d(A)_{10} \cdot 2d(T)_{10}$, and $d(C^+{}_3T_4C^+{}_3) \cdot d(G_3A_4G_3) \cdot d(C_3T_4C_3)$ triple helices. Nucleic Acids Res. 18, 5743–5750.

Pilch, D.S., C. Levenson, and R.H. Shafer (1990b) Structural analysis of the $(dA)_{10} \cdot 2(dT)_{10}$ triple helix. Proc. Natl. Acad. Sci. USA 87, 1942–1946.

Pilch, D.S., C. Levenson, and R.H. Shafer (1991) Structure, stability, and thermodynamics of a short intermolecular purine-purine-pyrimidine triple helix. Biochemistry 30, 6081–6087.

Pilch, D.S., M.J. Waring, J.S. Sun, M. Rougée, C.H. Nguyen, E. Bisagni, T. Garestier, and C. Hélène (1993) Characterization of a triple helix-specific ligand. BePI {3-methoxy-7H-8-methyl-11-[3′-amino)propylamino]-benzo[e] pyrido[4,3-b]indole} intercalates into both double-helical, and triple-helical DNA. J. Mol. Biol. 232, 926–946.

Piriou, J.M., C. Ketterlé, J. Gabarro-Arpa, J.A.H. Cognet, and M. Le Bret (1994) A database of 32 DNA triplets to study triple helices by molecular mechanics, and dynamics. Biophys. Chem. 50, 323–343.

Pitot, H.C. (1993) The molecular biology of carcinogenesis. Cancer 72, 962–970.

Plum, G.E., and K.J. Breslauer (1995a) Thermodynamics of an intramolecular DNA triple helix: a calorimetric, and spectroscopic study of the pH, and salt dependence of thermally induced structural transitions. J. Mol. Biol. 248, 679–695.

Plum, G.E., Y.W. Park, S.F. Singleton, P.B. Dervan, and K.J. Breslauer (1990) Thermodynamic characterization of the stability, and the melting behavior of a DNA triplex: a spectroscopic, and calorimetric study. Proc. Natl. Acad. Sci. USA 87, 9436–9440.

Plum, G.E., D.S. Pilch, S.F. Singleton, and K.J. Breslauer (1995b) Nucleic acid hybridization: triplex stability, and energetics. Annu. Rev. Biophys. Biomol. Struct. 24, 319–350.

Pochon, F., and A.M. Michelson (1965) Polynucleotides. VI. Interaction between polyguanylic acid, and polycytidylic acid. Proc. Natl. Acad. Sci. USA 53, 1425–1430.

Pochon, F., and A.M. Michelson (1967) Polynucleotides. IX. Methylation of nucleic acids, homopolynucleotides, and complexes. Biochim. Biophys. Acta 149, 99–106.

Podust, L.M., S.A. Gaidamakov, T.V. Abramova, V.V. Vlassov, and V.V. Gorn (1989) Sequence specificity modification of double-stranded DNA with an alkylating derivative of oligodeoxyribonucleotide $pT(CT)_6$. Bioorg. Khim. 15, 363–369.

Pohl, F.M., and T.M. Jovin (1972) Salt-induced co-operative conformational change of a synthetic DNA: equilibrium, and kinetic studies with poly(dG-dC). J. Mol. Biol. 67, 375–396.

Porschke, D., and M. Eigen (1971) Co-operative non-enzymic base recognition. III. Kinetics of the helix-coil transition of the oligoribouridylic · oligoriboadenylic · acid system, and of oligoriboadenylic acid alone at acidic pH. J. Mol. Biol. 62, 361–381.

Porter, A.C.G., Y. Chernajovsky, T.C. Dale, C.S. Gilbert, G.R. Stark, and I.M. Stark (1988) Interferone response element of the human 6–16 gene. EMBO J. 7, 85–92.

Porumb, H., C. Dagneaux, R. Letellier, C. Malvy, and E. Taillandier (1994) Triple-helices targeted to the polypurine tract of a murine retrovirus. Gene 149, 101–107.

Postel, E.H. (1992) Modulation of c-*myc* transcription by triple helix formation. Ann. NY Acad. Sci. 660, 57–63.

Postel, E.H., S.E. Mango, and S.J. Flint (1989) A nuclease-hypersensitive element of the human c-*myc* promoter interacts with a transcription initiation factor. Mol. Cell. Biol. 9, 5123–5133.

Postel, E.H., S.J. Flint, D.J. Kessler, and M.E. Hongan (1991) Evidence that a triplex-forming oligodeoxyribonucleotide binds to the c-*myc* promoter in HeLa cells, thereby reducing c-*myc* mRNA levels. Proc. Natl. Acad. Sci. USA 88, 8227–8231.

Potaman, V.N., Y.A. Bannikov, and L.S. Shlyakhtenko (1980) Sedimentation of DNA in water-ethanol solutions during the B-A transition. Nucleic Acids Res. 8, 635–642.

Potaman, V.N., and V.N. Soyfer (1994) Divalent metal cations upon coordination to the N7 of purines differentially stabilize the PyPuPu DNA triplex due to unequal Hoogsteen-type hydrogen bond enhancement. J. Biomol. Struct. Dyn. 11, 1035–1040.

Povsic, T.J., and P.B. Dervan (1989) Triple helix formation by oligonucleotides on DNA extended to the physiological pH range. J. Am. Chem. Soc. 111, 3059–3061.

Povsic, T.J., and P.B. Dervan (1990) Sequence-specific alkylation of double-helical DNA by oligonucleotide-directed triple-helix formation. J. Am. Chem. Soc. 112, 9428–9430.

Povsic, T.J., S.A. Strobel, and P.B. Dervan (1992) Sequence-specific double-strand alkylation, and cleavage of DNA mediated by triple-helix formation. J. Am. Chem. Soc. 114, 5934–5941.

Prakash, G., and E.T. Kool (1992) Structural effects in the recognition of DNA by circular oligonucleotides. J. Am. Chem. Soc. 114, 3523–3527.

Praseuth, D., T. Le Doan, M. Chassignol, J.L. Decout, N. Habhoub, J. Lhomme, N.T. Thuong, and C.Hélène (1988a) Sequence-targeted photosensitized reactions in nucleic acids by oligo-α-deoxynucleotides, and oligo-β-deoxynucleotides covalently linked to proflavin. Biochemistry 27, 3031–3038.

Praseuth. D., L. Perrouault, T. Le Doan, M. Chassignol, N.T. Thuong, and C. Hélène (1988b) Sequence-specific binding, and photocrosslinking of α, and β oligodeoxynucleotides to the major groove of DNA via triple-helix formation. Proc. Natl. Acad. Sci. USA 85, 1349–1353.

Price, M.A., and T.D. Tullius (1993) How the structure of an adenine tract depends on sequence context: a new model for the structure of T_nA_n DNA sequences. Biochemistry 32, 127–136.

Priestley, E.S., and P.B. Dervan (1995) Sequence composition effects on the energetics of triple helix formation by oligonucleotides containing a designed mimic of protonated cytosine. J. Am. Chem. Soc. 117, 4761–4765.

Printz, M.P., and P.J. von Hippel (1965) Hydrogen exchange studies of DNA structure. Proc. Natl. Acad. Sci. USA 53, 362–369.

Puhl, H.L., S.R. Gudibande, and M.J. Behe (1991) Poly[d(A · T)], and other synthetic polydeoxynucleotides containing oligoadenosine tracts from nucleosomes easily. J. Mol. Biol. 222, 1149–1160.

Pulleyblank, D.E., D.B. Haniford, and A.R. Morgan (1985) A structural basis for S1 nuclease sensitivity of double-stranded DNA. Cell 42, 271–280.

Pullman, B., P. Claverie, and J. Caillet (1967) Interaction energies in hydrogen-bonded purine-pyrimidine triplets. Proc. Natl. Acad. Sci. USA 57, 1663–1669.

Raae, A.J., and K. Kleppe (1978) T4 polynucleotide ligase catalyzed joining on triple-stranded nucleic acids. Biochemistry 17, 2939–2942.

Radding, C.M. (1991) Helical interactions in homologous pairing, and strand exchange driven by RecA protein. J. Biol. Chem. 266, 5355–5358.

Radhakrishnan, I., C. de los Santos, and D.J. Patel (1993a) Nuclear magnetic resonance structural studies of A · AT base triple alignments in intramolecular purine · purine · pyrimidine DNA triplexes in solution. J. Mol. Biol. 234, 188–197.

Radhakrishnan, I., C. de los Santos, and D.J. Patel (1991a) Nuclear magnetic resonance structural studies of intramolecular purine · purine · pyrimidine DNA triplexes in solution – base triple pairing alignments, and strand direction. J. Mol. Biol. 221, 1403–1418.

Radhakrishnan. I., X. Gao. C. de los Santos. D. Live, and D. J. Patel (1991b) NMR amino proton, and nitrogen markers of G · TA base triple formation. Biochemistry 30, 9022–9030.

Radhakrishnan, I., and D.J. Patel (1993a) Solution structure of an intramolecular purine · purine · pyrimidine DNA triplex. J. Am. Chem. Soc. 115, 1615–1617.

Radhakrishnan, I., and D.J. Patel (1993b) Solution structure of a purine · purine · pyrimidine DNA triplex containing G · GC, and T · AT triples. Structure 1, 135–152.

Radhakrishnan, I., and D.J. Patel (1994a) Solution structure of a pyrimidine · purine · pyrimidine DNA triplex containing T · AT, C+ · GC, and G · TA triples. Structure 2, 17–32.

Radhakrishnan, I., and D.J. Patel (1994b) Hydration sites in purine · purine · pyrimidine, and pyrimidine · purine · pyrimidine DNA triplexes in aqueous solution. Structure 2, 395–405.

Radhakrishnan, I., and D.J. Patel (1994c) Solution structure, and hydration patterns of a pyrimidine · purine · pyrimidine DNA triplex containing a novel T · CG base-triple. J. Mol. Biol. 241, 600–619.

Radhakrishnan, I., and D.J. Patel (1994d) DNA triplexes: solution structures, hydration sites, energetics, interactions,, and function. Biochemistry 33, 11405–11416.

Radhakrishnan, I., D.J. Patel, and X. Gao (1991c) NMR assignment strategy for DNA protons through three-dimensional proton-proton connectivities. Application to an intramolecular DNA triplex. J. Am. Chem. Soc. 113, 8542–8544.

Radhakrishnan, I., D.J. Patel, and X. Gao (1992a) Three-dimensional homonuclear NOESY-TOCSY of an intramolecular pyrimidine · purine · pyrimidine DNA triplex containing a central G · TA triple: nonexchangeable assignments, and structural implications. Biochemistry 31, 2514–2523.

Radhakrishnan, I., D.J. Patel, E.S. Priestley, H.M. Nash, and P.B. Dervan (1993b) NMR structural studies of a nonnatural deoxyribonucleoside which mediates recognition of GC base pairs in pyrimidine · purine · pyrimidine DNA triplexes. Biochemistry 32, 11228–11234.

Radhakrishnan, I., D.J. Patel, J.M. Veal, and X. Gao (1992b) Solution conformation of a G · TA triple in an intramolecular pyrimidine · purine · pyrimidine DNA triplex. J. AM. Chem. Soc. 114, 6913–6915.

Raghu, G., S. Tevosian, S. Anant, K.N. Subramanian, D.L. George, and S.M. Mirkin (1994) Transcriptional activity of the homopurine-homopyrimidine repeat of the c-Ki-*ras* promoter is independent of its H-forming potential. Nucleic Acids Res. 22, 3271–3279.

Raghunathan, G., H.T. Miles, and V. Sasisekharan (1993) Symmetry, and molecular structure of a DNA triple helix: $d(T)_n \cdot d(A)_n \cdot d(T)_n$. Biochemistry 32, 455–462.

Rahmouni, A.R., and R. D. Wells (1989) Stabilization of Z DNA in vivo by local supercoiling. Science 246, 358–363.

Rahn, R.O. (1992) Photochemistry of halogen pyrimidines: iodine release studies. Photochem. Photobiol. 56, 9–15.

Rainen, L.C., and B.D. Stollar (1977) Antisera to poly(A)-poly(U)-poly(I) contain antibody subpopulations specific for different aspects of the triple helix. Biochemistry 3, 2003–2007.

Rajagopal, P., and J. Feigon (1989a) Triple-strand formation in the homopurine: homopyrimidine DNA oligonucleotides $d(G\text{-}A)_4$ and $d(T\text{-}C)_4$. Nature 339, 637–640.

Rajagopal, P., and J. Feigon (1989b) NMR studies of triple-strand formation from the homopurine-homopyrimidine deoxyribonucleotides $d(GA)_4$ and $(TC)_4$. Biochemistry 28, 7859–7870.

Ramanathan, M., M. Lantz, R.D. MacGregor, M.R. Garovoy, and C.A. Hunt (1994) Characterization of the oligodeoxynucleotide-mediated inhibition of interferon-γ-induced major histocompatibility complex Class I, and intercellular adhesion molecule-1. J. Biol. Chem. 269, 24564–24574.

Ramsing, R.B., and T.M. Jovin (1988) Parallel-stranded DNA. Nucleic Acids Res. 16, 6659–6676.

Rando, R.F., L. DePaolis, R.H. Durland, K. Jayaraman, D.J. Kessler, and M.E. Hogan (1994) Inhibition of T7, and T3 RNA polymerase directed transcription elongation in vitro. Nucleic Acids Res. 22, 678–685.

Rao, B.J., and C.M. Radding (1994) Formation of base triplets by non-Watson-Crick bonds mediates homologous recognition in RecA recombination filaments. Proc. Natl. Acad. Sci. USA 91, 6161–6165.

Rao, B.S. (1994) Pausing of simian virus 40 DNA replication fork movement in vivo by $(dG\text{-}dA)_n \cdot (dT\text{-}dC)_n$ tracts. Gene 140, 233–237.

Rao, B.S., H. Manor, and R.G. Martin (1988) Pausing in simian virus 40 DNA replication by a sequence containing $(dG\text{-}dA)_{27} \cdot (dT\text{-}dC)_{27}$. Nucleic Acids Res. 16, 8077–8094.

Rao, T.S., R.H. Durland, D.M. Seth, M.A. Myrick, V. Bodepui, and G.R. Revankar (1995) Incorporation of 2′-deoxy-6-thioguanosine into G-rich oligodeoxyribonucleotides inhibits G-tetrad formation, and facilitates triplex formation. Biochemistry 34, 765–772.

Rau, D.C., and V.A. Parsegian (1992) Direct measurement of the intermolecular forces between counterion-condensed DNA double helices. Evidence for long range attractive hydration forces. Biophys. J. 61, 246–259.

Raudaschl-Sieber, G., H. Schollhorn, U. Thewalt, and B. Lippert (1985) Simultaneous binding of Pt (II) to three different sites (N7, N1, N3) of a guanine nucleobase. J. Am. Chem. Soc. 107, 3591–3595.

Reaban, M.E., and J.A. Griffin (1990) Induction of RNA-stabilized DNA conformers by transcription of an immunoglobulin switch region. Nature 348, 342–344. (Erratum published in Nature 351, 447–448).

Reaban, M.E., J. Lebowitz, and J.A. Griffin (1994) Transcription induces the formation of a stable RNA · DNA hybrid in the immunoglobulin α switch region. J. Biol. Chem. 269, 21850–21857.

Record, M.T., Jr. (1975) Effect of sodium, and magnesium ions on the helix-coil transition of DNA. Biopolymers 14, 2137–2158.

Record, M.T., Jr., C.F. Anderson, and T.M. Lohmam (1978) Thermodynamic analysis of ion effects on the binding, and conformational equilibriums of proteins, and nucleic acids: the roles of ion association or release, screening, and ion effects on water activity. Q. Rev. Biophys. 11, 103–178.

Reily, M.D., and L.G. Marzilli (1986) Novel, definitive NMR evidence for N(7), α-PO_4 chelation of 6-oxopurine nucleotide monophosphate to platinum anticancer drugs. J. Am. Chem. Soc. 108, 8299–8300.

Rentzeperis, D., D.W. Kupke, and L.A. Marky (1992) Differential hydration of homopurine sequences relative to alternating purine/pyrimidine sequences. Biopolymers 32, 1065–1075.

Reynolds, M.A., L.J. Arnold, Jr., M.T. Almazan, T.A. Beck, R.I. Hogrefe, M.D. Metzler, S.R. Stoughton, B.Y. Tseng, T.L. Trapane, P.O.P. Ts'o, and T.M. Woolf (1994) Triple-strand-forming methylphosphonate oligodeoxynucleotides targeted to mRNA efficiently block protein synthesis. Proc. Natl. Acad. Sci. USA 91, 12433–12437.

Rhodes, D. (1979) Nucleosome cores reconstituted from poly (dA · dT), and the octamer of histones. Nucleic Acids Res. 6, 1805–1816.

Rhodes, D., and A. Klug (1980) Helical periodicity of DNA determined by enzyme digestion. Nature 286, 573–578.

Rich, A. (1958a) Formation of two-, and three-stranded helical molecules by polyinosinic acid, and polyadenylic acid. Nature 181, 521–525.

Rich, A. (1958b) The molecular structure of polyinosinic acid. Biochim. Biophys. Acta 29, 502–509.

Rich, A. (1960) A hybrid helix containing both deoxyribose, and ribose polynucleotides, and its relation to the transfer of information between the nucleic acids. Proc. Natl. Acad. Sci. USA 46, 1044–1053.

Rich, A., A. Nordheim, and A.H. Wang (1984) The chemistry, and biology of left-handed Z-DNA. Annu. Rev. Biochem. 53, 791–846.

Richard, H., J.P. Schreiber, and M. Daune (1973) Interaction of metallic ions with DNA. V. DNA renaturation mechanism in the presence of copper(2+) ions. Biopolymers 12, 1–10.

Richardson, J.P. (1975) Initiation of transcription by *Escherichia coli* RNA polymerase from supercoiled, and non-supercoiled bacteriophage PM2 DNA. J. Mol. Biol. 91, 477–487.

Rigas, B., A.A. Welcher, D.C. Ward, and S.M. Weisman (1986) Rapid plasmid library screening using RecA-coated biotinylated probes. Proc. Natl. Acad. Sci. USA 83, 9591–9595.

Riley, M., B. Maling, and M.J. Chamberlin (1966) Physical, and chemical characterization of two-, and three-stranded adenine-thymine, and adenine-uracil homopolymer complexes. J. Mol. Biol. 20, 359–389.

Rippe, K., V. Fritsch, E. Westhof, and T.M. Jovin (1992) Alternating d(G-A) sequences form a parallel-stranded DNA homoduplex. EMBO J. 11, 3777–3786.

Rippe, K., and T.M. Jovin (1992) Parallel-stranded duplex DNA. Methods Enzymol. 211, 199–220.

Robert-Nicoud, M., D.J. Arndt-Jovin, D.A. Zarling, and T.M. Jovin (1984) Immunological detection of left-handed Z DNA in isolated polytene chromosomes. Effects of ionic strength, pH, temperature, and topological stress. EMBO J. 3, 721–731.

Roberts, R.W., and D.M. Crothers (1991) Specificity, and stringency in DNA triplex formation. Proc. Natl. Acad. Sci. USA 88, 9397–9401.

Roberts, R.W., and D.M. Crothers (1992) Stability, and properties of double, and triple helices: dramatic effects of RNA, and DNA backbone composition. Science 258, 1463–1466.

Robinson, H., G.A. van der Marel, J.H. van Boom, and A.H.-J. Wang (1992) Unusual DNA conformation at low pH revealed by NMR: parallel-stranded DNA duplex with homo base pairs. Biochemistry 31, 10510–10517.

Roig, V., R. Kurfurst, and N.T. Thuong (1993) Oligo-β-deoxyribonucleotides, and oligo-α-deoxyribonucleotides involving 2-aminopurine, and guanine for triple-helix formation. Tetrahedron Lett. 34, 1601–1604.

Rooney, S.M., and P.D. Moore (1995) Antiparallel, intramolecular triplex DNA stimulates homologous recombination in human cells. Proc. Natl. Acad. Sci. USA 92, 2141–2144.

Ross, C., M. Samuel, and S.L. Broitman (1992) Transcriptional inhibition of the bacteriophage T7 early promoter region by oligonucleotide triple helix formation. Biochem. Biophys. Res. Commun. 189, 1674–1680.

Ross, P., and R.L. Scruggs (1965) Heat of the reaction forming the three-stranded poly(A+2U) complex. Biopolymers 3, 491–496.

Rougée, M., B. Faucon, J.L. Mergny, F. Barcelo, C. Giovannangéli, T. Garestier, and C. Hélène (1992) Kinetics, and thermodynamics of triple-helix formation: effects of ionic strength, and mismatches. Biochemistry 31, 9269–9278.

Roy, C. (1993) Inhibition of gene transcription by purine rich triplex forming oligodeoxyribonucleotides. Nucleic Acids Res. 21, 2845–2852.

Roy, C. (1994) Triple-helix formation interferes with the transcription, and hinged DNA structure of the interferon-inducible 6–16 gene promoter. Eur. J. Biochem. 220, 493–503.

Roy, C., and B. Lebleu (1991) DNA protein interactions at the interferon-responsive promoter elements: potential for an H-DNA conformation. Nucleic Acids Res. 19, 517–524.

Rubin, C.M., and C.W. Schmid (1980) Pyrimidine-specific chemical reactions useful for DNA sequencing. Nucleic Acids Res. 8, 4613–4619.

Rubin, E., T.L. McKee, and E.T. Kool (1993) Binding of two different DNA sequences by conformational switching. J. Am. Chem. Soc. 115, 361–362.

Ruiz-Carillo, A. (1984) The histone H5 gene is flanked by S1 hypersensitive structures. Nucleic Acids Res. 12, 6473–6492.

Ruskin, B., and M.R. Green (1985) The role of the 3′ splice site consensus sequence in mammalian pre-mRNA splicing. Nature 317, 732–734.

Saenger, W. (1984) Principles of Nucleic Acid Structure. Springer-Verlag, New York.

Sage, E. (1993) Distribution, and repair of photolesions in DNA: genetic consequences, and the role of sequence context. Photochem. Photobiol. 57, 163–174.

Salunkhe, M., T.F. Wu, and R.L. Letsinger (1992) Control of folding, and binding of oligoncleotides by use of a nonucleotide linker. J. Am. Chem. Soc. 114, 8768–8772.

Samadashwily, G.M., A. Dayn, and S.M. Mirkin (1993) Suicidal nucleotide sequences for DNA polymerization. EMBO J. 12, 4975–4983.

Samadashwily, G.M., and S.M. Mirkin (1994) Trapping DNA polymerases using triplex-forming oligodeoxyribonucleotides. Gene 149, 127–136.

Sambrook, J., E.F. Fritsch, and T. Maniatis (1989) Molecular Cloning, A Laboratory Manual. 2nd edition. Cold Spring Harbor Laboratory Press, Cold Spring Harbor, New York.

Sancar, A., and M.S. Tang (1993) Nucleotide excision repair. Photochem. Photobiol. 57, 905–921.

Sandor, Z., and A. Bredberg (1994) Repair of triple helix directed psoralen adducts in human cells. Nucleic Acids Res. 22, 2051–2056.

Sanger, F., S. Nicklen, and A.R. Coulson (1977) DNA sequencing with chain terminating inhibitors. Proc. Natl. Acad. Sci. USA 74, 5463–5467.

Santra, M., K.G. Danielson, and R.V. Iozzo (1994) Structural, and functional characterization of the human decorin gene promoter. A homopurine-homopyrimidine S1 nuclease-sensitive region is involved in transcriptional control. J. Biol. Chem. 269, 579–587.

Sarai, A., S. Sugiura, H. Torigoe, and H. Shindo (1993) Thermodynamic, and kinetic analyses of DNA triplex formation: application of filter-binding assay. J. Biomol. Struct. Dyn. 11, 245–252.

Sarhan, S., and N. Seiler (1989) On the subcellular localization of polyamines. Biol. Chem. Hoppe-Seyler 370, 1279–1284.

Sarkar, P.K., and J.T. Yang (1965a) Optical rotatory dispersion, and conformation of polyadenylic, and polyuridylic acids. J. Biol. Chem. 240, 2088–2093.

Sarkar, P.K., and J.T. Yang (1965b) Optical activity, and the conformation of polyinosinic acid, and several other polynucleotide complexes. Biochemistry 4, 1238–1244.

Sarkar, P.S., and S.K. Brahmachari (1992) Intramolecular triplex potential sequence within a gene down regulates its expression in vivo. Nucleic Acids Res. 20, 5713–5718.

Sarocchi, M.T., Y. Courtois, and W. Guschlbauer (1970) Protonated polynucleotide structures. Specific complex formation between polycytidylic acid, and guanosine or guanylic acid. Eur. J. Biochem. 14, 411–421.

Sasisekharan, V., and P.B. Sigler (1965) X-ray diffraction study of poly(A+U). J. Mol. Biol. 12, 296–298.

Satchwell, S.C., H.R. Drew, and A.A. Travers (1986) Sequence periodicities in chicken nucleosome core DNA. J. Mol. Biol. 191, 659–675.

Saucier, J.M., and J.C. Wang (1972) Angular alteration of the DNA helix by *Escherichia coli* RNA polymerase. Nature (New Biol). 239, 167–170.

Scagiante, B., C. Morassutti, G. Tolazzi, A. Michelutti, M. Baccarani, and F. Quadrifoglio (1994) Effect of unmodified triple helix-forming oligodeoxyri-

bonucleotide targeted to human multidrug-resistance gene *mdr*1 in MDR cancer cells. FEBS Lett. 352, 380–384.
Scaria, P.V., and R.H. Shafer (1991) Binding of ethidium bromide to a DNA triple helix: evidence for intercalation. J. Biol. Chem. 266, 5417–5423.
Scaria, P.V., S. Will, C. Levenson, and R.H. Shafer (1995) Physicochemical studies of the $d(G_3T_4G_3)$*$d(G_3A_4G_3) \cdot d(C_3T_4C_3)$ triple helix. J. Biol. Chem. 270, 7295–7303.
Schildkraut, C., and S. Lifson (1965) Dependence of the melting temperature of DNA on salt concentration. Biopolymers 3, 195–208.
Schleif, R. (1992) DNA looping. Annu. Rev. Biochem. 61, 199–223.
Schmid, N., and J.-P. Behr (1991) Location of spermine, and other polyamines on DNA as revealed by photoaffinity cleavage with polyaminobenzenediazonium salts. Biochemistry 30, 4357–4361.
Schon, E., T. Evans, J. Welsh, and A. Efstratiadis (1983) Conformation of promoter DNA: fine mapping of S1-hypersensitive sites. Cell 35, 837–848.
Schwartz, A.M., and G.D. Fasman (1979) Thermal denaturation of chromatin, and lysine copolymer-DNA complexes. Effects of ethylene glycol. Biopolymers 18, 1045–1063.
Seela, F., J. Ott, and D. Franzen (1982) Poly(adenylic acids) containing the antibiotic tubercidin — base pairing, and hydrolysis by nuclease S1. Nucleic Acids Res. 10, 1389–1397.
Sekharudu, C.Y., N. Yathindra, and M. Sundaralingam (1993) Molecular dynamics investigations of DNA triple helical models: unique features of the Watson-Crick duplex. J. Biomol. Struct. Dyn. 11, 225–244.
Semerad, C.L., and L.J. Maher, III (1994) Exclusion of RNA strands from a purine motif triple helix. Nucleic Acids Res. 22, 5321–5325.
Sen, D., and W. Gilbert (1990) A sodium-potassium switch in the formation of four-stranded G4-DNA. Nature 344, 410–414.
Shaw, J.P., J.F. Milligan, S.H. Krawczyk, and M. Matteucci (1991) Specific high-efficiency, triple-helix-mediated cross-linking to duplex DNA. J. Am. Chem. Soc. 113, 7765–7766.
Shchyolkina, A.K., Y.P. Lysov, I.A. Il'ichova, A.A. Chernyi, Y.B. Golova, B.K. Chernov, B.P. Gottikh, and V.L. Florentiev (1989) Parallel stranded DNA with AT base pairing. FEBS Lett. 244, 39–42.
Shchyolkina, A.K., O.K. Mamaeva, O.F. Borisova, Y.P. Lysov, E.N. Timofeev, I.A. Il'icheva, B.P. Gottikh, and V.L. Florent'ev (1994) Three-stranded "clip" of the oligonucleotide $5'$-$(dT)_{10}pO(CH_2CH_2O)_3p(dT)_{10}pO(CH_2CH_2O)_3p(dA)_{10}$-$3'$. Antisense Res. Dev. 4, 27–33.
Shea, R.G., P. Ng, and N. Bischofberger (1990) Thermal denaturatioin profiles, and gel mobility shift analysis of oligodeoxyribonucleotide triplexes. Nucleic Acids Res. 18, 4859–4866.
Sheldrick, W.S., and B. Gunther (1988) Preparation, and structural characterization of dicarbonyl-rhodium(I) complexes of 8-azaguanine, and allopurinol derivatives. Inorg. Chim. Acta 152, 223–226.
Shen, C.K.J. (1983) Superhelicity induces hypersensitivity of a human polypyrimidine · polypurine DNA sequence in the human $\alpha2$–$\alpha1$ globin intergenic region to S1 nuclease digestion-high resolution mapping of the clustered cleavage sites. Nucleic Acids Res. 11, 7899–7910.

Shifrin, S., and C.L. Parrott (1975) Influence of glycerol, and other polyhydric alcohols on the quaternary structure of an oligomeric protein. Arch. Biochem. Biophys. 166. 426–432.

Shimizu, M., J.C. Hanvey, and R.D. Wells (1989) Intramolecular DNA triplexes in supercoiled plasmids. I. Effect of loop size on formation, and stability. J. Biol. Chem. 264, 5944–5949.

Shimizu, M., J.C. Hanvey, and R.D. Wells (1990) Multiple non-B-DNA conformations of polypurine · polypyrimidine sequences in plasmids. Biochemistry 29, 4704–4713.

Shimizu, M., H., Inoue, and E. Ohtsuka (1991) Design of oligonucleotides for sequence-specific triple-helix formation. Properties of oligonucleotides containing 2′-deoxyxanthosine. Nucleic Acids Symp. Ser. 25, 141–142.

Shimizu, M., H. Inoue, and E. Ohtsuka (1994) Detailed study of sequence-specific DNA cleavage of triplex-forming oligonucleotides linked to 1,10-phenathroline. Biochemistry 33, 606–613.

Shimizu, M., T. Koizumi, H. Inoue, and E. Ohtsuka (1994) Effects of 5-methyl substitution in 2′-O-methyloligo(pyrimidine)nucleotides on triple-helix formation. Bioorg. Med. Chem. Lett. 4, 1029–1032.

Shimizu, M., A. Konishi, Y. Shimada, H. Inoue, and E. Ohtsuka (1992) Oligo(2′-O-methyl)ribonucleotides. Effective probes for duples DNA. FEBS Lett. 302, 155–158.

Shimizu, M., K. Kubo, U. Matsumoto, and H. Shindo (1994) The loop sequence plays crucial roles for isomerization of intramolecular DNA triplexes in supercoiled plasmids. J. Mol. Biol. 235, 185–197.

Shin, Y.A., and G.L. Eichhorn (1968) Interactions of metal ions with polynucleotides, and related compounds. XI. The reversible unwinding, and rewinding of deoxyribonucleic acid by zinc(II) ions through temperature manipulations. Biochemistry 7, 1026–1032.

Shindo, H., H. Torigoe, and A. Sarai (1993) Thermodynamic, and kinetic studies of DNA triplex formation of an oligohompyrimidine, and a matched duplex by filter binding assay. Biochemistry 32, 8963–8969.

Shull, M.M. D.G. Pugh, and J.B. Lingrel (1989) Characterization of the human Na, K-ATPase α2 gene, and identification of intragenic restriction fragment length polymorphisms. J. Biol. Chem. 264, 17532–17543.

Shvarts, V.S., A.V. Kant, and N.M. Frolova (1990) Basic factors in the stability of RNA-duplexes. The contribution of dehydration. Mol. Biol. (Moscow) 24, 1484–1494.

Siegfried, E., G.H. Thomas, U.M. Bond, and S.C.R. Elgin (1986) Characterization of a supercoil-dependent S1 sensitive site 5′ to the *Drosophila melanogaster hsp*26 gene. Nucleic Acids Res. 14, 9425–9444.

Sigel, H. (1989) Metal-nucleotide interactions. In (T.D. Tullius, Ed.) Metal-DNA Chemistry, ACS Symposium Series 402, American Chemical Society, Washington, DC, pp. 159–204.

Sigler, P.B., D.R. Davis, and H.T. Miles (1962) A displacement reaction between a polynucleotide helix, and a random coil. J. Mol. Biol. 5, 709–717.

Sigman, D.S. (1990) Chemical nucleases. Biochemistry 29, 9097–9105.

Simpson, R.T., and P. Kunzler (1979) Chromatin, and core particles formed from inner histones, and synthetic polydeoxyribonucleotides of defined sequence. Nucleic Acids Res. 6, 1387–1415.

Sinden, R.R. (1994) DNA Structure, and Function, Chapter 6. Academic Press, San Diego, California, pp. 217–258.

Sinden, R.R., and D.E. Pettijohn (1981) Chromosomes in living *Escherichia coli* cells are segregated into domains of supercoiling. Proc. Natl. Acad. Sci. USA 78, 224–228.

Sinden, R.R., and D.W. Ussery (1992) Analysis of DNA structure in vivo using psoralen photobinding: measurement of supercoiling, topollogical domains, and DNA-protein interactions. Methods Enzymol. 212, 319–335.

Sinden, R.R., and R.D. Wells (1992) DNA structure, mutations, and human genetic disease. Curr. Opin. Biotechnol. 3, 612–622.

Sinden, R.R., G. Zheng, R.G. Brankamp, and K.N. Allen (1991) On the deletion of inverted repeated DNA in *E. coli*: effects of length, thermal stability, and cruciform formation in vivo. Genetics 129, 991–1005.

Singleton, S.F., and P.B. Dervan (1992a) Thermodynamics of oligodeoxyribonucleotide-directed triple-helix formation: an analysis using quantitative affinity cleavage titration. J. Am. Chem. Soc. 114, 6957–6965.

Singleton, S.F., and P.B. Dervan (1992b) Influence of pH on the equilibrium association constants for oligodeoxyribonucleotide-directed triple helix formation at single DNA sites. Biochemistry 31, 10995–11003.

Singleton, S.F., and P.B. Dervan (1993) Equilibrium association constants for oligonucleotide-directed triple helix formation at single DNA sites: linkage to cation valence, and concentration. Biochemistry 32, 13171–13179.

Singleton, S.F., and P.B. Dervan (1994) Temperature dependence of the energetics of oligonucleotide-directed triple-helix formation at a single DNA site. J. Am. Chem. Soc. 116, 10376–10382.

Sklenar, V., and J. Feigon (1990) Formation of a stable triplex from a single DNA strand. Nature 345, 836–838.

Skoog, J.U., and L.J. Maher, III (1993a) Repression of bacteriophage promoters by DNA, and RNA oligonucleotides. Nucleic Acids Res. 21, 2131–2138.

Skoog, J.U., and L.J. Maher, III (1993b) Relief of triple-helix-mediated promoter inhibition by elongating RNA polymerases. Nucleic Acids Res. 21, 4055–4058.

Skuratovskii, I.Y., and V.N. Bartenev (1978) Investigation of the structure of magnesium, and lithium salts of T2 phage DNA by the method of X-ray diffraction. The possible mechanisms of the participation of cations in the structural transformation of double-stranded DNA. Mol. Biol. (Moscow) 12, 1359–1376.

Slonczewski, J.L., B.P. Rosen, J.R. Alger, and R.M. Macnab (1981) pH homeostasis in *Escherichia coli*: measurement by ^{31}P nuclear magnetic resonance of methylphosphonate, and phosphate. Proc. Natl. Acad. Sci. USA 78, 6271–6275.

Smith, K.C., and P. Hanawalt (1969) Molecular Photobiology. Academic Press, New York.

Solomon, L.R., L.R. Massom, and H.W. Jarrett (1992) Enzymatic synthesis of DNA-silicas using DNA polymerase. Anal. Biochem. 203, 58–69.

Soyfer, V.N. (1969) Molecular Mechanisms of Mutagenesis. Nauka Publishers, Moscow.

Soyfer, V.N. (1975) Chemical Basis of Mutation. In (T.G. Dobzhansky, M.K. Hecht, and W.C. Steere, Eds.) Evolutionary Biology. Plenum Press, New York, pp. 121–235.

Soyfer, V.N. (Soifer, W.N.) (1976) Molekulare Mechanismen der Mutagenese und Reparatur. Akademie Verlag, Berlin.

Soyfer, V.N., V.V. Demidov, and N.I. Soyfer (1994) Site-directed DNA modification through the formation of triplexes with photosensitizing oligonucleotides. Abstracts of the First International Antisense Conference of Japan, Kyoto, December 4–7, 1994, p. 135.

Soyfer, V.N., O.N. Voloshin, V.A. Malkov, and M.D. Frank-Kamenetskii (1992) Photofootprinting of inter-, and intramolecular DNA triplexes. In (R.H. Sarma, and M.H. Sarma, Eds.) Structure, and Function, Vol. 1: Nucleic Acids. Adenine Press, New York, pp. 29–41.

Spencer, C.A., and M. Groudine (1990) Transcription elongation, and eukaryotic gene regulation. Oncogene 5, 777–785.

Spitzner, J.R., I.K. Chung, and M.T. Muller (1995) Determination of 5′, and 3′ DNA triplex interference boundaries reveals the core DNA binding sequence for topoisomerase II. J. Biol. Chem. 270, 5232–5243.

Sprous, D., and S.C. Harvey (1992) A three-dimensional model for nodule DNA. J. Biol. Chem. 267, 5502.

Srinivasan, A.R., and W.K. Olson (1993) Molecular modeling of triple helical nucleic acid structures. J. Biomol. Struct. Dyn. 10, a185.

Stasiak, A. (1992) Three-stranded DNA structure; is this the secret of DNA homologous recognition? Mol. Microbiol. 6, 3267–3276.

Staubli, A.B., and P.B. Dervan (1994) Sequence specificity of the non-natural pyrido[2,3-d]pyrimidine nucleoside in triple helix formation. Nucleic Acids Res. 22, 2637–2642.

Steiner, R.F., and R.F. Beers (1959) Polynucleotides. VI. The influence of various factors upon the structural transition of polyriboadenylic acid at acid pH. Biochim. Biophys. Acta 32, 166–176.

Stettler, U.H., H. Weber, T. Koller, and H. Weissman (1979) Preparation and characterization of Form V DNA, the duplex DNA resulting from association of complementary circular single-stranded DNA. J. Mol. Biol. 131, 21–40.

Stevens, C.L., and G. Felsenfeld (1964) The conversion of two-stranded poly(A+U) to three strand poly(A+2U), and poly A by heat. Biopolymers 2, 293–314.

Stilz, H.U., and P.B. Dervan (1993) Specific recognition of CG base pairs by 2-deoxynebularine within the purine · purine · pyrimidine triple-helix motif. Biochemistry 32, 2177–2185.

Stokrova, J., M. Vojtiskova, and E. Palecek (1989) Electron microscopy of supercoiled pEJ4 DNA containing homopurine · homopyrimidine sequences. J. Biomol. Struct. Dyn. 6, 891–898.

Stollar, B.D. (1992) Immunochemical analyses of nucleic acids. In (W.E. Cohn, and K. Moldave, Eds.) Progress in Nucleic Acids Research, and Molecular Biology, Vol. 42, Academic Press, San Diego, California, pp. 40–77.

Stollar, B.D., and V. Raso (1974) Antibodies recognize specific structures of triple-helical polynucleotides built on poly(A) or poly(dA). Nature 250, 231–234.

Stonehouse, T.J., and K.R. Fox (1994) DNase I footprinting of triple helix formation at polypurine tracts by acridine-linked oligopyrimidines: stringency, structural changes, and interaction with minor groove binding ligands. Biochim. Biophys. Acta 1218, 322–330.

Stonington, O., and D. Pettijohn (1971) The folded genome of *Escherichia coli* isolated in a protein-DNA-RNA complex. Proc. Natl. Acad. Sci. USA 68, 6–9.
Strobel, S.A, and P.B. Dervan (1989) Cooperative site-specific binding of oligonucleotides to duplex DNA. J. Am. Chem. Soc. 111, 7286–7287.
Strobel, S.A., and P.B. Dervan (1990) Site-specific cleavage of a yeast chromosome by oligonucleotide-directed triple-helix formation. Science 249, 73–75.
Strobel, S.A., and P.B. Dervan (1991) Single-site enzymatic cleavage of yeast genomic DNA mediated by triple helix formation. Nature 350, 172–174.
Strobel, S.A., and P.B. Dervan (1992) Triple helix-mediated single-site enzymatic cleavage of megabase genomic DNA. Methods Enzymol. 216, 309–321.
Strobel, S.A., L.A. Doucette-Stamm, L. Riba, D.E. Housman, and P.B. Dervan (1991) Site-specific cleavage of human chromosome 4 mediated by triple helix formation. Science 254, 1639–1642.
Strobel, S.A., H.E. Moser, and P.B. Dervan (1988) Double-strand cleavage of genomic DNA at a single site by triple-helix formation. J. Am. Chem. Soc. 110, 7927–7929.
Struhl, K. (1985) Naturally occurring poly(dA-dT) sequences are upstream promoter elements for constitutive transcription in yeast. Proc. Natl. Acad. Sci. USA 82, 8419–8423.
Sugimoto, N., Y. Shintani, and M. Sasaki (1991) Effect of the third-strand length on the formation of DNA triple helix. Nucleic Acids Res. Symp. Ser. 25, 183–184.
Sugimoto, N., Y. Shintani, A. Tanaka, and M. Sasaki (1992) Thermodynamics of triple-helix, and double-helix formations by octamers of deoxyriboadenylic, and deoxyribothymidylic acids. Bull. Chem. Soc. Japan 65, 535–540.
Sugiyama, H., Y. Tsutsumi, K. Fujimoto, and I. Saito (1993) Photoinduced deoxyribose C2′ oxidation in DNA. Alkali-dependent cleavage of erythrose-containing sites via a retroaldol reaction. J. Am. Chem. Soc. 115, 4443–4448.
Sugiyama, H., Y. Tsutsumi, and I. Saito (1990) Highly sequence selective photoreaction of 5-bromouracil-containing deoxyhexanucleotides. J. Am. Chem. Soc. 112, 6720–6721.
Sun, J.-S., T. De Bizemont, G. Duval-Valentin, T. Montenay-Garestier, and C. Hélène (1991a) Extension of the range of recognition sequences for triple helix formation by oligonucleotides containing guanines, and thymines. C.R. Acad. Sci. Ser. III 313, 585–590.
Sun, J.-S., J.C. François, T. Montenay-Garestier, T. Saison-Behmoaras, V. Roig, N.T. Thuong, and C. Hélène (1989) Sequence-specific intercalating agents: intercalation at specific sequences on duplex DNA via major groove recognition by oligodeoxynucleotide-intercalator conjugates. Proc. Natl. Acad. Sci. USA 86, 9198–9202.
Sun, J.-S., C. Giovannangéli, J.C. François, R. Kurfurst, T. Montenay-Garestier, U. Asseline, T. Saison-Behmoaras, N.T. Thuong, and C. Hélène (1991b) Triple-helix formation by α oligodeoxynucleotides, and α oligodeoxynucleotide-intercalator conjugates. Proc. Natl. Acad. Sci. USA 88, 6023–6027.
Sun, J.S., and C. Hélène (1993) Oligonucleotide-directed triple-helix formation. Curr. Opin. Struct. Biol. 3, 345–356.
Sun, J.S., and R. Lavery (1992) Strand orientation of [α]-oligodeoxynucleotides in triple helix structures: dependence on nucleotide sequence. J. Mol. Recognit. 5, 93–98.

Sun, J.S., R. Lavery, J. Chomilier, K. Zakrzewska, T. Montenay-Garestier, and C. Hélène (1991c) Theoretical study of ethidium intercalation in triple-stranded DNA, and at triplex-duplex junctions. J. Biomol. Struct. Dyn. 9, 425–436.

Sun, J.S., J.L. Mergny, R. Lavery, T. Montenay-Garestier, and C. Hélène (1991d) Triple helix structures: sequence dependence, flexibility, and mismatch effects. J. Biomol. Struct. Dyn. 9, 411–424.

Sundquist, W.I., and A. Klug (1989) Telomeric DNA dimerizes by formation of guanine tetrads between hairpin loops. Nature 342, 825–829.

Suwalsky, M., W. Traub, and U. Shmueli (1969) An X-ray study of the interaction of DNA with spermine. J. Mol. Biol. 42, 363–373.

Svinarchuk, F.P., Y.V. Lavrovskii, D.A. Konevets, and V.V. Vlassov (1993) Nuclear peptides specifically binding oligonucleotides. Mol. Biol. (Moscow) 27, 187–180 (Engl Transl).

Svinarchuk, F., J.R. Bertrand, and C. Malvy (1994) A short purine oligonucleotide forms a highly stable triple helix with the promoter of the murine c-*pim*-1 proto-oncogene. Nucleic Acids Res. 22, 3742–3747.

Svoboda, D., J.-S. Taylor, J.E. Hearst, and A. Sancar (1993) DNA repair by eukaryotic nucleotide excision nuclease. J. Biol. Chem. 268, 1931–1936.

Sykes, K., and R. Kaufman (1990) A naturally occurring gamma globin gene mutation enhances Sp1 binding activity. Mol. Cell. Biol. 10, 95–102.

Szybalski, W., H. Kubinski, and P. Sheldrick (1966) Pyrimidine clusters on the transcribing strand of DNA, and their possible role in the initiation of RNA synthesis. Cold Spring Harbor Symp. Quant. Biol. 31, 123–127.

Tabor, H. (1962) The protective effect of spermine, and other polyamines against heat denaturation of deoxyribonucleic acid (DNA). Biochemistry 1, 496–501.

Tabor, C.W., and Tabor, H. (1976) 1,4-Diaminobutane (putrescine), spermidine, and spermine. Annu. Rev. Biochem. 45, 285–306.

Tabor, C.W., and Tabor, H. (1984) Polyamines. Annu. Rev. Biochem. 53, 749–790.

Taillandier, E., and J. Liquier (1992) Infrared spectroscopy of DNA. Methods Enzymol. 211, 307–335.

Takabatake, T., K. Asada, Y. Uchimura, M. Ohdate, and N. Kusukawa (1992) The use of purine-rich oligonucleotides in triplex-mediated DNA isolation, and generation of unidirectional deletions. Nucleic Acids Res. 20, 5853–5854.

Takasugi, M., A. Guendouz, M. Chassignol, J.L. Decout, J. Lhomme, N.T. Thuong, and C. Hélène (1991) Sequence-specific photo-induced cross-linking of the two strands of double-helical DNA by a psoralen covalently linked to a triple helix-forming oligonucleotide. Proc. Natl. Acad. Sci. USA 88, 5602–5606.

Tang, M.S., H. Htun, Y. Cheng, and J.E. Dahlberg (1991) Suppression of cyclobutane, and <6–4> dipyrimidines formation in triple-stranded H-DNA. Biochemistry 30, 7021–7026.

Tao, N.J., S.M. Lindsay, and A. Rupprecht (1989) Structure of DNA hydration shells studied by Raman spectroscopy. Biopolymers 28, 1019–1030.

Tarui, M., M. Doi, T. Ishida, M. Inoue, S. Nakaike, and K. Kitamura (1994) DNA-binding characterization of a novel anti-tumor benzo[a]phenazine derivative NC-182: spectroscopic, and viscometric studies. Biochem. J. 304, 271–279.

Thiele, D., and W. Guschlbauer (1968) Evidence for a three-stranded complex between poly I, and poly C. FEBS Lett. 1, 173–175.

Thiele, D., and W. Guschlbauer (1969) Protonated polynucleotides. VII. Thermal transitions between various polyinosinic, and polycytidylic acid complexes in acid media. Biopolymers 8, 361–378.

Thiele, D., and W. Guschlbauer (1971) Protonated polynucleotide structures. IX. Disproportionation of poly(G) · poly(C) in acid medium. Biopolymers 10, 143–157.

Thomas, B.J., and R. Rothstein (1989) Elevated recombination rates in transcriptionally active DNA. Cell 56, 619–630.

Thomas, T.J., and V.A. Bloomfield (1984) Ionic, and structural effects on the thermal helix-coil transition of DNA complexed with natural, and synthetic polyamines. Biopolymers 23, 1295–1306.

Thomas, T.J., V.A. Bloomfield, and Z.N. Canellakis (1985) Differential effects on the B-to-Z transition of poly(dG-me5dC) · poly(dG-me5dC) produced by N1-, and N8-acetyl spermidine. Biopolymers 24, 725–729.

Thomas, T., and T.J. Thomas (1993) Selectivity of polyamines in triplex DNA stabilization. Biochemistry 32, 14068–14074.

Thrierr, J.-C., and M. Leng (1972) A study of complexes between polyriboadenylic acid, and polyribouridylic acid: formation, and structure. Biochim. Biophys. Acta 272, 238–251.

Thuong, N.T., and C. Hélène (1993) Sequence-specific recognition, and modification of double-helical DNA by oligonucleotides. Angew. Chem. Int. Ed. Engl. 32, 666–690.

Tibanyenda, N., S.H. De Bruin, C.A.G. Haasnot, G.A. Van der Marel, J.H. Van Boom, and C.W. Hilbers (1984) The effect of single base-pair mismatches on the duplex stability of d(G-T-T-G-A-A-C-T-A-T-A-A-T-T-A-T) · d(C-A-A-C-T-T-G-A-T-A-T-T-A-A-T-A). Eur. J. Biochem. 139, 19–27.

Timofeev, V.P., A.I. Arutiunian, and A.I. Petrov (1990) Segmental flexibility of single-, double-, and triple-stranded polyribonucleotides from the data of spin label method. Formation of the triple-stranded poly(A · A · U) polynucleotide helix. Mol. Biol. (Moscow) 24, 140–155.

Torrence, P.F., and B. Witkop (1975) Polynucleotide duplexes based on poly(7-deazaadenylic acid). Biochim. Biophys. Acta 395, 56–66.

Trapane, T.L., M.S Christopherson, C.D. Roby, P.O.P. Ts'o, and D. Wang (1994) DNA triple helices with C-nucleosides (deoxypseudouridine) in the second strand. J. Am. Chem. Soc. 116, 8412–8413.

Tripathi, J., and S.K. Brahmachari (1991) Distribution of simple repetitive $(TG/CA)_n$, and $(CT/AG)_n$ sequences in human, and rodent genomes. J. Biomol. Struct. Dyn. 9, 387–397.

Tsao, Y.P., H.Y. Wu, and L.F. Liu (1989) Transcription-driven supercoiling of DNA: direct biochemical evidence from in vitro studies. Cell 56, 111–118.

Ts'o, P.O.P., L. Aurelian, E. Chang, and P.S. Miller (1992) Nonionic oligonucleotide analogs (Matagen) as anticodic agents in duplex and triplex formation. Ann. NY Acad. Sci. 660, 159–177.

Tsuboi, M. (1964) On the melting temperature of nucleic acid in solution. Bull. Chem. Soc. Japan, 37, 1514–1522.

Tsukahara, S., S.G. Kim, and H. Takaku (1993) Inhibition of restriction endonuclease cleavage site via triple helix formation by homopyrimidine phosphorothioate oligonucleotides. Biochem. Biophys. Res. Commun. 196, 990–996.

Tuerk, C., and L. Gold (1990) Systematic evolution of ligands by exponential enrichment: RNA ligands to bacteriophage T4 DNA polymerase. Science 249, 505–510.

Tullius, T.D. (1989) Physical studies of protein-DNA complexes by footprinting. Annu. Rev. Biophys. Biophys. Chem. 18, 213–237.

Tung, C.H., K.J. Breslauer, and S. Stein (1993) Polyamine-linked oligonucleotides for DNA triple helix formation. Nucleic Acids Res. 21, 5489–5494.

Tunis-Schneider, M.J.-B., and M.F. Maestre (1970) Circular dichroism spectra of oriented and unoriented deoxyribonucleic acid films. Preliminary studies. J. Mol. Biol. 52, 521–541.

Turner, D.H., N. Sugimoto, S.D. Dreiker, and S.M. Freier (1988) Hydrogen bonding and stacking contributions to nucleic acid stability. In (R.H. Sarma, and M.H. Sarma, Eds.) Structure and Expression, Adenine Press, New York, pp. 249–259.

Uhlmann, E., and A. Peyman (1990) Antisense oligonucleotides: a new therapeutic principle. Chem. Rev. 90, 544–584.

Ulrich, M.J., W.J. Gray, and T.J. Ley (1992) An intramolecular DNA triplex is disrupted by point mutations associated with hereditary persistence of fetal hemoglobin. J. Biol. Chem. 267, 18649–18658.

Ulyanov, N.B., K.D. Bishop, V.I. Ivanov, and T.L. James (1994) Tertiary base interactions in slipped loop-DNA: an NMR and model building study. Nucleic Acids Res. 22, 4242–4249.

Umemoto, K., M.H. Sarma, G. Gupta, J. Luo, and R.H. Sarma (1990) Structure and stability of a DNA triple helix in solution: NMR studies on $d(T)_6 \cdot d(A)_6 \cdot d(T)_6$ and its complex with a minor groove binding drug. J. Am. Chem. Soc. 112, 4539–4545.

Usdin, K., and A.V. Furano (1989a) The structure of the guanine-rich polypurine · polypyrimidine sequence at the right end of the L1 (LINE) element. J. Biol. Chem. 264, 15681–15687.

Usdin, K., and A.V. Furano (1989b) Insertion of L1 elements into sites that can form non-B DNA. Interaction of non-B DNA-forming sequences. J. Biol. Chem. 264, 20736–20743.

Ussery, D.W., R.W. Hoepfner, and R.R. Sinden (1992) Probing DNA structure with psoralen in vitro. Methods Enzymol. 212, 242–262.

Ussery, D.W., and R.R. Sinden (1993) Environmental influences of the in vivo level of intramolecular triplex DNA in *Escherichia coli*. Biochemistry 32, 6206–6213.

Vallette, F., E. Mege, A. Reiss, and M. Adesnik (1989) Construction of mutant and chimeric genes using the polymerase chain reaction. Nucleic Acids Res. 17, 723–733.

van de Sande, J.H., N.B. Ramsing, M.W. Germann, W. Elhorst, B.W. Kalisch, E. von Kitzing, R.T. Pon, R.C. Clegg, and T.M. Jovin (1988) Parallel stranded DNA. Science 241, 551–557.

van Genderen, M.H.P., M.P. Hilbers, L.H. Koole, and H.M. Buck (1990) Peptide-induced parallel DNA duplexes for oligopyrimidines. Stereospecificity in complexation for oligo(L-lysine) and oligo(L-ornitine). Biochemistry 29, 7838–7845.

van Holde, K. (1989) Chromatin. Springer-Verlag, New York.

van Holde, K., and J. Zlatanova (1994) Unusual DNA structures, chromatin and transcription. BioEssays 16, 59–68.

van Meervelt, L., D. Vliege, A. Dautant, B. Gallois, G. Precigoux, and O. Kennard (1995) High-resolution structure of a DNA helix forming (C · G)*G base triplets. Nature 374, 742–744.

van Vlijmen, H.W.T., G.L. Ramé, and B.M. Pettitt (1990) A study of model energetics and conformational properties of polynucleotide triplexes. Biopolymers 30, 517–532.

Varma, R.S. (1993) Synthesis of oligonucleotide analogs with modified backbones. Synlett. No. 9, 621–637.

Vary, C.P. (1992) Triple-helical capture assay for quantitation of polymerase chain reaction products. Clin. Chem. 38, 687–694.

Verspieren, P., N. Loreau, N.T. Thuong, D. Shire, and J.J. Toulmé (1990) Effect of RNA secondary structure and modified bases on the inhibition of trypanosomatid protein synthesis in cell free extracts by antisense oligodeoxynucleotides. Nucleic Acids Res. 18, 4711–4717.

Vertino, P.M., R.J. Bergeron, P.F. Cavanaugh, Jr., and C.W. Porter (1987) Structural determinants of spermidine-DNA interactions. Biopolymers 26, 691–703.

Veselkov, A.G., V.A. Malkov, M.D. Frank-Kamenetskii, and V.N. Dobrynin (1993) Triplex model of chromosome ends. Nature 364, 496.

Vinograd, J., J. Lebowitz, R. Radloff, R. Watson, and P. Laipis (1965) The twisted circular form of polyoma viral DNA. Proc. Natl. Acad. Sci. USA 53, 1104–1111.

Vlassov, V.V., S.A. Gaidamakov, V.F. Zarytova, D.G. Knorre, A.S. Levina, A.A. Nikonova, L.M. Podust, and O.S. Fedorova (1988) Sequence-specific chemical modification of double-stranded DNA with alkylating oligodeoxyribonucleotide derivatives. Gene 72, 313–322.

Vogt, V.M. (1980) Purification, and properties of S1 nuclease from *Aspergillus*. Methods Enzymol. 65, 248–255.

Vojtiskova, M., S. Mirkin, V. Lyamichev, O. Voloshin, M. Frank-Kamenetskii, and E. Palecek (1988) Chemical probing of the homopurine · homopyrimidine tracts in supercoiled DNA at single-nucleotide resolution. FEBS Lett. 234, 295–299.

Vojtiskova, M., and E. Palecek (1987) Unusual protonated structure in the homopurine · homopyrimidine tract of supercoiled and linearized plasmids recognized by chemical probes. J. Biomol. Struct. Dyn. 5, 283–296.

Völker, J., D.P. Botes, G.G. Lindsey, and H.H. Klump (1993) Energetics of a stable intramolecular DNA triple helix formation. J. Mol. Biol. 230, 1278–1290.

Völker, J., and H.H. Klump (1994) Electrostatic effects in DNA triple helices. Biochemistry 33, 13502–13508.

Volkmann, S., J. Dannull, and K. Moelling (1993) The polypurine tract, PPT, of HIV as target for antisense and triple-helix-forming oligonucleotides. Biochimie 75, 71–78.

Volkmann, S., J. Jendis, A. Frauendorf, and K. Moeling (1995) Inhibition of HIV-1 reverse transcription by triple-helix forming oligonucleotides with viral RNA. Nucleic Acids Res. 23, 1204–1212.

Vologodskii, A.V. (1992) Topology, and Physics of Circular DNA. CRC Press, Boca Raton, Florida.

Voloshin, O.N., S.M. Mirkin, V.I. Lyamichev, B.P. Belotserkovskii, and M.D. Frank-Kamenetskii (1988) Chemical probing of homopurine-homopyrimidine mirror repeats in supercoiled DNA. Nature 333, 475–476.

Voloshin, O.N., A.G. Veselkov, B.P. Belotserkovskii, O.N. Danilevskaya, M.N. Pavlova, V.N. Dobrynin, and M.D. Frank-Kamenetskii (1992) An eclectic DNA structure adopted by human telomeric sequence under superhelical stress and low pH. J. Biomol. Struct. Dyn. 9, 643–652.

Votavova, H., D. Kucerova, J. Felsberg, and J. Sponar (1986) Changes in conformation, stability and condensation of DNA by univalent and divalent cations in methanol-water mixtures. J. Biomol. Struct. Dyn. 4, 477–489.

Vu, H., T.S. Hill, and K. Jayaraman (1994) Synthesis and properties of cholesteryl-modified triple-helix forming oligonucleotides containing a triglycyl linker. Bioconjugate Chem. 5, 666–668.

Wagner, R.W., M.D. Matteucci, J.G. Lewis, A.J. Gutierrez, C. Moulds, and B.C. Froehler (1993) Antisense gene inhibition by oligonucleotides containing C-5 propyne pyrimidines. Science 260, 1510–1513.

Wang, A.H.-J., R. Gessner, G. van der Marel, J.H. Van Boom, and A. Rich (1985) A crystal structure of Z-DNA without an alternating purine-pyrimidine sequence. Proc. Natl. Acad. Sci. USA 82, 3611–3615.

Wang, A.H.-J., G.J. Quigley, F.J. Kolpak, J. Crawford, J.H. van Boom, G. van der Marel, and A. Rich (1979) Molecular structure of left-handed double helical DNA fragment at atomic resolution. Nature 282, 680–686.

Wang, C.-I., and J.-S. Taylor (1992) In vitro evidence that UV-induced frameshift and substitution mutations at T tracts are the result of misalignment-mediated replication past a specific thymine dimer. Biochemistry 31, 3671–3681.

Wang, E., S. Malek, and J. Feigon (1992) Structure of a G · T · A triplet in an intramolecular DNA triplex. Biochemistry 31, 4838–4846.

Wang, G., D.D. Levi, M.M. Seidman, and P.M. Glaser (1995) Targeted mutagenesis in mammalian cells mediated by intracellular triple helix formation. Mol. Cell. Biol. 15, 1759–1768.

Wang, J.C. (1985) DNA topoisomerases. Annu. Rev. Biochem. 54, 665–697.

Wang, J.C., P.R. Caron, and R.A. Kim (1990) The role of DNA topoisomerases in recombination and genome stability: a double-edged sword? Cell 62, 403–406.

Wang, J., C. Jones, M. Norcross, E. Bohnlein, and A. Razzaque (1994) Identification and characterization of a human herpesvirus 6 gene segment capable of transactivating the human immunodeficiency virus type 1 long terminal repeat in an Sp1 binding site-dependent manner. J. Virol. 68, 1706–1713.

Wang, J.C., and A.S. Lynch (1993) Transcription and DNA supercoiling. Curr. Opin. Genet. Dev. 3, 764–768.

Wang, L., P. Pancoska, and T.A. Keiderling (1994) Detection and characterization of triple-helical pyrimidine-purine-pyrimidine nucleic acids with vibrational circular dichroism. Biochemistry 33, 8428–8435.

Wang, Q., S. Tsukahara, H. Yamakawa, K. Takai, and H. Takaku (1994) pH-independent inhibition of restriction endonuclease cleavage via triple helix formation by oligonucleotides containing 8-oxo-2′-deoxyadenosine. FEBS Lett. 355, 11–14.

Wang, S., M.A. Booher, and E.T. Kool (1994) Stabilities of nucleotide loops bridging the pyrimidine strands in DNA pyrimidine-purine-pyrimidine triplexes: special stability of the CTTTG loop. Biochemistry 33, 4639–4644.

Wang, S., and E.T. Kool (1994a) Circular RNA oligonucleotides. Synthesis, nucleic acid binding properties, and a comparison with circular DNAs. Nucleic Acids Res. 22, 2326–2333.

Wang, S., and E.T. Kool (1994b) Recognition of single-stranded nucleic acids by triplex formation: the binding of pyrimidine-rich sequences. J. Am. Chem. Soc. 116, 8857–8858.

Wang, S., and E.T. Kool (1995a) Relative stabilities of triple helices composed of combinations of DNA, RNA and 2′-O-methyl-RNA backbones: chimeric circular oligonucleotides as probes. Nucleic Acids Res. 23, 1157–1164.

Wang, S., and E.T. Kool (1995b) Origins of large differences in stability of DNA and RNA helices: C-5 methyl and 2′-hydroxyl effects. Biochemistry 34, 4125–4132.

Wang, S.Y., Ed. (1976) Photochemistry and Photobiology of Nucleic Acids, Vol. 1, Academic Press, New York.

Wang Y., and D.J. Patel (1995) Bulge defects in intramolecular pyrimidine · purine · pyrimidine DNA triplexes in solution. Biochemistry 34, 5696–5704.

Wang, Z.Y., X.H. Lin, M. Nobuyoshi, Q.Q. Qui, and T.F. Deuel (1992) Binding of single-stranded oligonucleotides to a non-B-form DNA structure results in loss of promoter activity of the platelet-derived growth factor A-chain gene. J. Biol. Chem. 267, 13669–13674.

Ward, J.F. (1988) DNA damage produced by ionizing radiation in mammalian cells: identities, mechanisms of formation, and repairability. Progr. Nucleic Acids Res. Mol. Biol. 35, 95–125.

Waring, M.J. (1974) Stabilization of two-stranded ribohomopolymer helices and destabilization of a three-stranded helix by ethidium bromide. Biochem. J. 143, 483–486.

Washbrook, E., and K.R. Fox (1994a) Alternate-strand DNA triple-helix formation using short acridine-linked oligonucleotides. Biochem. J. 301, 569–575.

Washbrook, E., and K.R. Fox (1994b) Comparison of antiparallel A · AT, and T · AT triplets within an alternate-strand DNA triple helix. Nucleic Acids Res. 22, 3977–3982.

Watson, J.D., and F.H.C. Crick (1953) Molecular structure of nucleic acids: a structure for deoxynucleic acid. Nature 171, 737–738.

Watson, J.D., N.H. Hopkins, J.W. Roberts, J.A. Steitz, and A.M. Weiner (1989) Molecular Biology of the Gene. Benjamin-Cummings, Menlo Park, CA.

Weerasinghe, S., P.E. Smith, V. Mohan, Y.-K. Cheng, and B.M. Pettitt (1995) Nanosecond dynamics, and structure of a model DNA triple helix in salt-water solition. J. Am. Chem. Soc. 117, 2147–2158.

Weinreb, A., D.A. Collier, B.K. Birshtein, and R.D. Wells (1990) Left-handed Z-DNA and intramolecular triplex formation at the site of an unequal sister chromatid exchange. J. Biol. Chem. 265, 1352–1359.

Weintraub, H. (1983) A dominant role for DNA secondary structure in forming hypersensitive structures in chromatin. Cell 32, 1191–1203.

Wells, R.D., D.A. Collier, J.C. Hanvey, M. Shimizu, and F. Wohlrab (1988) The chemistry and biology of unusual DNA structures adopted by oligopurine · oligopyrimidine sequences. FASEB J. 2, 2939–2949.

Wells, R.D., and S.C. Harvey, Eds. (1988) Unusual DNA Structures. Springer-Verlag, New York.

Wells, R.D., and R.R. Sinden (1993) Defined ordered sequence DNA, DNA structure, and DNA-directed mutation. In Genome Analysis, Vol. 7: Genome Rearrangement and Stability. Cold Spring Harbor Laboratory Press, Cold Spring Harbor, New York, pp. 107–138.

Westhof, E., P. Dumas, and D. Moras (1985) Crystallographic refinement of yeast aspartic acid transfer RNA. J. Mol. Biol. 184, 119–145.

Westhof, E., P. Romby, P.J. Romaniuk, J.-P. Ebel, C. Ehresmann, and B. Ehresmann (1989) Computer modeling from solution data of spinach chloroplast and of *Xenopus laevis* somatic and oocyte 5S rRNAs. J. Mol. Biol. 207, 417–431.

White, A.P., and J.W. Powell (1995) Observation of the hydration-dependent conformation of the $(dG)_{20} \cdot (dG)_{20}(dC)_{20}$ oligonucleotide triplex using FTIR spectroscopy. Biochemistry 34, 1137–1142.

Wiegand, R.C., K.L. Beattie, W.K. Holloman, and C.M. Radding (1977) Uptake of homologous single stranded fragments by superhelical DNA. III. The product and its enzymic conversion to a recombinant molecule. J. Mol. Biol. 116, 805–824.

Williamson, J.R., M.K. Raghuraman, and T.R. Cech (1990) Monovalent cation-induced structure of telomeric DNA: the G-quartet model. Cell 59, 871–880.

Wilson, R.W., and V.A. Bloomfield (1979) Counterion-induced condensation of deoxyribonucleic acid. A light-scattering study. Biochemistry 18, 2192–2196.

Wilson, W.D., H.P. Hopkins, S. Mizan, D.D. Hamilton, and G. Zon (1994) Thermodynamics of DNA triplex formation in oligomers with and without cytosine bases: influence of buffer species, pH, and sequence. J. Am. Chem. Soc. 116, 3607–3608.

Wilson, W.D., F.A. Tanious, S. Mizan, S. Yao, A.S. Kiselyov, G. Zon, and L. Strekowski (1993) DNA triple-helix specific intercalators as antigene enhancers: unfused aromatic cations. Biochemistry 32, 10614–10621.

Wilson, W.D., S. Mizan, F.A. Tanious, S. Yao, and G. Zon (1994) The interaction of intercalators and groove-binding agents with DNA triple-helical structures: the influence of ligand structure, DNA backbone modifications and sequence. J. Mol. Recogn. 7, 89–98.

Wohlrab, F., M.J. McLean, and R.D. Wells (1987) The segment inversion site of herpex simplex virus type 1 adopts a novel DNA structure. J. Biol. Chem. 262, 6407–6416.

Wolffe, A. (1992) Chromatin Structure, and Function. Academic Press, San Diego, California.

Wong, A.K., H.A. Yee, J.H. van de Sande, and J.B. Rattner (1990) Distribution of CT-rich tracts is conserved in vertebrate chromosomes. Chromosoma 99, 344–351.

Worcel, A., and E. Burgi (1972) On the structure of the folded chromosome of *Escherichia coli*. J. Mol. Biol. 71, 127–147.

Wu, H.-Y., J.C. Shyy, J.C. Wang, and L.F. Liu (1988) Transcription generates positively, and negatively supercoiled domains in the template. Cell 53, 433–450.

Wu, S.M., D.W. Stafford, and J. Ware (1990) Deduced amino sequence of mouse blood-coagulation factor IX. Gene 86, 275–278.

Xiang, G., W. Soussou, and L.W. McLaughlin (1994) A new pyrimidine nucleoside ($m^{5ox}C$) for the pH-independent recognition of G-C base pairs by oligonucleotide-directed triplex formation. J. Am. Chem. Soc. 116, 11155–11156.

Xodo, L.E., M. Alunni-Fabbroni, and G. Manzini (1994a) Effect of 5-methylcytosine on the structure and stability of DNA. Formation of triple-stranded concatenamers by overlapping oligonucleotides. J. Biomol. Struct. Dyn. 11, 703–720.

Xodo, L.E., M. Alunni-Fabbroni, G. Manzini, and F. Quadrifoglio (1993) Sequence-specific DNA-triplex formation at imperfect homopurine-homopyrimidine sequences within a DNA plasmid. Eur. J. Biochem. 212, 395–401.

Xodo, L.E., M. Alunni-Fabbroni, G. Manzini, and F. Quadrifoglio (1994b) Pyrimidine phosphorothioate oligonucleotides form triple-stranded helices and promote transcription inhibition. Nucleic Acids Res. 22, 3322–3330.

Xodo, L.E., G. Manzini, and F. Quadrifoglio (1990) Spectroscopic and calorimetric investigation on the DNA triplex formed by d(CTCTTCTTTCTTT-TCTTTCTTCTC) and d(GAGAAGAAAGA) at acidic pH. Nucleic Acids Res. 18, 3557–3564.

Xodo, L.E., G. Manzini, F. Quadrifoglio, G. van der Marel, and J. van Boom (1991) Effect of 5-methylcytosine on the stability of triple-stranded DNA — a thermodynamic study. Nucleic Acids Res. 19, 5625–5631.

Xu, Q., R.K. Shoemaker, and W.H. Braunlin (1993) Induction of B-A transitions of deoxyoligonucleotides by multivalent cations in dilute aqueous solution. Biophys. J. 65, 1039–1049.

Yagil, G. (1991) Paranemic structures of DNA and their role in DNA unwinding. Crit. Rev. Biochem. Mol. Biol. 26, 475–559.

Yancopoulos, G.D., and F.W. Alt (1985) Developmentally controlled and tissue-specific expression of unrearranged V_H gene segments. Cell 40, 271–281.

Yang, L., and T.A. Keiderling (1993) Vibrational CD study of the thermal denaturation of poly(rA) · polg(rU) Biopolymers 33, 315–327.

Yang, M., S.S. Ghosh, and G.P. Millar (1994) Direct measurement of thermodynamic and kinetic parameters of DNA triple helix formation by fluorescence spectroscopy. Biochemistry 33, 15329–15337.

Yavachev, L.P., O.I. Georgiev, E.A. Braga, T.A. Avdonina, A.E. Bogomolova, V.B. Zhurkin, V.V. Nosikov, and A.A. Hadjiolov (1986) Nucleotide sequence analysis of the spacer regions flanking the rat tRNA transcription unit and identification of repetitive elements. Nucleic Acids Res. 14, 2799–2810.

Ye, X., K. Kimura, and D.J. Patel (1993) Site-specific intercalation of an anthracycline antitumor antibiotic into a Y · RY triplex through covalent adduct formation. J. Am. Chem. Soc. 115, 9325–9326.

Yee, H.A., A.K.C. Wong, J.H. van de Sande, and J.B. Rattner (1991) Identification of novel single-stranded $d(TC)_n$ binding proteins in several mammalian species. Nucleic Acids Res. 19, 949–953.

Yoon, K., C.A. Hobbs, J. Koch, M. Sardaro, R. Kutny, and A.L. Weis (1992) Elucidation of the sequence-specific third strand recognition of four Watson–Crick base pairs in a pyrimidine triple-helix motif: T · AT, C · GC, T · CG, and G · TA. Proc. Natl. Acad. Sci. USA 89, 3840–3844.

Yoon, K., C.A. Hobbs, A.E. Walter, and D.H. Turner (1993) Effect of 5′-phosphate on the stability of triple helix. Nucleic Acids Res. 21, 601–606.

Young, S.L., S.H. Krawczyk., M.D. Matteucci, and J.J. Toole (1991) Triple helix formation inhibits transcription elongation in vitro. Proc. Natl. Acad. Sci. USA 88, 10023–10026.

Zakian, V.A. (1989) Structure and function of telomeres. Annu. Rev. Genet. 23, 579–604.

Zamecnik, P.C., J. Goodchild, Y. Taguchi, and P.S. Sarin (1986) Inhibition of replication and expression of human T-cell lymphotropic virus type III in cultured cells by exogenic synthetic oligonucleotides complementary to viral DNA. Proc. Natl. Acad. Sci. USA 83, 4143–4146.

Zarytova, V.F., E.M. Ivanova, and G.G. Karpova (1981) Complementary addressed alkylation of poly(A)-fragments of mRNA in Krebs ascitic tumor cells. Bioorg. Khim. 7, 1515–1522.

Zarytova, V.F., A.S. Levina, L.M. Khalimskaya, and Z.A. Sergeeva (1992) Sequence-specific modification of nucleic acids by oligonucleotide derivative containing alkylating groups in the C-5-position of deoxyuridine. Bioorg. Khim. 18, 640–645.

Zendegui, J.G., K.M. Vasquez, J.H. Tinsley, D.J. Kessler, and M.E. Hogan (1992) In vivo stability and kinetics of absorption and disposition of 3′ phosphoprophyl amineoligonucleotides. Nucleic Acids Res. 20, 307–314.

Zheng, G., T. Kochel, R.W. Hoepfner, S.E. Timmons, and R.R. Sinden (1991) Torsionally tuned cruciform and Z-DNA probes for measuring unrestrained supercoiling at specific sites in DNA of living cells. J. Mol. Biol. 221, 107–129.

Zhurkin, V.B. (1985) Sequence-dependent bending of DNA and phasing of nucleosomes. J. Biomol. Struct. Dyn. 2, 785–804.

Zhurkin, V.B., G. Ragunathan, N.B. Ulyanov, R.D. Camerini-Otero, and R.L. Jernigan (1994a) Recombination triple helix, R-form DNA. A stereochemichal model for recognition and strand exchange. In (R.H. Sarma, and M.H. Sarma, Eds.) Structural Biology: The State of the Art 1993, Adenine Press, New York, pp. 43–66.

Zhurkin, V.B., G. Ragunathan, N.B. Ulyanov, R.D. Camerini-Otero, and R.L. Jernigan (1994b) A parallel DNA triplex as a model for the intermediate in homologous recombination. J. Mol. Biol. 239, 181–200.

Zimmer, C., G. Luck, and H. Triebel (1974) Conformation and reactivity of DNA. IV. Base binding ability of transition metal ions to native DNA and effect on helix conformation with special reference to DNA–zinc(II) complex. Biopolymers 13, 425–453.

Zuidema, D., F.M. Van den Berg, and R.A. Flavell (1978) The isolation of duplex DNA fragments containing (dG · dC) clusters by chromatography on poly(rC)-Sephadex. Nucleic Acids Res. 5, 2471–2483.

Index

E